STUDENT STUDY GUIDE
FOR CAMPBELL'S

BIOLOGY

Martha R. Taylor, Ph.D.

Cornell University

The Benjamin/Cummings Publishing Company, Inc.
Menlo Park, California • Reading, Massachusetts
Don Mills, Ontario • Wokingham, U.K. • Amsterdam • Sydney
Singapore • Tokyo • Madrid • Bogota • Santiago • San Juan

Sponsoring Editor: Robin Williams
Production Coordinator: The Book Company: George Calmenson
Designer: The Book Company: Wendy Calmenson
Copyeditor: Lyn Dupré
Composition: First Image

About the cover: Floral radiograph of squash blossom. © Albert G. Richards.

ISBN 0-8053-1842-9

DEFGHIJ-AL-898

The Benjamin/Cummings Publishing Company, Inc.
2727 Sand Hill Road
Menlo Park, California 94025

CONTENTS

PREFACE

When I first thought about writing a student study guide for Neil Campbell's *Biology*, I wanted to call it a structuring guide. The purpose of this book is to help you to structure and organize your developing knowledge of biology and to create your own personal understanding of the topics covered in the text. The four components of each chapter of this study guide are designed to help you structure your knowledge. The *Framework* identifies the overall picture in each chapter. It provides a conceptual framework into which the chapter information fits. The *Chapter Summary* condenses the major concepts of each chapter. The *Structure Your Knowledge* section directs you to organize and relate these concepts by completing tables, answering essay questions, and constructing concept maps. In the *Test Your Knowledge* section, you are provided with a battery of objective questions to test your understanding. Suggested answers to the Structure Your Knowledge and Test Your Knowledge questions are provided at the end of each unit.

A *concept map* is a diagram or map that shows the relationship between concepts. Developing a concept map for a group of concepts requires that you evaluate the relative importance of the concepts (which are most inclusive and important; which are less important and subordinate to other concepts), arrange the concepts, and draw connections between them in a manner that helps you make their meanings explicit. The *structure* of a concept map is a hierarchically organized cluster of concepts, enclosed in boxes or circles and connected with lines that are labeled to explain the relationships between the concepts. The *function* of a concept map is to help you structure your understanding of a topic. The *value* of concept mapping is in the process of thinking and evaluating that you must do in order to create a map. (Additional information about concept mapping may be found in *Learning How to Learn*, Joseph D. Novak and D. Bob Gowin, Cambridge University Press, 1984.)

This book uses concept maps in several ways. A map of a chapter may be presented in the Framework section, to show how I have organized the key concepts in that chapter. More often, you will develop a concept map on a certain subset of ideas from the chapter. The Answers section will have a *suggested* concept map that shows one way of organizing the material.

A concept map is an individual picture of your understanding at the time you made the map. Meanings change and grow as you gain more experience in an area, enabling you to make more connections between concepts. Concept maps are also context dependent. I can take the same group of concepts from Chapter 5 and organize them differently depending on whether my focus is enzyme action or protein structure. Do not look to the answer section for the "right" concept map. *After* you have organized your own thoughts, look at the suggested map to make sure you have included the key concepts (although you may have added more), to check that the connections you have made are reasonable, and perhaps to see another way to organize the information.

In the multiple choice questions presented in each chapter, you are asked to choose the best answer. Some answers may be partly correct; almost all choices have been written to test your ability to think and discriminate. Make sure you understand why the other choices are incorrect, as well as why the given answer is correct.

Biology is a fascinating and broad subject. Campbell's *Biology* is filled with terminology and facts that are useful for understanding the major themes of modern biology. This *Student Study Guide* is intended to help you learn and recall information, and, most importantly, to encourage and guide you as you develop your own understanding of and appreciation for biology.

CHAPTER 1

INTRODUCTION: THEMES IN THE STUDY OF LIFE

FRAMEWORK

This chapter outlines seven themes that unify the study of biology and describes the scientific construction of biological knowledge. A course in biology is neither a vocabulary course nor a classification exercise for the diverse forms of life. Biology is a collection of facts and concepts that are structured within theories and organizing principles. Recognizing the common themes within biology will help you to structure your knowledge of this fascinating and challenging study of life.

CHAPTER SUMMARY

Biology is the scientific study of life, an extension of what E. O. Wilson terms *biophilia*, an innate affinity for life in its diverse forms. The scope of biology is immense, spanning from the submicroscopic realm of molecules to the complex webs of ecosystems, from the present back through 4 billion years of evolutionary history. Recent advances in research methods, coupled with the large number of biologists spread throughout the many subfields of biology, have led to an explosion of information. A beginning student can make sense of this expanding universe of knowledge by focusing on a few enduring themes that unify the study of biology.

A Hierarchy of Structural Organization

Biological organization is based on a hierarchy of structural levels, ranging from atoms, through biological molecules, organelles, cells, tissues, organs, organ systems, organisms, populations, communities, to ecosystems. The study of biology includes understanding the various levels of biological organization.

Emergent Properties

The interactions among the components of a particular level of organization lead to the emergence of novel properties at the next level: the whole is often greater than the sum of its parts. The emergent properties of life do not support the doctrine of vitalism, which claims that life is driven by forces that defy explanation. The principles of physics and chemistry can be used to unravel and understand the properties of life that arise from the hierarchy of structural organization.

The characteristics of life, which emerge from this highly ordered and complex organization, include the ability to take in energy and transform it to do work and to maintain an ordered state; the ability to respond to stimuli from the environment; the reproduction, growth and development of organisms as directed by heritable programs; and the evolution of a population as it accumulates adaptations to its specific environment.

Holism studies the higher organizational levels of life in order to comprehend the properties that arise from the interactions of the system's parts. Reductionism, in contrast, breaks down complex systems to simple components that are more manageable to study. The reductionist approach to biology has been a powerful strategy for understanding the most basic components of this complex phenomenon called life. The exposition of one level of organization gives insight into the emergent properties of the next level.

The Cellular Theme of Life

The cell is the simplest level of structure capable of performing all the activities of life; it is life's basic unit of structure and function. Hooke first described and named cells in 1665, when he observed a slice of cork with a simple microscope. Leeuwenhoek developed lenses that permitted him to discover the world of unicellular organisms. In 1839, Schleiden and Schwann concluded that all living things consist of cells. The recent advent of the electron microscope has revealed the complex structural organization of cells.

Correlation of Structure and Function

A study of the form of a biological structure gives information on its functioning, and a study of functions provides insight into structural organization. The principle that form fits function applies to the many structural levels of biological organization.

Heritable Programs

The biological instructions for the development and functioning of organisms are coded in symbolic form in DNA molecules and transmitted from parents to offspring in the units of inheritance called genes. A gene has precise information encoded in the specific sequential arrangement of the four nucleotides that make up DNA. A DNA molecule exists as a double helix, two spiralling chains with complementary nucleotide sequences, providing for the exact copying of the genetic material every time a cell divides. All forms of life use the same genetic code to write and transcribe their heritable script.

Unity in Diversity

The diversity of life is estimated at about 30 to 40 million species. Taxonomy, the branch of biology that names and classifies species, groups organisms into hierarchical classes that reflect their relationships. The diverse forms of life are organized into five kingdoms: Monera, Protoctista, Plantae, Fungi, and Animalia.

The kingdom Monera, which contains bacteria, is distinguished on the basis of the simple prokaryotic cell type. All other kingdoms have eukaryotic cells. Protoctists are mostly unicellular, or simple multicellular forms, and include the protozoa. The other three kingdoms are multicellular eukaryotic groups that are characterized to a large extent by their mode of nutrition. Plants are photosynthetic, able to convert light energy into the chemical energy of sugar. Fungi are mostly decomposers that absorb their nutrients from dead organisms or organic wastes. Animals obtain their food by ingestion.

The diversity of life forms is unified by the universal genetic code, similarities in metabolic pathways, and commonalities in cell structure.

Evolution: The Core Theme

Evolution is the one biological theme that connects all of life by common ancestry. The history of living forms extends back over 3 billion years to the ancient prokaryotes. Each species is the tip of an evolutionary branch that connects closely related and similar species with their common ancestors.

In *The Origin of Species*, published in 1859, Charles Darwin presented his case for "descent with modification," the theory of the evolution of present forms from a succession of ancestral forms as a result of a mechanism called natural selection. Darwin synthesized the concept of natural selection from three generalized observations: (1) individuals in a population of a species vary in many heritable traits; (2) many more young are produced within a population than can be supported by the environment; and (3) individuals with traits best suited for the environment leave a larger proportion of offspring than do less fit individuals. This differential reproductive success results in the gradual accumulation of adaptations within a population.

According to Darwin, new species originate as populations are separated by geographic barriers and unique traits and combinations of traits are selected for by differing environments. Descent with modification makes sense of both the unity and diversity of life; evolution is the core theme of biology.

Biology as Science

Science emerges from our curiosity and drive to understand ourselves, the world, and the universe. Asking questions about nature and believing that those questions are answerable provide the basis of science. Few scientists rigidly follow the prescribed steps of the "scientific method," but researchers do focus on gathering evidence in experiments or by observations to test hypotheses—tentative answers to specific questions. Scientific studies usually include a control, an unaltered system or group of organisms that permits the testing of the effect of a single variable in the experimental group.

Science is characterized as progressive and self-correcting. Scientists build on the work done by others, refining or refuting previous ideas. Scientists working on the same problem both cooperate and compete.

Facts, in the form of observations and experimental results, are prerequisites of science, but new ways of organizing and relating those facts advance science.

Newton, Darwin, and Einstein stand out as scientists because they synthesized theories with great explanatory power. A theory is broader in scope and more widely accepted than is a hypothesis. Good theories generate testable hypotheses.

Biology is a demanding science—partly because living systems are so complex and partly because biology incorporates concepts from chemistry, physics, and math. There is a wealth of information presented in this book. The basic themes of biology will serve to help you understand, appreciate, and structure your growing knowledge of biology.

STRUCTURE YOUR KNOWLEDGE

1. This chapter presents seven unifying themes of biology. Briefly describe each of these themes.

    A hierarchy of structural organization
    Emergent properties
    The cellular theme of life
    Correlation of structure and function
    Heritable programs
    Unity in diversity
    Evolution

2. Biology is the scientific study of life. How would you characterize a scientific approach to developing biological knowledge?

ANSWER SECTION

CHAPTER 1: INTRODUCTION: THEMES IN THE STUDY OF LIFE

Suggested Answers to Structure Your Knowledge

1.   a. Living things exhibit a hierarchy of structural levels of organization, ranging all the way from the organization of particles in atoms, through molecules, macromolecules, cellular organelles, cells, tissues, organs, organ systems, whole organisms, populations, communities, to ecosystems.

b. Unique properties emerge at each structural level as a result of the interactions and organization of components.

c. All living forms are composed of cells, the basic unit of structure and function that exhibits the characteristics of life.

d. Within each level of organization, there is a correlation between structure and function; that is, between the physical arrangement of parts and the function they serve.

e. The structures and processes of life are coded for by DNA, the universal language of inheritance.

f. The diversity of life, as illustrated by the five-kingdom classification of organisms, is unified by the universal genetic code, similar metabolic pathways, and common cellular characteristics.

g. Evolution, the connection of all life forms through common ancestry and the origin of new life forms and the accumulation of adaptations to specific environments through natural selection, is the core theme of biology.

2.   The scientific study of life rests on the belief that natural phenomena have natural causes and that questions about life are answerable. Scientists develop hypotheses, which are tentative answers to questions, that are then tested through the accumulation of evidence from experiments and observations. Evidence and conclusions are shared, and scientists repeat, reinforce, and refute each other's ideas. Theories are broad explanations of various phenomena that serve to organize biological knowledge, generate new questions, and direct future studies.

Unit I

The Chemistry of Life

CHAPTER **2**

ATOMS, MOLECULES, AND CHEMICAL BONDS

FRAMEWORK

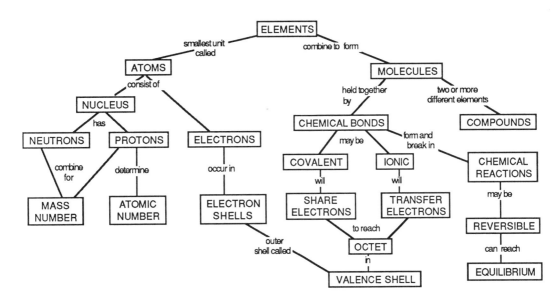

This chapter considers the basic principles of chemistry that explain the behavior of atoms and molecules and form the basis for our modern understanding of biology. The emergent properties associated with each new level of structural organization are evident even as subatomic particles are organized into atoms and atoms into molecules. The above concept map sketches out the relationships among the key concepts included in this chapter.

CHAPTER SUMMARY

The study of the phenomenon of life necessitates an understanding of the structure and interaction of the atoms and molecules that constitute living organisms. A reductionist approach to the study of biology focuses on lower and basic levels of organization. Understanding the principles of chemistry that govern how atoms and molecules behave lays the foundation for the consideration of the successive structural levels of

the organization of life and the emergent properties that accompany those levels.

Matter: Elements and Compounds

Chemistry is the study of *matter*, anything that takes up space and has mass. Mass is a measure of the amount of matter within an object. Weight is the measure of how strongly a mass is pulled by gravity. On earth, weight is proportional to the quantity of matter or the mass in an object.

The basic constituents of matter are *elements*, substances that cannot be broken down by ordinary chemical means. Chemists have identified 92 naturally occurring elements, each symbolized by one or two letters. About 25 elements are essential to life. Of these, carbon (C), oxygen (O), hydrogen (H), and nitrogen (N) make up 99% of living matter. The remaining 1% is composed of phosphorus (P), sulfur (S), calcium (Ca), and a few other elements. Some elements, such as iron (Fe) and iodine (I), are needed in extremely minute

7

quantities and are called *trace elements*.

A *compound* is made up of two or more elements combined in a fixed ratio. A compound usually has characteristics quite different from its constituent elements, an example of the emergence of novel properties in higher levels of organization.

The Structure and Behavior of Atoms

An *atom* is the smallest unit of an element retaining the physical and chemical properties of that element. Each element has its own unique atomic structure, made up of a characteristic number of subatomic particles. Uncharged *neutrons* and positively charged *protons* are packed tightly together to form the nucleus of an atom. Negatively charged *electrons* orbit rapidly about the nucleus, electrically attracted to the nucleus. The number of electrons is always equal to the number of protons; thus atoms have no net charge.

Protons and neutrons have similar masses of almost exactly 1 dalton apiece. A *dalton* is the atomic mass measurement unit. Electrons have negligible mass. The *mass number* of an element is equal to the number of protons and neutrons in its nucleus and approximates the mass of an atom in daltons. The term *atomic weight* is often used to refer to the mass of an atom.

The *atomic number* of an atom refers to the number of protons in its nucleus. Each element has a characteristic and consistent atomic number. A subscript to the left of the symbol for an element indicates its atomic number; a superscript to the left indicates its mass number. The number of neutrons in an atom is equal to the difference between the mass number and the atomic number.

Although the number of protons is constant, the number of neutrons can vary within the atoms of an element, creating different *isotopes*. An element can occur in nature as a mixture of its isotopes. Some isotopes are unstable or *radioactive*; their nuclei spontaneously decay, giving off particles and energy that can be measured with electronic instruments. The characteristic half-life for radioactive isotopes (the length of time it takes 50% of the radioactive atoms in a sample to decay) provides a basis for dating fossils. Radioactive isotopes are important tools in biological research and medicine because they can be detected in such minute quantities. Molecules can be labeled with radioactive isotopes and followed through the steps of metabolic pathways, traced throughout an organism, or incorporated into DNA for molecular genetics research. Radiation from decaying isotopes poses a significant hazard to life. Thus, the development of nuclear weapons and the disposal of radioactive wastes from weapon manufacture, research, and power plants are important public issues.

The nucleus of an atom is very small compared to the large area in which the electrons orbit. The chemical behavior of atoms—the type of interactions they have with other atoms, is determined by the number and location of the electrons.

Energy is defined as the ability to do work. *Potential energy* is energy stored in matter as a consequence of the relative position of masses or charges. Matter naturally tends to move toward a lower level of potential energy (which is a more stable state), and requires the input of energy to return to a higher potential energy. The negatively charged electrons are attracted to the positively charged nucleus; their potential energy increases as their distance from the nucleus increases. Energy must be absorbed to move an electron farther from the nucleus.

Electrons can orbit in several different potential energy states, called *energy levels* or *electron shells*, surrounding the nucleus. The three-dimensional space or volume within which an electron is most likely to be found is called an *orbital*. No more than two electrons can occupy an orbital. The first electron shell or energy level can contain two electrons in a single spherical orbital, called the 1*s* orbital. The second electron shell has four orbitals, each with the potential of containing two electrons. They include a 2*s* spherical orbital and three dumbbell-shaped *p* orbitals located along the x, y, and z axes. Higher electron shells contain additional orbitals, but the outermost energy shell of an atom never contains more than eight electrons, located in its *s* and three *p* orbitals.

The chemical behavior of an atom is a function of its electron configuration—in particular, of the number of electrons in its outermost energy shell, or *valence shell*. As the number of outer electrons, or *valence electrons*, increases, the orbitals are filled in the following order: two in the *s* orbital, one in each of the *p* orbitals, then a second in each *p* orbital. The *octet rule* states that a valence shell of eight electrons is complete, resulting in an unreactive or inert atom. Atoms with incomplete valence shells are chemically reactive.

The periodic table of elements is arranged in order of the sequential filling of electron orbitals. Atoms with the same number of electrons in their valence shell have similar chemical properties.

Chemical Bonds and Molecules

Atoms with incomplete valence shells can either completely transfer electrons from or share electrons with other atoms such that each atom is able to complete its valence shell. These interactions usually result in attractions, called *chemical bonds*, that hold the atoms together. A *molecule* consists of two or more atoms held together by chemical bonds. A *covalent bond* is formed by the sharing of a pair of valence electrons by two atoms. An *ionic bond* is formed when one or more electrons is transferred from one atom to another.

A *structural formula* represents the atoms and bonding within a molecule. Thus, H-H indicates a hydrogen molecule of two atoms of hydrogen held together by a covalent bond. O=O represents an oxygen molecule in which two pairs of valence electrons are shared between oxygen atoms, forming a *double covalent bond*. A *molecular formula* (such as O_2) indicates the kinds and numbers of atoms in a molecule, but not the bonding between them. A *compound* consists of molecules formed from more than one element.

The *valence* or bonding capacity of an atom is a measure of the number of covalent bonds it must form in order to complete its outer shell. The valences of the four most common elements of living matter are hydrogen 1, oxygen 2, nitrogen 3, and carbon 4.

Molecules have characteristic sizes and shapes. The function of a molecule is often dependent on its structure or geometry; emergent properties derive from the arrangement or organization of constituents. The *s* and *p* orbitals in the valence shell are rearranged in the formation of a covalent bond; they hybridize to form four teardrop-shaped orbitals extending from the nucleus in a pyramidal, three-dimensional array. A methane molecule (CH_4) has the shape of a tetrahedron with the carbon atom in the center and each of the four hydrogen atoms sharing a pair of electrons at the corners of the carbon's valence orbitals. A water molecule (H_2O) is in the shape of a V, two hydrogen atoms sharing electrons at the corners of two of oxygen's valence orbitals, with oxygen's remaining two pairs of electrons found in the other two orbitals.

Electronegativity is the attraction of an atom for electrons. The more electronegative an atom, the more strongly it pulls the shared electrons of a covalent bond toward itself. If the atoms in a molecule have similar electronegativities, the electrons remain equally shared between the two nuclei, and the covalent bond is said to be *nonpolar*. If one atom is more electronegative, it pulls the shared electrons closer to itself, creating a *polar* covalent bond. This unequal sharing of electrons results in a partial negative charge associated with the more electronegative atom and a partial positive charge associated with the atom from which the electron is pulled.

If two atoms are very different in their attraction for the shared electrons, the more electronegative atom may completely transfer an electron from another atom, forming an *ionic bond*. This transfer of a negatively charged electron from one atom to another results in the formation of charged atoms called *ions*. The atom that lost the electron is positively charged and called a *cation*. The atom that gained the electron is negatively charged and called an *anion* (an Anion Adds

an electron). The transfer of electrons allows atoms to achieve complete valence shells. The atoms are held together because of the attraction of their opposite charges. A salt is an ionic compound in which ions form a three-dimensional crystalline lattice arrangement held together by electrical attraction.

The sharing of electrons between atoms in the completion of their valence shell can fall on a continuum from nonpolar covalent bonds in which electrons are equally shared, through polar covalent bonds, to ionic bonds in which electrons are shared so unequally as to be actually transferred from one atom to another. Covalent bonds are strong; they are hard to break. Ionic bonds are strong in a dry salt crystal, but are easily broken apart when the salt dissolves. In water, ionic bonds are weak; anions and cations easily separate from each other.

Hydrogen bonds, van der Waals interactions, and hydrophobic interactions are weak chemical bonds within and between molecules that are important to cellular chemistry. They determine the spatial arrangements of molecules and thus influence the functioning, interactions, and emergent properties of these cellular molecules. *Hydrogen bonds* are formed when a hydrogen atom (with a partial positive charge) that is covalently bonded with one electronegative atom is also attracted to another electronegative atom. Hydrogen bonds are responsible for many of the unusual properties of water. *Van der Waals interactions* occur between atoms and molecules in the form of weak attractions between transient regions of negative and positive charge that result from the unequal distribution of orbiting electrons. Hydrophobic molecules do not dissolve in water. *Hydrophobic interactions* between molecules result in the clumping or coalescing of hydrophobic molecules due to their repulsion or exclusion from water.

Chemical Reactions

Chemical reactions involve the making or breaking of chemical bonds in the transformation of matter into different forms. Matter is conserved in chemical reactions; the same number and kind of atoms are present in both *reactants* and *products*, although the rearrangement of electrons and atoms causes the properties of these molecules to be different. Some reactions go to completion, but most are reversible—the products of the forward reaction can become reactants in the reverse reaction. The rate of a reaction is speeded by increasing concentrations of reactants. More collisions occur between reactants and more products are formed. Eventually, *chemical equilibrium* may be reached when the forward and reverse reactions proceed at the same rate and the relative concentrations of reactants and products remain fixed.

These relative concentrations will vary depending on the reaction; chemical equilibrium does not mean that reactants and products are equal in concentration.

STRUCTURE YOUR KNOWLEDGE

Take the time to write out or discuss your answers to the following questions. Then refer to the suggested answers at the end of the unit.

1. Describe an atom. Include the concepts of particle, mass, charge, and spatial arrangement.

2. Atoms can have various numbers associated with them. Explain the following: atomic number, mass number, atomic weight, valence. Which of these numbers is most related to the chemical behavior of an atom?   Explain.

3. Explain what is meant by the statement that there is no distinct dividing line between covalent bonds and ionic bonds.

4. Linus Pauling was quoted as saying:

   I, myself, have confidence that all of the properties of living organisms could ultimately be discovered by the process of attempting to reduce the organism in our minds to a combination of the different parts: essentially the molecules that make up the organism.

   Do you agree with this reductionist approach to biology? Why is it important to understand the most basic units of living matter, and why is it also important to keep moving up through organizational levels to study life?

TEST YOUR KNOWLEDGE

MULTIPLE CHOICE: *Choose the one best answer.*

1. Each element has its own characteristic atom in which
   a. the atomic weight is constant.
   b. the atomic number is constant.
   c. the mass number is constant.
   d. two of the above are correct.

2. Isotopes can be used in studies of metabolic pathways because
   a. their half-life allows one to time an experiment.
   b. they are more reactive.
   c. the cell does not recognize the extra protons in the nucleus, so isotopes are readily used by the cell.
   d. only electrons are involved in interactions.

3. At equilibrium in a reaction
   a. the forward and reverse reactions are occurring at the same rate.
   b. the reactants and products are in equal concentration.
   c. the forward reaction has gone further than the reverse reaction.
   d. both a and b are correct.

4. Oxygen has eight electrons. You would expect these to be found arranged in the orbitals in the following way:
   a. Two in $1s$, two in $2s$, two in $2px$, two in $2py$, zero in $2pz$.
   b. Two in $1s$, two in $2s$, two in $2px$, one in $2py$, one in $2pz$.
   c. Two in $2s$, two in $2px$, two in $2py$, two in $2pz$.
   d. Two in $1s$, two in $2px$, two in $2py$, two in $2pz$.

5. A covalent bond is likely to be polar if
   a. one of the atoms sharing electrons is much more electronegative than the other.
   b. the two atoms sharing electrons are equally electronegative.
   c. the two atoms sharing electrons are of the same element.
   d. the bond is part of a tetrahedrally shaped molecule.

6. Of the three weak bonds, which is unique to compounds with polar covalent bonds?
   a. hydrogen bonds
   b. van der Waals interactions
   c. hydrophobic interactions
   d. None are found in association with polar compounds.

7. Which of these classes of substances would be least soluble in the polar compound water?
   a. ionic compounds
   b. polar compounds
   c. hydrophobic compounds
   d. salts

8. Which of the following inquiries represents the most reductionist approach?
   a. study of the absorption of radioactively labeled sugars by a yeast cell
   b. study of van der Waals interactions
   c. study of the clinical effects of sickle-cell anemia
   d. study of the effect of vitamin C on the common cold

9. The octet rule states that
   a. a valence shell with eight electrons is stable and complete.
   b. no energy level can contain more than eight electrons.
   c. a valence shell with eight electrons produces a neutral atom.
   d. no more than eight electrons can be shared between atoms.

10. The most accurate structural formula for water is
    a. /O\
       H   H
    b. H-O-H
    c. H_2O
    d. $H^+ O = H^+$

11. A triple covalent bond
    a. would be very polar.
    b. involves the bonding of three atoms.
    c. involves the bonding of four atoms.
    d. involves the sharing of six electrons.

12. It is difficult to speak of a molecule of the salt NaCl because
    a. each sodium ion is attracted to four chloride ions.
    b. salt occurs as a crystalline lattice of many sodium and chloride ions.
    c. the ratio of sodium and chlorine atoms may vary.
    d. the bonds in a salt crystal are weak and break easily.

13. A cation
    a. has gained an electron.
    b. can easily form hydrogen bonds.
    c. is hydrophobic.
    d. has a positive charge.

The six elements most common in living organisms are:

$^{12}_{6}C$ $^{16}_{8}O$ $^{1}_{1}H$ $^{14}_{7}N$ $^{32}_{16}S$ $^{31}_{15}P$

Use this information to answer the following questions.

14. How many electrons does phosphorus have in its valence shell?
    a. 15
    b. 5
    c. 7
    d. 8

15. Which of these atoms does not undergo *sp3* hybridization when it forms covalent bonds?
    a. C
    b. O
    c. H
    d. P

16. A rare, radioactive isotope of phosphorus has the mass number 32.
    How many neutrons does this isotope have?
    a. 32
    b. 1
    c. 16
    d. 17

17. Based on electron configuration, which of these elements would have chemical behavior most like that of oxygen?
    a. C
    b. N
    c. S
    d. P

18. How many covalent bonds is a phosphorus atom most likely to form?
    a. 1
    b. 2
    c. 3
    d. 4

19. What is the valence of sulfur?
    a. 1
    b. 2
    c. 3
    d. 4

20. How many of these elements are found beside each other on the periodic chart?
    a. one group of two
    b. two groups of two
    c. one group of two and one group of three
    d. all of them

CHAPTER 3

WATER AND THE FITNESS OF THE ENVIRONMENT

FRAMEWORK

In this chapter you will be introduced to how the chemistry and emergent properties of water contribute to the biological fitness of the external and internal environment of living organisms. The significance of the *polarity* of the water molecule, the *hydrogen bonding* among water molecules, and the *dissociation* of water into hydrogen and hydroxide ions will be examined.

CHAPTER SUMMARY

Water is the biological medium that makes life possible. It makes up 70% to 95% of the content of the cells of living organisms and covers 75% of the earth's surface. The extraordinary properties of water significantly contribute to the fitness of the external environment for life and the fitness of an organism's internal environment for the chemical and physical processes of life. The properties of water can be traced to the structure and interactions of its molecules.

Water Molecules and Hydrogen Bonding

Water molecules consist of two hydrogen atoms covalently bonded to an oxygen atom. The molecule has a tetrahedral or pyramidal shape. The two hydrogen atoms and the two valence orbitals of oxygen that each contain a pair of unshared electrons form the four corners. The electronegative oxygen nucleus pulls the shared electron away from the hydrogen nucleus, creating a polar covalent bond and a partial positive charge on the hydrogen atom. The two unshared orbitals possess a partial negative charge. This asymmetric shape and the polarity of the molecule create the potential for the formation of *hydrogen bonds*, in which a hydrogen atom covalently bonded to an oxygen atom is shared with an oxygen atom of another molecule. This bonding orders water molecules into a higher level of structural organization and accounts for the emergent properties of this extraordinary substance.

Extraordinary Properties of Water

Liquid water is unusually cohesive due to the hydrogen bonds that tend to hold the molecules together. This *cohesion* creates a more structured organization to the liquid and produces a high surface tension at the interface between water and air. Small insects are able to "walk on water" due to this surface tension. Water's hydrogen bonds and polarity also result in *adhesion*, in which water molecules cling to *hydrophilic* (water-loving) substances. The combination of cohesion and adhesion allows water to rise in thin tubes of hydrophilic material by *capillary action*. *Imbibition* is the soaking of water by capillary action into a porous hydrophilic substance such as wood or the coat of a germinating seed.

The ability of water to stabilize air temperatures is related to the high specific heat of water. A large amount of heat is absorbed or released during a slight change in water temperature. *Heat* is the total quantity of *kinetic energy*, the energy created by the movement of atoms and molecules. *Temperature* measures the average kinetic energy of the molecules in a substance.

Temperature is measured by a *Celsius scale*. Water freezes at 0° C and boils at 100° C. Heat is measured by the *calorie*. One calorie is the amount of heat energy it takes to raise 1 gram of water 1 degree Celsius. A *kilocalorie* (kcal, also designated as C) is 1000 calories, the amount of heat required to raise 1 kilogram of water 1 degree Celsius.

Specific heat is the amount of heat absorbed or lost when 1 gram of a substance changes temperature by 1 degree Celsius. Water's specific heat of 1 cal/gm/° C is unusually high compared with that of other common substances. Water must absorb or lose a relatively large quantity of heat in order for its temperature to change. This property arises from the hydrogen bonding of water molecules. Heat must be absorbed to break hydrogen bonds, and heat is released when hydrogen bonds form. Heat energy must be used to disrupt hydrogen bonds before water molecules can move faster and temperature can rise. Likewise, as the temperature of water drops slightly, many hydrogen bonds form and release a considerable amount of heat energy. The high proportion of water in the environment and within organisms, coupled with water's high specific heat, keeps temperature fluctuations within limits that permit life.

The transformation from a liquid to a gas is called vaporization or evaporation. Molecules with sufficient kinetic energy overcome their attraction to other molecules and enter the air as a gas. The addition of heat increases the rate of evaporation by increasing the kinetic energy of molecules. The *heat of vaporization* is the quantity of heat a liquid must absorb for 1 gram of a substance to be converted from liquid to gaseous state. Water has a high heat of vaporization (540 cal/gm) because a large amount of heat is needed to break the hydrogen bonds holding water molecules together.

As a substance vaporizes, the liquid left behind loses the kinetic energy of the escaping molecules and cools down. *Evaporative cooling* helps to protect terrestrial organisms from overheating and contributes to the stability of temperatures in and around lakes and ponds.

As water cools below 4° C, it expands. By 0° C, each water molecule becomes hydrogen bonded to four other molecules, creating a crystalline lattice and spacing the molecules farther apart. Ice is about 10% less dense than is liquid water at 4° C and therefore floats. The floating ice insulates the liquid water below, allowing it to be at a temperature warmer than the air above it. The layer of ice protects bodies of water from freezing solidly from the bottom upward in the winter.

The formation and melting of ice serve to temper the transitions between seasons. Heat is released into the air when hydrogen bonds form as water solidifies into ice and snow. The melting of ice and snow absorbs heat as hydrogen bonds are broken. This release and absorption of heat make temperature fluctuations within the environment less abrupt.

Water is the most versatile solvent known. A *solution* is a homogeneous mixture of two or more substances; the dissolving agent is called the *solvent*, and the substance that is dissolved is the *solute*. An *aqueous solution* is one in which water is the solvent. The positive and negative regions of water molecules are attracted to oppositely charged ions or charged regions of polar molecules. Thus, solute molecules become surrounded by water molecules and dissolve into solution. Nonpolar compounds, however, will not dissolve in water.

Aqueous Solutions

A *mole* is the amount of a substance that has a mass in grams numerically equivalent to its molecular weight (sum of the weight of all atoms in the molecule) in daltons. A mole of any substance has exactly the same number of molecules—6.02×10^{23}, called *Avogadro's number*. The *molarity* of a solution (abbreviated M) refers to the number of moles of a solute dissolved in 1 liter of solution.

A water molecule can *dissociate* into a *hydrogen ion*, H^+ (which is transferred to another water molecule to form a *hydronium ion*, H_3O^+), and a hydroxide ion (OH^-). Although reversible and statistically rare, this dissociation is very important in the chemistry of life. In pure water, the concentrations of H^+ and OH^- ions are the same; both are equal to 10^{-7} M. When acids or bases dissolve in water, the H^+ and OH^- balance shifts. An *acid* adds H^+ to a solution, whereas a *base* reduces H^+ in a solution by either accepting hydrogen ions or adding hydroxide ions. A strong acid or strong base is a substance that dissociates completely when mixed with water. A weak acid dissociates reversibly to release or reaccept H^+. A weak base reversibly binds and releases H^+.

A solution with a higher concentration of H^+ than of OH^- is considered acidic. A basic solution has a higher concentration of OH^- than H^+. In any solution, the product of the H^+ and OH^- concentrations is constant at 10^{-14} M. In a neutral solution, $[H^+] = 10^{-7}$ and $[OH^-] = 10^{-7}$ M; so the product is 10^{-14} M. Brackets, [ ], indicate molar concentration. If the $[H^+]$ increases, then the $[OH^-]$ decreases, due to the tendency of excess hydrogen ions to combine with the hydroxide ions in solution and form water. Likewise, an increase in $[OH^-]$ causes a decrease in $[H^+]$. The product of the concentrations remains constant at 10^{-14}; an increase in the concentration of one ion results in an equivalent decrease in the other.

The hydrogen and hydroxide ion concentrations can vary by many orders of magnitude. The range of concentrations is compressed, through the use of logarithms, in the *pH scale*. The pH of a solution is defined as the negative log (base 10) of the $[H^+]$: pH $= -\log_{10} [H^+]$. For a neutral solution, $[H^+]$ is 10^{-7} M, and the pH value equals 7. As the $[H^+]$ increases in an acidic solution, the *pH value* decreases. A pH value below 7 indicates an acidic solution; one above 7 denotes a basic solution. Each pH unit represents a ten-fold difference in ion concentration. A slight change in pH reflects a substantial change in [H⁺] and [OH⁻].

The chemical reactions of a cell often produce acids and bases, yet a shift in pH within a cell can be harmful. *Buffers* within the cell serve to minimize the changes in concentrations of hydrogen and hydroxide ions, and thus maintain a constant pH. A buffer is a substance that accepts H^+ ions when they are in excess and donates H^+ ions to a solution when their concentration decreases. The carbonic acid/bicarbonate buffering system is an important biological buffer. Carbonic acid acts as an acid to donate H^+ ions when the pH tends to increase. When the pH falls, the excess H^+ ions are accepted by the bicarbonate ion, which acts like a base. An acid and base in equilibrium with each other is typical of most buffering systems.

The environmental problem of *acid rain* emphasizes the sensitivity of life to pH. Acid rain, with a pH value lower than the normal pH value (5.6) of rain, is due to the reaction of water in the atmosphere with the sulfur oxides and nitrogen oxides released by the combustion of fossil fuels.

Acid rain has harmful effects on both terrestrial and freshwater ecosystems. Lowering the pH of the soil solution affects the solubility of minerals needed by plants. A lowered pH of lakes and ponds affects many species of fishes, amphibians, and aquatic invertebrates. Nearly one-half of the lakes in the higher elevations of the western Adirondacks of New York are now more acidic than pH 5.0, with the result that fishes have disappeared from almost all of them. The reduction of acid rain depends on the development of industrial controls and antipollution devices.

STRUCTURE YOUR KNOWLEDGE

There are two major clusters of concepts that you need to be able to relate in order to have a good understanding of water and its contribution to the fitness of the environment. Suggested concept maps are included at the end of this unit, but remember that your concept map should represent your own understanding. The value of this exercise is your process of organizing these concepts for yourself .

1. The tendency of water to form hydrogen bonds between its polar molecules has a profound effect on temperature regulation within the environment. Select the key concepts that relate to this important property of water and create a concept map showing how the breaking and formation of hydrogen bonds are related to temperature.

2. To become proficient in the use of the concepts relating to pH, develop a concept map to organize your understanding of the following terms: pH, [H⁺], [OH⁻], acidic, basic, neutral, buffer, 1-14, acid-base pair. Remember to label connecting lines and add additional concepts as you need them.

TEST YOUR KNOWLEDGE

MULTIPLE CHOICE: *Choose the one best answer.*

1. Water contributes to the fitness of the environment because
   a. plants need water to grow.
   b. life evolved in water.
   c. the making and breaking of H bonds helps regulate temperature.
   d. the surface tension of water creates a niche for water organisms.

2. The polarity of the water molecule
   a. promotes the making of hydrogen bonds.
   b. helps water to dissolve nonpolar solutes.
   c. lowers the heat of vaporization.
   d. makes water most dense at 4° C.

3. Capillary action is the result of
   a. adhesion of water molecules to hydrophobic substances.
   b. hydrogen bonds creating cohesion and adhesion.
   c. imbibition.
   d. the pull of gravity on a column of water.

4. A low specific heat would mean that
   a. little heat must be absorbed or released to effect a temperature change.
   b. breaking hydrogen bonds releases a small amount of heat.
   c. breaking hydrogen bonds absorbs a small amount of heat.
   d. boiling temperature would probably be high.

5. Temperature is a measure of
   a. specific heat.
   b. average kinetic energy of molecules.
   c. total kinetic energy of molecules.
   d. Celsius degrees.

6. Evaporative cooling is a result of
   a. a low heat of vaporization.
   b. a high heat of melting.
   c. a reduction in the average kinetic energy of the liquid remaining after molecules enter the gaseous state.
   d. sweating on a humid day.

7. Ice floats because
    a. air is trapped in the crystalline lattice.
    b. the formation of hydrogen bonds releases heat; warmer objects float.
    c. it has a larger surface area than liquid water.
    d. the maximum number of H bonds are formed, spacing the molecules farther apart and creating a less dense structure than water has.

8. The molarity of a solution is equal to
    a. Avogadro's number of molecules in 1 liter of solvent.
    b. number of moles of a solute in 1 liter of solution.
    c. molecular weight of solute in 1 liter of solution.
    d. number of solute particles in 1 liter of solvent.

9. A solution with pH 2, compared to a solution with pH 4,
    a. is one-half as acidic.
    b. is 100 times more acidic.
    c. is 1000 times more acidic.
    d. has two times more $[OH^-]$.

10. A buffer
    a. maintains pH at 7.
    b. absorbs excess H^+.
    c. releases H^+.
    d. is often a weak acid-base pair.

11. Which of the following is least soluble in water?
    a. polar compounds
    b. nonpolar compounds
    c. ionic compounds
    d. hydrophillic molecules

12. What factor accounts for the movement of water up xylem vessels in a plant?
    a. capillary action
    b. hydrogen bonding
    c. adhesion
    d. all of the above

13. What bonds must be broken for water to vaporize?
    a. polar covalent bonds
    b. nonpolar covalent bonds
    c. hydrogen bonds
    d. all of the above

14. How would you make a 0.1 M solution of acetic acid ($C_2H_4O_2$)? The mass numbers for these elements are C - 12, O - 16, H - 1.
    a. Mix 60 g acetic acid with enough water to yield 1 liter of solution.
    b. Mix 2 g carbon, 4 g hydrogen, and 2 g oxygen in 1 liter of water.
    c. Mix 6 g acetic acid with enough water to make 1 liter of solution.
    d. Mix 0.1 mol acetic acid with 1 mol water.

15. How many molecules of acetic acid would be in the solution in question 14?
    a. 0.1
    b. 6
    c. 0.6×10^{23}
    d. 60

TRUE OR FALSE: *Mark T or F; then change the false statements so that they are true.*

1. _____ Heat is a measure of the total kinetic energy of the molecules within a substance or body.

2. _____ The high surface tension of water is a result of adhesion.

3. _____ The formation of ice in winter slows the transition to cold weather because the formation of hydrogen bonds absorbs cold.

4. _____ Nonpolar substances dissolve in water because the water molecules form a hydration shell around them.

5. _____ Kinetic energy is the energy of motion.

6. _____ A calorie is the amount of heat it takes to raise the temperature of 1 gram of a substance by $1°$ C.

7. _____ The high heat of vaporization of water is a result of the heat needed to break hydrogen bonds.

8. _____ A base accepts H^+ ions or donates OH^- ions to a solution.

9. _____ The breaking down of the crystalline structure when ice melts releases heat to the environment.

10. _____ The dissociation into H^+ and OH^- ions makes water a good solvent for biological materials.

FILL IN THE BLANKS: *Complete the following table on pH.*

$[H^+]$	$[OH^-]$	pH	Acidic, basic, or neutral?
	10^{-11}	3	acidic
10^{-8}		8	
10^{-12}			
	10^{-5}		
		1	
	10^{-7}		

CARBON AND MOLECULAR DIVERSITY

FRAMEWORK

This chapter presents the basics of *organic chemistry*, the study of carbon-containing compounds. Carbon can form *four covalent bonds* and bind with itself to make straight or branching chains and rings. These *carbon skeletons* serve as the basis for complex organic molecules. The diversity of molecules is increased due to the existence of *isomers*—molecules with the same molecular formula but different architecture, and thus different properties. An array of *functional groups* may be bound to carbon skeletons, conferring specific properties to these organic compounds.

CHAPTER SUMMARY

The ability of carbon to form large and complex molecules is central to the existence of life. The critical molecules associated with living matter—proteins, DNA, sugars, and fats—all have carbon atoms, bound to each other and other atoms, at the foundation of their structures. The incredible diversity and functional capacity of carbon-based molecules can be shown to emerge from the structure and interactions of these molecules.

Foundations of Organic Chemistry

Organic chemistry specializes in the study of carbon-containing organic molecules. *Vitalism* was the underlying philosophy of early organic chemistry. Because chemists could not synthesize the complex organic molecules found in living organisms, they attributed the existence of life and the formation of the molecules associated with living things to a life force outside the jurisdiction of physical and chemical laws. As chemists

learned to synthesize organic compounds from inorganic substances in the 1800s, the foundation of vitalism was questioned. *Mechanism* is the belief that physical and chemical laws and explanations are sufficient to account for all natural phenomena, even the processes of life. This philosophy replaced vitalism as the basis for biological inquiry.

Versatility of Carbon in Molecular Architecture

Carbon has six electrons, with four in its valence shell. To complete its valence shell, carbon forms *four covalent bonds* with other molecules. This covalent bonding capacity is at the center of carbon's ability to form large and complex molecules with characteristic three-dimensional shapes and properties. When carbon forms covalent bonds, its four electron orbitals hybridize into four teardrop-shaped orbitals angling out into the shape of a *tetrahedron*. The spatial arrangement of atoms can determine the function of a molecule in a cell.

The four covalent bonds of carbon can include single, double, or even triple bonds. Carbon atoms readily bond with each other, producing chains or rings of carbon atoms. These molecular backbones can vary in length, branching, placement of double bonds, and location of atoms of other elements. The simplest organic molecules are *hydrocarbons*, consisting of only carbon and hydrogen.

Isomers are molecules with the same molecular formula but different structural arrangements and thus, different properties. *Structural isomers* differ in the arrangement of atoms and often in the location of double bonds. Ethyl alcohol and dimethyl ether (Figure 4.1a) have the same number and kinds of atoms but a different bonding sequence and very different properties.

Geometric isomers have the same sequence of covalent bonds but differ in their spatial arrangement due to the inflexibility of double bonds. There is free rotation of atoms about a single covalent bond. Double bonds restrict that rotation and thus fix the geometry of a molecule to a specific spatial arrangement. Maleic acid and fumaric acid (Figure 4.1b) are geometric isomers. The difference in spatial arrangements of geometric isomers can dramatically affect the biological properties of organic molecules.

Enantiomers are isomers that are mirror images of each other. An *asymmetric* carbon is one that is covalently bound to four different atoms or groups of atoms. Due to the tetrahedral shape of the asymmetric carbon, the four groups can be attached in spatial arrangements that are not superimposable on each other. Enantiomers are left- and right-handed versions of each other, and they differ greatly in their biological activity. Although *l*- and *d*-lactic acid look similar in a flat representation of their structures, they are not superimposable due to the tetrahedral arrangement of carbon bonds (Figure 4.1c). Cells can differentiate between left- and right-handed versions of a molecule; usually, only one form is biologically active.

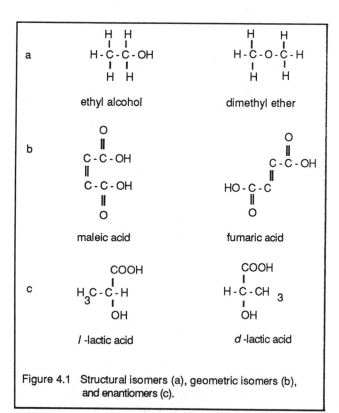

Figure 4.1   Structural isomers (a), geometric isomers (b), and enantiomers (c).

Functional Groups

The properties of organic molecules depend not only on the arrangement of the carbon skeleton, but also on specific groups of atoms, known as functional groups, bonded to the carbon backbone.

These *functional groups* are usually involved in the chemical reactions within a cell and behave consistently from one molecule to another.

Molecules can be separated from each other in a mixture using the technique of *chromatography*. The relative solubility in a mobile solvent compared to the attraction or binding to a stationary phase (paper, gel on glass plate, column of beads or resin) affects the movement of molecules through a chromatographic system. The polarity or ionic state of the molecule is usually the basis for differential solubility and binding.

There are six important functional groups in the chemistry of life. The *hydroxyl group* consists of an oxygen atom and a hydrogen atom (-OH) covalently bonded to the carbon skeleton. The hydroxyl group is polar, due to the electronegative oxygen drawing the hydrogen electron closer to itself. This polarity makes the molecule soluble in water. Organic molecules with hydroxyl groups are called alcohols, and their specific names often end in *ol*.

Carbonyl groups consist of a carbon double bonded to an oxygen (-C=O). If the carbonyl group is at the end of the carbon skeleton, the compound is called an aldehyde. If it is not at the end, the compound is called a ketone. This group is very polar and reactive. Sugars have both carbonyl and hydroxyl groups.

A *carboxyl group* consists of a carbon double bonded to an oxygen and also attached to a hydroxyl group (-COOH). Compounds with a carboxyl group are called carboxylic acids or organic acids and tend to dissociate to release an H^+ and act as weak acids.

An *amino group* consists of a nitrogen atom bonded to two hydrogens ($-NH_2$). Compounds with an amino group, called amines, can act as bases. The nitrogen, with its pair of unshared electrons, can attract a hydrogen ion, becoming $-NH_3^+$. Amino acids, the building blocks of proteins, contain both an amino and a carboxyl group (hence the name *amino acid*). Amino acids can exist as *zwitterions*, molecules with cationic and anionic groups. Such amino acids can act as buffers, losing H^+ or binding H^+ as the pH value in the cell varies.

The *sulfhydryl group* consists of a sulfur atom bound to a hydrogen atom (-SH). Compounds with sulfhydryl groups are called thiols. The cross-linkings of sulfhydryl groups help to stabilize protein structure.

A *phosphate group* binds to the carbon skeleton by an oxygen atom attached to a phosphorus atom bound to three other oxygen atoms ($-OPO_3H_2$). One of these oxygens is double bonded to the phosphorus. A hydrogen ion can dissociate from each of the other two oxygens, producing a phosphate anion. Organic compounds with phosphate groups can store energy and pass it to another molecule by the transfer of a phosphate group.

Elements of Life: A Review

Carbon, oxygen, hydrogen, nitrogen, and smaller quantities of sulfur and phosphorus, all capable of forming strong covalent bonds, are combined into the architecture of the complex organic molecules of living matter. The versatility of carbon in forming four covalent bonds, linking readily with itself to produce chains and rings, and binding with other elements and functional groups, makes possible the incredible diversity of organic molecules.

STRUCTURE YOUR KNOWLEDGE

1. One characteristic of organic compounds is that they can exist in the form of isomers— molecules with the same molecular formula but different spatial arrangements of atoms and different properties. Construct a concept map that illustrates your understanding of isomers, their significance, and the characteristics of the three types of isomers.

2. Fill in the following table on the functional groups.

TEST YOUR KNOWLEDGE

MULTIPLE CHOICE: *Choose the one best answer.*

1. A zwitterion is
   a. an enantiomer.
   b. a high-energy phosphate ion.
   c. a molecule with both carbonyl and hydroxyl groups.
   d. a molecule with both cationic and anionic groups.

2. A hydrocarbon and an alcohol molecule could be separated by chromatography because
   a. a nonpolar solvent would carry the hydrocarbon farther in the system.
   b. a nonpolar solvent would carry the alcohol farther in the system.
   c. they could be separated on the basis of color.
   d. the alcohol is a heavier molecule.

3. Which functional group forms cross-links that stabilize protein structure?
   a. amino   c. phosphate
   b. carboxyl   d. sulfhydryl

4. Which of the following is not true of an asymmetric carbon atom?
   a. It is attached to four different atoms or groups.
   b. It creates geometric isomers.
   c. It creates enantiomers.
   d. Its bonds are in the shape of a tetrahedron.

5. A reductionist approach to considering the structure and function of organic molecules would be based on
   a. mechanism.   c. determinism.
   b. holism.   d. vitalism.

NAME OF GROUP	MOLECULAR FORMULA	CHARACTERISTICS CONFERRED TO ORGANIC COMPOUND
	- OH	
		Aldehyde or ketone, polar and reactive group
Carboxyl		
	- NH_2	
		Forms crosslinks in 3-D protein conformation
Phosphate group		

6. The functional group that confers basic
   properties on organic molecules is
   a. -COOH.
   b. -OH.
   c. -SH.
   d. -NH$_2$.

7. The functional group that confers acidic
   properties on organic molecules is
   a. -COOH.
   b. -OH.
   c. -SH.
   d. -NH$_2$.

8. Which is not true about geometric isomers?
   a. They have different chemical properties.
   b. They have the same molecular formula.
   c. Their atoms and bonds are arranged in
       different sequences.
   d. The rigidity of the double carbon bond
       restricts movement.

9. The tetrahedral shape of the four covalent bonds
   of a carbon atom is due to
   a. the hybridization of its *sp* ^{3}orbitals.
   b. the difference in electronegativity
       of the bonds.
   c. the asymmetric nature of the carbon atom.
   d. geometric isomerism.

10. Which of the following is not true of
    amino acids?
    a. They can act as buffers.
    b. They are the building blocks of
        nucleic acids.
    c. They can be zwitterions.
    d. They are named for their amine and
        carboxylic
        functional groups.

11. How many asymmetric carbons are there
    in galactose?
    a. 2
    b. 4
    c. 5
    d. 6

12. Which functional groups can act as acids?
    a. amine and sulfhydryl
    b. carbonyl and carboxyl
    c. carboxyl and phosphate
    d. alcohol and aldehyde

MATCHING: *Match the formulas to the correct
descriptions. Choices may be used more than once; more
than one right choice may be available.*

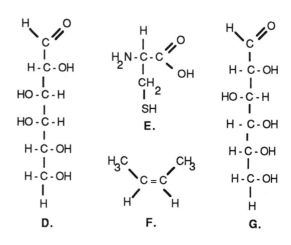

1.____&____ structural isomers

2.____&____ geometric isomers

3.____&____ enantiomers

4._____ carboxylic acid

5._____ thiol

6._____ alcohol

7._____ aldehyde

8._____ amino acid

9._____ organic phosphate

10._____ hydrocarbon

11._____ amine

12._____ ketone

CHAPTER 5

STRUCTURE AND FUNCTION OF MACROMOLECULES

FRAMEWORK

CLASS	MONOMERS	LINKAGE	FUNCTIONS
Carbohydrates	Monosaccharides	Glycosidic	Energy, raw materials, energy storage, structural compounds
Lipids	Glycerol, fatty acids, phosphates, carbon rings	Ester	Energy storage, membranes, hormones
Proteins	Amino acids	Peptide	Enzymes, structural compounds, movement, transport, hormones
Nucleic Acids	Nucleotides (pentose, nitrogenous bases, phosphate group)	Phosphodiester	Information coding, heredity

This chapter introduces the major classes of macromolecules that constitute living organisms. Two ideas are central to the chapter: the great diversity of macromolecules, created by combining a small number of monomers or subunits into unique sequences and three-dimensional structures, forms the basis for the diversity of life; and the variety of functions performed by these macromolecules is directly related to their specific structure. The chart above briefly summarizes the major characteristics of the four classes of macromolecules.

CHAPTER SUMMARY

The four classes of large molecules found in living matter are carbohydrates, lipids, proteins, and nucleic acids. These giant molecules, called *macromolecules*, represent another level in the hierarchy of biological organization. The functions of these molecules are related to their complex and unique architecture.

Macromolecules Are Polymers

Polymers are large molecules formed from the linking together of many similar or identical small molecules, called *monomers*. Macromolecules are constructed from about 40 to 50 common *monomers* and a few rarer molecules. The limitless variety of polymers arises from the almost infinite number of possibilities in the sequencing and arrangements of these basic subunits. The molecular basis for the diversity of life rests on the ordering of common small molecules into distinctive and unique macromolecules.

Macromolecules of all four classes are formed by the same chemical process. Monomers are joined by *dehydration synthesis*, the removal of a molecule of water in a condensation reaction. One monomer loses a hydroxyl (-OH) and the other contributes a hydrogen (-H) to the water molecule, and a covalent bond between the monomers is formed. Energy is required to join monomers and the process is facilitated by enzymes.

Hydrolysis is the breaking of bonds between monomers

through the addition of water molecules. A hydroxyl is joined to one monomer while a hydrogen is bonded with the other. The disassembling of polymers is necessary in the digestion of food so that small molecules can be absorbed and transported through the blood stream. Enzymes also control hydrolysis.

Carbohydrates

Carbohydrates include sugars and their derivatives. *Monosaccharides* are simple sugars; polysaccharides are polymers of many sugars joined together by dehydration synthesis. Monosaccharides have the general formula of $(CH_2O)_n$. The number of these units forming a sugar varies from three to seven, with hexoses $(C_6H_{12}O_6)$, trioses $(C_3H_6O_3)$, and pentoses $(C_5H_{10}O_5)$ found most commonly. Glucose, a hexose, is a molecule of key importance in the energy transformations central to life. It is formed by green plants in photosynthesis and broken down to yield energy in respiration in almost all living organisms.

Sugars are aldehydes or ketones, depending on which of their carbon atoms is bound to the carbonyl group (-O). Because the other carbon atoms characteristically all bear hydroxyl groups (-OH), sugars are alcohols as well. Additional diversity among sugar molecules is provided by the spatial arrangement of parts around asymmetric carbons. For each asymmetric carbon in a molecule, there are two mirror-image isomeric forms (enantiomers). The hexoses glucose and galactose, for instance, differ only in the arrangement around one asymmetric carbon. These small structural differences affect the recognition and interaction of molecules within cells.

In aqueous solutions, most monosaccharides are in a ring structure. In glucose, an oxygen forms the bridge between carbons number one and five. The carbon skeleton is numbered from the end closest to the carbonyl group in the open chain form.

Monosaccharides may have additional functional groups attached, as in acid sugars, amino sugars, and sugar phosphates. Sugars serve as the main fuel for cellular work, the raw materials for synthesis of other small organic molecules such as amino acids and fatty acids, and monomers that are synthesized into polysaccharides that serve as storage or structural molecules.

Glycosidic linkages are the bonds formed by dehydration synthesis between two monosaccharides. They are named by the carbons they join. Thus maltose, a *disaccharide*, has a 1 4 linkage between the number one carbon of one glucose and the number four carbon of another glucose. Diversity in disaccharides is increased by the location of the glycosidic linkage and also by the monomers used. Sucrose, a common disaccharide, is formed from the joining of a glucose and fructose molecule.

Polysaccharides are macromolecules made from a few hundred to a few thousand monosaccharides. *Starch*, a storage molecule in plants, is a polymer made of glucose molecules joined by 1 4 linkages. Starch has a helical shape resulting from the angle of these bonds. Amylose is an unbranched form of starch; amylopectin, a more complex starch, adds 1 6 glycosidic linkages to form a branching polymer. Plants store sugar for later use in starch molecules. Most animals have enzymes to hydrolyze plant starch into glucose, providing nutrients for the cell. Animals produce glycogen, a highly branched polymer of glucose, as their food storage form.

Cellulose, the major component of plant cell walls, is the most abundant organic compound on earth. It differs from starch by the configuration, called alpha (α) or beta (ß), of the ring form of glucose. In cellulose, the glucose monomers are in the ß configuration, and the glycosidic bonds are called ß 1 4 linkages. In starch, there are α 1 4 linkages between the α glucose molecules. The geometry of the glycosidic bonds is responsible for the different three-dimensional shapes and properties of these plant polysaccharides. Enzymes that digest starch are unable to hydrolyze the ß linkages of cellulose. Only a few organisms (some bacteria, microorganisms, and fungi) have enzymes that can digest cellulose.

In a plant cell wall, a thousand or more parallel cellulose molecules are held together by hydrogen bonds between hydroxyl groups to form a microfibril. Several microfibrils are in turn intertwined to form a cellulose fibril, which may supercoil with other fibrils to form strong structural cables.

Chitin is a structural polysaccharide formed from modified amino sugar monomers and found in the exoskeleton of arthropods and the cell walls of many fungi.

Lipids

Fats, phospholipids, and steroids are a diverse assemblage of macromolecules, classed together as lipids because they are all insoluble in water. Fats are composed of fatty acids attached to the three-carbon alcohol, glycerol. A fatty acid consists of a long carbon skeleton (often 16 or 18 carbons in length) with a carboxyl group at the "head" end. The long hydrocarbon tail is nonpolar and thus hydrophobic. This hydrophobic nature causes fats to be insoluble in water.

Fatty acids are linked to glycerol in a dehydration synthesis in which a hydroxyl from glycerol combines with a hydrogen of the fatty acid's carboxyl group, leaving the glycerol and fatty acid bound with an *ester linkage*. Diglycerides are formed when two fatty acids are attached to a glycerol. In a triglyceride, or fat, three fatty acids attach by dehydration synthesis to glycerol.

Fatty acids with double bonds in their carbon skeletons are called *unsaturated*. The double bonds create a kink in the shape of the molecule and prevent the fat molecules from packing close together and becoming solidified at room temperature. *Saturated* fats have fatty acids with no double bonds in their carbon skeletons; each carbon atom is "saturated" with hydrogen. Most animal fats are saturated and solid at room temperature.

Fats are excellent energy storage molecules. A gram of fat stores more than twice as much energy as a gram of starch. Adipose tissue, made of fat storage cells, also cushions organs and insulates the body.

Phospholipids consist of a glycerol linked to two fatty acids and a negatively charged phosphate group. Other small molecules may be attached to the phosphate group. The phosphate head of this molecule is hydrophilic and water soluble, whereas the two fatty acid chains produce a nonpolar and hydrophobic tail. This unique structure of phospholipids makes them ideal constituents of cell membranes. Arranged in a bilayer, the hydrophilic heads face toward the aqueous solutions inside and outside the cell, and the hydrophobic tails interact in the center to hold the molecules of the membrane together.

Steroids are a class of lipids distinguished by four fused carbon rings with various functional groups attached. *Cholesterol* is an important steroid that is a common component of animal cell membranes and a precursor for most other steroids, including many hormones.

Proteins

Proteins are crucial macromolecules, serving such diverse functions as structural support (fibers), storage (of amino acids and energy), transport (hemoglobin), internal coordination (hormones), movement (actin and myosin in muscle), defense (antibodies), and, perhaps most important, control of metabolism (enzymes). This wide range of functions is made possible by the highly structured and unique three-dimensional shapes of these macromolecules. Twenty *amino acids* constitute the monomers from which the huge variety of protein molecules is composed.

Most amino acids are composed of an asymmetric carbon (called the α carbon) bonded to a carboxyl group, an amino group, a hydrogen, and a variable *side chain* called the "R" group. The R group confers the unique physical and chemical properties of each amino acid. Side chains may be nonpolar and hydrophobic; polar or charged and thus hydrophilic; acidic or basic.

With their asymmetric carbon, amino acids can exist in two isomeric forms (enantiomers) considered to be left- or right-handed. Usually only the *l*-form of an amino acid is used by cells to make proteins.

A dehydration synthesis joins the carboxyl group of one amino acid with the amino group of another in a linkage known as a *peptide bond*. A polymer of amino acids is called a *polypeptide chain*, with a free carboxyl group at one end (the C-terminus) and a free amino group at the other (the N-terminus). The other amino and carboxyl groups are linked into a polypeptide backbone held together by peptide bonds. The side chains of each amino acid extend out from this chain and participate in interactions that help create the three-dimensional structure of proteins. Polypeptides may vary in length from a few monomers to a thousand or more amino acids. Each specific polypeptide has its own unique sequence of amino acids.

Proteins have unique three-dimensional shapes, or *conformations*, created by the twisting or folding of one or more polypeptide chains. Structural proteins tend to be fibrous, whereas enzymes are more spherical or globular. The unique conformation of a protein enables it to recognize and bind specifically to another molecule.

There are three superimposed structural levels of architecture in the conformation of a protein. A fourth level may be present when a protein consists of more than one polypeptide chain. *Primary structure* is the unique sequence of amino acids within a protein, coded for by genetic information. Even a slight deviation from the sequence of amino acids can severely affect a protein's function by altering the protein's conformation.

In the early 1950s, Sanger determined the primary structure of insulin through the laborious process of hydrolyzing the protein into small peptide chains, using *chromatography* to separate the small pieces, determining their sequences of amino acids, and then overlapping the sequences of the small fragments to reconstruct the whole polypeptide. Most of these steps are now automated, and the primary structures of hundreds of proteins have been determined.

Secondary structure involves the twisting or folding of the polypeptide backbone, stabilized by hydrogen bonds between peptide linkages. An *alpha helix* is a delicate coil produced by hydrogen bonding between every fourth peptide bond. Linus Pauling and Robert Corey first described the alpha helix in 1951. Some fibrous proteins have alpha helices along their entire length. Globular proteins are more likely to have regions of alpha helix alternating with nonhelical regions where the side chains are too bulky to fit into the coil.

A *beta sheet* is a secondary structure in which the polypeptide chain folds back and forth. Hydrogen bonds between oppositely running polypeptide segments create this zigzag structure. Beta sheets are found in the dense core of many globular proteins and may dominate some fibrous proteins.

Interactions between the various side chains of the

constituent amino acids produce *tertiary structure.* Hydrophobic interactions between nonpolar side groups in the center of the molecule, hydrogen bonds, and ionic bonds between negatively and positively charged side chains produce a stable and unique shape to the protein. Strong covalent bonds, called *disulfide bridges,* may occur between the sulfhydryl side groups of cysteine monomers that have been brought close together by the folding of the polypeptide.

Quaternary structure occurs in proteins that are composed of more than one polypeptide chain. The individual polypeptide chains, called subunits, are held together in a precise structural arrangement. Collagen is a fibrous protein composed of three helical *subunits* supercoiled together. Hemoglobin is a globular protein consisting of four polypeptide chains and an iron-containing heme group held in the center of the molecule.

The specific function of a protein is an emergent property developing from its intricate three-dimensional architecture. This conformation arises spontaneously and is dependent on the interactions among the amino acids making up the polypeptide chain. These interactions can be disrupted by changes in pH value, salt concentration, temperature, or other aspects of the environment, and the protein may *denature,* losing its native conformation and thus its function. Any factor that disrupts the hydrogen bonding, hydrophobic interactions, ionic bonds or disulfide bridges within a protein molecule will cause denaturation. For example, excess heat may cause enough molecular agitation to disrupt these bonds. Some denaturation is reversible; if the protein does not precipitate out of solution, it can return to its three-dimensional conformation when returned to its normal environment. The shape of a protein ultimately emerges from its sequence of amino acids, its primary structure.

Nucleic Acids

Nucleic acids are macromolecules with the unique ability to reproduce themselves and to carry the code that directs all the cell's activities. *DNA, deoxyribonucleic acid,* is the genetic material that is inherited from one generation to the next and is reproduced in each cell of an organism. *RNA, ribonucleic acid,* reads the instructions coded in DNA and directs the synthesis of proteins, the ultimate enactors of the genetic program.

Nucleic acids are polymers of *nucleotides,* linked together by dehydration synthesis. Each nucleotide consists of a pentose (five-carbon) sugar covalently bonded to a phosphate group and to one of five nitrogenous bases. There are two families of nitrogenous bases. *Pyrimidines,* including cytosine (C),

thymine (T), and uracil (U), are characterized by six-membered rings of carbon and nitrogen atoms. Thymine is found only in DNA; uracil is found only in RNA. *Purines,* adenine (A) and guanine (G), add a five-membered ring to the pyrimidine ring. The pentose sugar is either *ribose,* in RNA, or *deoxyribose* (missing one oxygen atom), in DNA. *Nucleosides* consist of a nitrogenous base joined to a sugar. *Nucleotides* have a phosphate group attached to the number five carbon of the sugar.

Nucleotides can function as monomer subunits in nucleic acids, or as individual molecules. Adenosine triphosphate, ATP, is a nucleotide involved in energy transfer in cellular processes.

Nucleotides are linked together into polynucleotide chains by phosphodiester linkages, which join the phosphate of one nucleotide with the sugar of the next by a covalent bond. The nitrogenous bases extend from this repeating sugar-phosphate backbone.

DNA molecules consist of two such polynucleotide chains spiraling around an imaginary axis in a double helix. Watson and Crick, in 1953, first proposed the *double helix* arrangement, which consists of two sugar-phosphate backbones on the outside of the helix and their nitrogenous bases pairing and hydrogen bonding together in the inside of the helix. Adenine pairs with only thymine; guanine always pairs with cytosine. Thus, the sequences of nitrogenous bases on the two strands of DNA are complementary, predictable counterparts of each other. DNA can replicate itself, providing for the precise copying of genes for inheritance, because of this specific base-pairing property.

STRUCTURE YOUR KNOWLEDGE

1. For each of the four families of macromolecules—carbohydrates, lipids, proteins, and nucleic acids—create a concept map (or a brief table, chart, or summary, if you prefer) that emphasizes the structure and functions of that group. Include the monomers involved, the type of linkage, and the unique characteristics of the structure of the polymers that relate to their function. A few examples in each class probably would be helpful. First list the key characteristics that you can remember for each group and think about how they relate. Then go back to the Chapter Summary or textbook to check that you have the major points for each group of macromolecules. Finally, organize this information so that you can relate structure to function.

TEST YOUR KNOWLEDGE

MATCHING: *Match the molecule with its class of macromolecules.*

1. _____ glycogen   **A.** carbohydrate
2. _____ cholesterol   **B.** lipid
3. _____ ATP   **C.** protein
4. _____ collagen   **D.** nucleic acid
5. _____ hemoglobin
6. _____ a gene
7. _____ triglyceride
8. _____ enzyme
9. _____ cellulose
10. _____ chitin

MATCHING OF FORMULAS: *Match the chemical formulas with their description. Answers may be used more than once.*

1. _____ molecules that would combine to form a fat
2. _____ molecule that would be attached to other monomers by a peptide bond
3. _____ molecules or groups that would combine to form a DNA nucleotide
4. _____ molecules that are carbohydrates
5. _____ molecule that is a purine
6. _____ monomer of a protein
7. _____ phosphodiester bonds are formed between these two groups

MULTIPLE CHOICE: *Choose the one best answer.*

1. Dehydration synthesis is a process that
   a. creates bonds between amino acids in the formation of a peptide chain.
   b. involves the removal of a water molecule.
   c. is a condensation reaction.
   d. involves all of the above.

2. Which of the following is not true of pentoses?
   a. They are found in nucleic acids.
   b. They can occur in a ring structure.
   c. They have the formula $C_5H_{12}O_5$.
   d. They have hydroxyl and carbonyl groups.

3. Disaccharides can differ from each other in all of the following ways except
   a. in the number of their monosaccharides.
   b. in the existence of enantiomers.
   c. in the type of monomer involved.
   d. in the location of their glycosidic linkage.

4. Which of the following is not true of cellulose?
   a. It is the most abundant organic compound on earth.
   b. It differs from starch because of the ß configuration of glucose.
   c. It is a highly branched, strong structural component of cell walls.
   d. Few organisms have enzymes that hydrolyze its glycosidic linkages.

5. Plants store most of their energy as
   a. glucose.
   b. glycogen.
   c. starch.
   d. sucrose.

6. When a protein denatures, it
   a. loses its primary structure.
   b. loses its secondary and tertiary structure.
   c. becomes insoluble and precipitates.
   d. hydrolyzes into component amino acids.

7. The alpha helix of proteins is
   a. part of the tertiary structure and is stabilized by disulfide bridges.
   b. a double helix.
   c. stabilized by hydrogen bonds and commonly found in fibrous proteins.
   d. found in some regions of globular proteins and stabilized by hydrophobic interactions.

8. A fatty acid that has the formula $C_{16}H_{32}O_2$ is
   a. saturated.
   b. unsaturated.
   c. polyunsaturated.
   d. liquid at room temperature.

9. Three molecules of the fatty acid in question 8 are joined to a molecule of glycerol ($C_3H_8O_3$). The resulting molecule has the formula
   a. $C_{48}H_{96}O_6$.
   b. $C_{51}H_{104}O_9$.
   c. $C_{51}H_{102}O_8$.
   d. $C_{51}H_{98}O_6$.

10. The molecule formed in question 9 is
    a. a triglyceride.
    b. a lipid.
    c. a fat.
    d. all of the above.

11. Which of the following are hydrophobic?
    a. steroids
    b. chitin
    c. head ends of phospholipids
    d. polynucleotides

12. Beta sheets are characterized by
    a. disulfide bridges between cysteine amino acids.
    b. back-and-forth folds of the polypeptide chain held together by hydrophobic interactions.
    c. folds stabilized by hydrogen bonds between segments of polypeptide chains running in opposite directions.
    d. membrane sheets composed of phospholipids.

FILL IN THE BLANKS

1. The nitrogenous base absent in RNA is _____ .
2. Cytosine always pairs with _____ .
3. Adenine and guanine are _____ .
4. A nitrogenous base joined to a pentose sugar is called a _____ .
5. When a phosphate groups is added to question 4, it is called a _____ .
6. _____ structure is found in proteins with more than one peptide chain.
7. The conformation of a protein is determined by its _____ .
8. The energy storage molecule of animals is _____ .
9. Membranes are composed of a bilayer of _____ .
10. The linkages between amino acids in a protein are called _____ .

CHAPTER 6

INTRODUCTION TO METABOLISM

FRAMEWORK

This chapter considers metabolism, the totality of the chemical reactions that take place in living organisms. Two key topics of metabolism are considered: the energy transformations that underlie all chemical reactions and the role of enzymes in the "cold chemistry" of the cell.

The reactions in a cell either increase or release free energy. The following illustration summarizes some of the components of free energy changes within a cell.

Enzymes are biological catalysts that lower the activation energy of a reaction and thus greatly speed up and control metabolic processes. An enzyme is a three-dimensional protein molecule with an active site specific for its substrate. Intricate control and feedback mechanisms produce the metabolic integration necessary for life.

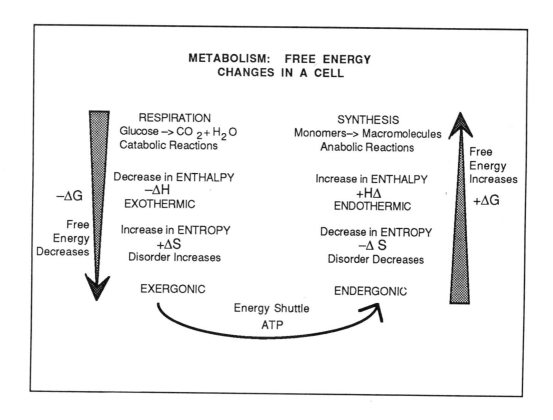

METABOLISM: FREE ENERGY CHANGES IN A CELL

RESPIRATION
Glucose $\rightarrow$ CO_2 + H_2O
Catabolic Reactions

SYNTHESIS
Monomers $\rightarrow$ Macromolecules
Anabolic Reactions

$-\Delta G$

Decrease in ENTHALPY
$-\Delta H$
EXOTHERMIC

Increase in ENTHALPY
$+H\Delta$
ENDOTHERMIC

Free Energy Increases
$+\Delta G$

Free Energy Decreases

Increase in ENTROPY
$+\Delta S$
Disorder Increases

Decrease in ENTROPY
$-\Delta S$
Disorder Decreases

EXERGONIC

Energy Shuttle
ATP

ENDERGONIC

CHAPTER SUMMARY

The Metabolic Map

Metabolism is the totality of an organism's chemical processes, involving the thousands of precisely coordinated, complex, efficient, and integrated chemical reactions in a cell. These reactions are ordered into metabolic pathways—sequenced and intricately branched routes controlled by enzymes. Through these pathways the cell creates and transforms the organic molecules that provide the material and energy for life. Metabolism is an emergent property arising from the organization and orderly interactions of cellular molecules.

Catabolic pathways release the energy stored in complex molecules through the breaking down or degradation of these molecules into simpler compounds. *Anabolic pathways* require energy to combine simpler molecules into complicated ones. This energy is often supplied through the coupling of catabolic and anabolic pathways in a cell. Energy is the basis of all metabolic processes. Thus, an understanding of energy transformations is essential to the study of metabolism.

Energy: Some Basic Principles

Energy has been defined as the capacity to do work, to move matter against an opposing force. *Kinetic energy* is the energy of motion, of matter that is moving. This matter does its work by transferring its motion to other matter. Heat and light are forms of kinetic energy due to the motion of molecules and photons. *Potential energy* is the capacity of matter to do work as a consequence of its location or arrangement. *Chemical energy* is a form of potential energy stored in the bonds holding atoms together in molecules.

Energy can be converted from one form to another: kinetic energy into potential energy and vice versa, chemical energy into heat, light into chemical energy. The study of the principles that govern the transformation of energy is called *thermodynamics*.

The *First Law of Thermodynamics* states that energy can be neither created nor destroyed. Energy can be transferred between matter and transformed from one kind to another, but the total energy of the universe is constant. According to this principle of the conservation of energy, chemical reactions that either require or produce energy are merely transforming a set amount of energy into a different form.

The *Second Law of Thermodynamics* states that every energy transformation or transfer results in an increasing disorder within the universe. Entropy is the term used as a quantitative measure of disorder or randomness. Thus, every process results in an increase in the entropy in the universe.

A system, such as a cell, may become more ordered, but it does so with an attendant increase in the entropy of its surroundings. A cell can use highly ordered organic molecules as a source of the energy needed to create its own highly ordered structure, but it returns heat and the simple molecules of carbon dioxide and water to the environment. In any energy transformation or transfer, some of the ordered form of energy is converted to heat, a less-ordered kinetic energy. No energy transfer is 100% efficient. The quantity of energy in the universe may be constant, but its quality is not. Every process results in an increasing disorder in energy in the form of the random molecular motion of heat.

Chemical Energy: A Closer Look

Chemical reactions rearrange atoms by the breaking and forming of chemical bonds. Energy is released when bonds form and is expended or absorbed in order to break bonds. *Bond energy*, usually expressed in kilocalories per mole of bonds formed or broken, is the quantity of energy involved in the forming or breaking of a particular bond.

The net energy released or consumed by the conversion of reactants into products is called the *heat of reaction*, signified by ΔH. ΔH represents the difference between the energy consumed when bonds break and that released when bonds form. *Enthalpy* is the heat content of a molecule, determined from the total potential energy stored in its bonds. The heat of reaction, ΔH, can be thought of as the difference between the enthalpy of the reactants and of the products, between the sum of the bond energies of the reactants and the sum of the bond energies of the products. A reaction that has a negative ΔH and thus releases heat does so because the products have less enthalpy than do the reactants. Such heat-releasing reactions are called exothermic. When the products of a reaction have more enthalpy than the reactants, the reaction is called endothermic, has a positive ΔH, and absorbs heat from the surroundings.

Spontaneous reactions occur without the addition of external energy. Energy-rich systems are intrinsically unstable, and tend to change so as to decrease their energy. Chemical reactions tend to be spontaneous if the energy stored in the molecules decreases. Exothermic reactions are usually spontaneous; endothermic reactions, requiring energy and increasing the chemical energy of molecules, are usually nonspontaneous.

Entropy, the trend toward disorder and randomization, also contributes to driving *spontaneous reactions*. The change in entropy (ΔS) of a reaction is represented by $\Delta S = S$ final state $- S$ initial state. Thus, a positive ΔS represents an increase in entropy and will contribute to a reaction being spontaneous. The increase in entropy may be enough to drive a reaction even though the reaction may be endothermic, as in the spontaneous evaporation of water. It is also possible for a reaction to represent a decrease in entropy but still proceed spontaneously due to the decrease in enthalpy, as in the formation of highly ordered snowflakes.

The balance between enthalpy and entropy can be represented by the concept of *free energy*. Free energy (G) in a system increases with an increase in enthalpy and a decrease in entropy. Free energy decreases with a decrease in enthalpy and an increase in entropy. The relationship among free energy, enthalpy, and entropy can be symbolized as $\Delta G = \Delta H - T\Delta S$. Temperature (T= °Kelvin= °C + 273) must be factored into the determination of entropy because a high temperature increases random molecular motion, disrupting order.

For a reaction to be spontaneous, the free energy of the system must decrease; ΔG must be negative. A reduction in free energy is dependent on the magnitude of the change in enthalpy ($-\Delta H$ is exothermic) and entropy ($+\Delta S$ increases entropy) and the temperature. At times, temperature may be the deciding factor in whether or not a reaction will proceed. For example, the denaturation of proteins is favored by an increase in entropy but retarded by the increase in enthalpy necessary to break the hydrogen bonds and other forces creating the shape of the molecule. An increase in temperature (usually above 60°C) will weight the entropy factor in favor of a net decrease in free energy so that denaturation will proceed.

Free energy is also a measure of how much work a spontaneous process can do. When temperature is constant, as it is in a cell, free energy is an indication of the total energy released from the system (enthalpy) minus the energy used in the cases when the system becomes more ordered (entropy).

Using the reference of free energy, reactions can be classified as *exergonic* or *endergonic*. An exergonic ($-\Delta G$) reaction proceeds with a net release of free energy and is spontaneous. The magnitude of ΔG indicates the maximum amount of work the reaction can do. Endergonic reactions ($+\Delta G$) are nonspontaneous; they must absorb free energy from the surroundings. The magnitude of ΔG indicates the minimum amount of work needed to drive the reaction. In metabolism, exergonic reactions are often coupled closely with endergonic reactions; the free energy released from the former is used to power the latter.

At equilibrium in a chemical reaction, the forward and backward reactions are proceeding at the same rate. The products and reactants are then present in a fixed ratio called the *equilibrium constant* (K_{eq}). At equilibrium, $\Delta G = 0$ because there is no net free energy change. When approaching equilibrium, the ΔG of a reaction is negative; when moving away from equilibrium, ΔG is positive. In order to drive a reaction away from its normal equilibrium constant, a cell must add free energy, usually by coupling the reaction with an exergonic reaction. ATP is used as the energy shuttle in cellular metabolism.

Energy for Cellular Work

A cell must perform several kinds of work: mechanical work involved in movement of the cell or parts of the cell, transport work in pumping molecules across membranes, and chemical work in driving endergonic reactions to synthesize cellular molecules. In most cases, the immediate source of the energy to perform this work comes from *adenosine triphosphate,* or *ATP.*

ATP is a nucleoside triphosphate, the purine base adenine bonded to the sugar ribose, connected to a chain of three phosphate groups. The bonds between the phosphate groups are unstable and can be broken by hydrolysis. Thus, ATP can be hydrolyzed to ADP (adenosine diphosphate) and an inorganic phosphate molecule, releasing 7.3 kilocalories of energy per mole of ATP. This quantity of free energy has been experimentally measured; the ΔG of the reaction in the cell is estimated to be closer to -10 to -12 kcal/mol.

Although the phosphate bonds in ATP are called high-energy bonds, they are actually weak bonds, easily hydrolyzed to produce stronger, more stable bonds. The transformation to a more stable state releases energy. In a cell, this energy can be used to transfer the phosphate group from ATP to another molecule (under the direction of specific enzymes), producing a phosphorylated, more reactive intermediate. The phosphorylation of other molecules by ATP forms the basis for almost all cellular work.

A cell regenerates ATP at a phenomenal rate; ten million molecules of ATP may be consumed and regenerated per second. The formation of ATP from ADP and inorganic phosphate is endergonic, with a ΔG of +7.3 kcal/mol. Cellular respiration (the catabolic processing of glucose and other organic molecules) provides the energy for the regeneration of ATP. Plants can also produce ATP using light energy.

Respiration is overwhelmingly exergonic. The stepwise degradation of glucose, with ATP capturing the small bursts of energy released at each step down the free energy gradient, ensures that the energy from respiration is available to the cell in the small doses required for its cellular work.

Respiration and other cellular chemical reactions are reversible and could reach equilibrium if the cell did not keep a steady supply of reactants and siphon off the products (as reactants for new processes or as waste products to be expelled). Chemical systems at equilibrium have a ΔG of 0 and can do no work. Respiration continues to produce ATP as long as the cell can provide glucose and expel CO_2.

Enzymes

Thermodynamics can indicate what reactions are spontaneous but not how fast those reactions occur. Some spontaneous reactions may take centuries to occur. *Enzymes* are used by the cell to speed and regulate metabolic reactions so that life is possible. Enzymes are biological catalysts, agents that change the speed of a reaction but are unchanged by the reaction.

Activation energy, E_a, is the energy that must be absorbed by reactants to reach the unstable transition state, in which bonds are more fragile and likely to break, and from which the reaction can proceed. This state can be reached by the addition of thermal energy from the surroundings, causing the reactant molecules to collide more often and more forcefully. Even in an exergonic reaction, in which ΔG is negative, energy must first be absorbed to reach the transition state.

The activation energy barrier is essential to life because it prevents the energy-rich macromolecules of the cell from decomposing spontaneously. In order for metabolism to proceed in a cell, however, E_a must be reached for selected reactions. Heat, a normal source of activation energy in reactions, would be harmful to the cell and would also speed metabolic reactions indiscriminately. Enzymes are able to lower E_a for specific reactions so that metabolism can proceed at cellular temperatures. Enzymes do not change ΔG for a reaction; they only speed up reactions that would otherwise occur very slowly.

Enzymes are proteins, macromolecules with characteristic three-dimensional shapes. The specificity of an enzyme for the particular *substrate* on which it works is determined by its unique shape. The substrate is temporarily bound to its enzyme at the *active site*, a pocket or groove, found on the surface of the enzyme molecule, that has a shape and charge arrangement complementary to the substrate molecule. When a substrate molecule enters the active site, the enzyme changes shape slightly, creating what is called an *induced fit* between substrate and active site, which enhances the ability of the enzyme to catalyze the chemical reaction.

The substrate is held in the active site by hydrogen or ionic bonds. The side chains (R groups) of the surrounding amino acids in the active site facilitate the conversion of substrate to product. The product then leaves the active site and the enzyme can bind with another substrate molecule. The conversion is extremely fast; an enzyme can catalyze 1000 reactions or more per second.

Enzymes can catalyze reactions involving the joining of two reactants by providing active sites in which the substrates are bound together closely and oriented properly. An induced fit can stretch or bend critical bonds in the substrate molecule and make them easier to break. An active site may provide a microenvironment that is necessary to a particular reaction. Enzymes may also actually participate in a reaction by forming brief covalent bonds with the substrate.

The rate at which an enzyme molecule works is partly dependent on the concentration of its substrate. The speed of a reaction will increase with increasing substrate concentration up to the point at which all enzyme molecules are saturated with substrate molecules and are working at full speed.

The activity of an enzyme is dependent on its three-dimensional shape, and this shape is dependent on certain environmental factors that affect the weak chemical bonds maintaining protein structure. The velocity of an enzyme-catalyzed reaction may increase with rising temperature up to the point at which increased thermal agitation of molecules begins to disrupt the hydrogen and ionic bonds that stabilize protein conformation. A change in pH value may denature an enzyme by disrupting the hydrogen bonding of the molecule. Enzymes are sensitive to salt concentration because inorganic ions may interfere with ionic bonds within the enzyme molecule. When the conformation of an enzyme changes through denaturation, its activity decreases.

Cofactors are small molecules that bind with enzymes and are necessary for enzyme function. They may be inorganic, such as various metal atoms, or organic molecules called *coenzymes*. Most vitamins are coenzymes or precursors of coenzymes.

Enzyme *inhibitors* selectively disrupt the action of enzymes either reversibly by binding with the enzyme with weak bonds, or irreversibly by attaching with covalent bonds. *Competitive inhibitors* compete with the substrate for the active site of the enzyme. Increasing the concentration of substrate molecules may overcome this type of inhibition as long as the inhibitor does not bind too strongly to the active site. *Noncompetitive inhibitors* bind to a part of the enzyme separate from the active site and change the conformation of the enzyme, thus impeding enzyme action. Metabolic poisons such as mercury, cyanide, and many pesticides are noncompetitive inhibitors of key enzymes. The cell itself uses selective inhibitors to control enzyme action and thus regulate metabolism.

The molecules that inhibit or activate enzyme activity may bind to an *allosteric* site, a receptor site not associated with the active site on the enzyme. An allosteric enzyme may oscillate between two confomational states depending on whether an activator or inhibitor is bound to an allosteric site. Allosteric enzymes are often complex molecules made of two or more polypeptide chains or subunits, each with its own active site. Allosteric sites may be located where subunits join. The subunits interact such that a conformational change in one is transmitted to the other(s). Through this phenomenon, called *cooperativity,* a single activator or inhibitor molecule bound to one allosteric site can affect the active sites of all subunits. Allosteric enzymes may be critical regulators of metabolic pathways.

Control of Metabolism

Feedback inhibition commonly regulates metabolic pathways. The product of a pathway can act as an inhibitor of an enzyme early in the pathway and shut down the process when the cell has produced sufficient endproduct.

Cellular metabolism would be nonfunctional without enzymes. The intricate ordering and control of the chemical pathways of metabolism are made possible through the production and regulation of enzymes. The complex internal structure of the cell serves to order metabolic pathways in space and time. Enzymes can be grouped into multienzyme complexes in which the enzymes regulating successive steps of a metabolic pathway are serially arranged. Specialized cellular compartments may contain high concentrations of the enzymes and substrates needed for a particular pathway. An efficient arrangement occurs when multienzyme complexes are located in sequential order in the membranes of cellular compartments. Thus the structural level of cellular organization serves to facilitate and control metabolism.

Emergent Properties: A Reprise

This unit has illustrated how life is organized into a hierarchy of structural levels with emergent properties associated with each new level of order. The structure and interactions of atoms, molecules, monomers, and macromolecules have been linked to metabolism, the orderly chemistry characteristic of life.

STRUCTURE YOUR KNOWLEDGE

This chapter introduces many complex ideas concerning the thermodynamics of metabolism. Take the time to organize your understanding of small "chunks" of this information and then try to integrate these pieces into your picture of the energy transformations taking place within the cells of living organisms.

1. Create a simple concept map concerning energy: its definition, examples, and the first two laws of thermodynamics.

2. Develop two separate concept maps, one for enthalpy and ΔH, and one for entropy and ΔS. Then try to combine these two maps under the broad umbrella of free energy and ΔG. The value in this exercise is for you to wrestle with and organize these concepts for yourself. Do not turn to the suggested concept map until you have worked on your own understanding. Remember that the concept map in the answer section is only one way of structuring these ideas—have confidence in your own organization.

3. Now take a break from concept mapping and answer these questions about proteins. (A map on proteins would, of course, be helpful.)

    a. Why are enzymes so critical to the existence of life?
    b. What characteristics of proteins make them good catalysts?
    c. Briefly describe the process by which an enzyme catalyzes a reaction.
    d. How does a cell control enzyme action and regulate metabolism?

TEST YOUR KNOWLEDGE

MULTIPLE CHOICE: *Choose the one best answer.*

1. When glucose is converted to CO_2 and H_2O, changes in enthalpy, entropy, and free energy are as follows:
    a. $-\Delta H, -\Delta S, -\Delta G$.
    b. $-\Delta H, +\Delta S, -\Delta G$.
    c. $-\Delta H, +\Delta S, +\Delta G$.
    d. $+\Delta H, +\Delta S, +\Delta G$.

2. When water evaporates spontaneously, the following changes apply:
    a. $+\Delta H, +\Delta S, +\Delta G$.
    b. $-\Delta H, +\Delta S, -\Delta G$.
    c. $+\Delta H, +\Delta S, -\Delta G$.
    d. $+\Delta H, -\Delta S, -\Delta G$.

3. When a protein forms from amino acids, the following changes apply:
    a. $+\Delta H, -\Delta S, +\Delta G$.
    b. $+\Delta H, +\Delta S, -\Delta G$.
    c. $-\Delta H, -\Delta S, +\Delta G$.
    d. $-\Delta H, +\Delta S, +\Delta G$.

4. Exothermic and exergonic reactions are similar in that
    a. they both release energy.
    b. they are always spontaneous.

    c. they both always result in an increase in entropy.

    d. all of the above are true.

5. A negative ΔG means that

    a. the quantity G of energy is available to do work.

    b. the reaction is spontaneous.

    c. the reactants have more free energy than the products.

    d. all of the above are true.

6. According to the first law of thermodynamics,

    a. for every action there is an equal and opposite reaction.

    b. every energy transfer results in an increase in disorder or entropy.

    c. the total amount of energy in the universe is conserved or constant.

    d. energy can be transferred or transformed, but disorder always increases.

7. Catabolic pathways

    a. combine molecules into more complex and energy-rich molecules.

    b. are usually coupled with anabolic pathways to which they supply energy in the form of ATP.

    c. involve endergonic reactions that break complex molecules into simpler ones.

    d. are spontaneous and do not need enzyme catalysis.

8. Chemical energy

    a. is a form of potential energy.

    b. is stored in the bonds between atoms in molecules.

    c. can be transferred from one molecule to another, but some of it will be transformed to heat energy.

    d. all of the above are true.

9. A spontaneous reaction

    a. proceeds rapidly.

    b. occurs without the addition of external energy (aside from E_a).

    c. does not need to be catalyzed by enzymes.

    d. all of the above are true.

10. The formation of ATP from ADP and inorganic phosphate

    a. is an endergonic process.

    b. transfers the phosphate to another intermediate that becomes more reactive.

    c. produces a strong, stable, high-energy bond that can drive cellular work.

    d. has a ΔG of -7.3 kcal/mol.

11. At equilibrium,

    a. $\Delta G = O$.

    b. the ratio between products and reactants equals the equilibrium constant (K_{eq}).

    c. the forward and backward reactions proceed at the same rate.

    d. all of the above are true.

12. An enzyme is capable of

    a. lowering the activation energy of a reaction.

    b. changing the K_{eq} for a reaction.

    c. increasing the free energy change for a reaction.

    d. all of the above.

13. In cooperativity,

    a. a multienzyme complex contains all the enzymes of a metabolic pathway.

    b. a product of a pathway serves as a competitive inhibitor of an early enzyme in the pathway.

    c. a molecule bound to an allosteric site affects the active site of several subunits.

    d. several substrate molecules can be catalyzed by the same enzyme.

14. Substrates are held in the active site of an enzyme by

    a. hydrogen and covalent bonds.

    b. the action of coenzymes and cofactors.

    c. the matching shape of the pocket or groove on the enzyme.

    d. the lowering of the activation energy.

15. According to the induced fit hypothesis,

    a. the binding of the substrate depends on the shape of the active site.

    b. a competitive inhibitor can outcompete the substrate for the active site.

    c. the binding of the substrate changes the shape of the enzyme slightly and can stress or bend substrate bonds.

    d. the active site creates a microenvironment ideal for the reaction.

FILL IN THE BLANKS

1. _____ is the totality of an organism's chemical processes.

2. _____ pathways require energy to combine molecules together.

3. _____ energy is the energy of motion.

4. _____ refers to the heat content or potential energy in the bonds of a molecule.

5. _____ is the term for the measure of disorder or randomness.

6. _____ is the energy that must be absorbed by molecules to reach the transition state.

7. _____ inhibitors change the enzyme's conformation by binding to an allosteric site.

8. _____ are organic molecules that bind to enzymes and are necessary for their functioning.

9. _____ is a regulatory device in which the product of a pathway binds to an enzyme early in the pathway.

10. _____ enzymes change between two conformations depending on whether an activator or inhibitor is bound to them.

ANSWER SECTION

CHAPTER 2: ATOMS, MOLECULES AND CHEMICAL BONDS

Suggested Answers to Structure Your Knowledge

1. An atom is the smallest unit of an element that maintains the physical and chemical properties of that element. Atoms are composed of subatomic particles, the largest and most stable of which are protons, neutrons, and electrons. Protons and neutrons are packed into the nucleus of an atom, whereas electrons travel at great speeds around the nucleus in orbitals within energy shells. Protons and neutrons both have a mass of 1 dalton; electrons have negligible mass. Neutrons do not carry a charge. Protons have a charge of +1; electrons have a charge of -1. Atoms that are not involved in chemical bonds are neutral because the number of electrons equals the number of protons. Ions are atoms that have gained or lost electrons, and their charge reflects the difference between the number of protons and electrons.

2. The atoms of each element have a characteristic number of protons in their nuclei, referred to as the atomic number. In a neutral atom, the *atomic number* also indicates the number of electrons.

The *mass number* is an indication of the approximate mass of an atom and is equal to the number of protons and neutrons in the nucleus.

The *atomic weight* refers to the atomic mass of an atom. It is equal to the mass number, and is measured in the atomic mass unit of daltons. Protons and neutrons both have a mass of approximately 1 dalton.

The *valence* is an indication of the bonding capacity of an atom. It is the number of covalent bonds that must be formed for an atom to satisfy the octet rule and complete its valence shell with eight electrons. The valence of an atom is most related to the chemical behavior of an atom because it is an indication of the number of bonds the atom will make, or the number of electrons the atom must share in order to reach a filled valence shell.

3. Ionic and nonpolar covalent bonds represent the two extremes along a continuum of electron sharing between atoms in a molecule. In ionic bonds the electrons are completely pulled away from one atom to the other, creating negatively and positively charged ions. In nonpolar covalent bonds the electrons are equally shared between two atoms. The middle ground of these two extremes is filled with polar covalent bonds in which one atom pulls the shared electrons closer to it, producing a partial negative charge associated with that portion of the molecule and a partial positive charge associated with the atom from which the electrons are pulled. Because electrons are rapidly orbiting around the nuclei and constantly changing positions, a very polar bond may vacillate between being covalent and ionic.

4. There are two compelling reasons for adopting a reductionist approach to the study of biology. First, the anatomy, physiology, history, diversity, and ecological relationships of living organisms are incredibly complex phenomena. These intricate and complicated systems are more readily studied and understood when they are broken down into smaller, simpler parts. The other reason for a focus on the level of molecular biology is that all the actions and reactions of living organisms, even those that are emergent properties from higher levels of organization, have their ultimate explanation in the chemical reactions of the atoms and molecules that make up living matter.

A strict reductionist approach to biology, however, would miss the consideration of the unique properties of living organisms that emerge at each level of structural organization as the result of the interaction and arrangement of the parts. A study of the parts is crucial, but not sufficient, for the understanding of the whole.

Answers to Test Your Knowledge

Multiple Choice:

1. b	6. b	11. d	16. d
2. d	7. c	12. b	17. c
3. a	8. b	13. d	18. c
4. b	9. a	14. b	19. b
5. a	10. a	15. c	20. c

CHAPTER 3: WATER AND THE FITNESS OF THE ENVIRONMENT

Suggested Answers to Structure Your Knowledge

1.

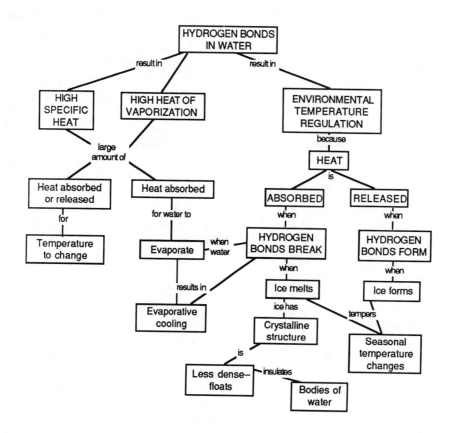

2.

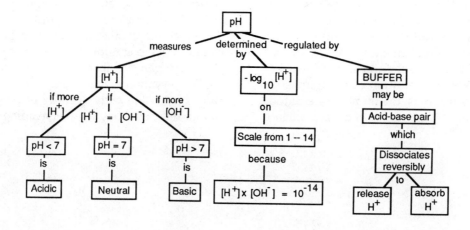

Answers to Test Your Knowledge

Multiple Choice:

1. c	6. c	11. b
2. a	7. d	12. d
3. b	8. b	13. c
4. a	9. b	14. c
5. b	10. d	15. c

True or False:

1. True
2. False, change *adhesion* to *hydrogen bonding*
3. False, change *absorbs cold* to *releases heat*
4. False, change *nonpolar substances* to *ions*
5. True
6. False, change *calorie* to *specific heat,* or change *substance* to *water*
7. True
8. True
9. False, change *releases* to *absorbs*, and *to* to *from*
10. False, change *dissociation into . . . ions* to *hydrogen bonding*

Fill in the Blanks:

$[H^+]$	$[OH^-]$	pH	Acidic, basic, or neutral?
10^{-3}	10^{-11}	3	acidic
10^{-8}	10^{-6}	8	basic
10^{-12}	10^{-2}	12	basic
10^{-9}	10^{-5}	9	basic
10^{-1}	10^{-13}	1	acidic
10^{-7}	10^{-7}	7	neutral

CHAPTER 4: CARBON AND MOLECULAR DIVERSITY

Suggested Answers to Structure Your Knowledge

1.

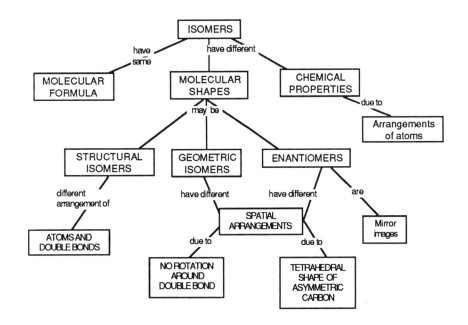

2.

FUNCTIONAL GROUPS TABLE

NAME OF GROUP	MOLECULAR FORMULA	CHARACTERISTICS CONFERRED TO ORGANIC COMPOUND
Hydroxyl	- OH	Polarity, solubility in water
Carbonyl	- C=O	Aldehyde or ketone, polar and reactive group
Carboxyl	- COOH	Acidic, can dissociate and release H^+
Amine	$- NH_2$	Basic, accepts H^+, becoming $-NH_3^+$
Sulfhydryl	- SH	Forms crosslinks in 3-D protein conformation
Phosphate group	$- OPO_3H_2$	Acidic, use in energy transfer

Answers To Test Your Knowledge

Multiple Choice:

1. d	5. a	19. a
2. a	6. d	10. b
3. d	7. a	11. b
4. b	8. c	12. c

Matching:

1. A & C	5. E	19. A,C
2. B & F	6. A,C,D,G	10. B,F
3. D & G	7. A,D,G	11. E
4. E	8. e	12. C

CHAPTER 5: STRUCTURE AND FUNCTION OF MACROMOLECULES

Suggested Answers to Structure Your Knowledge

1.

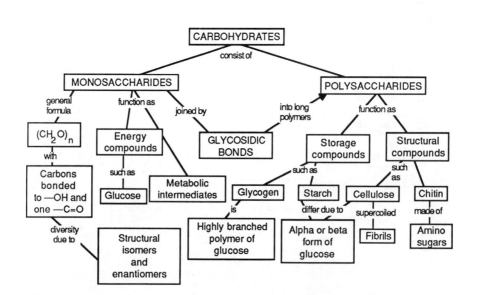

2.

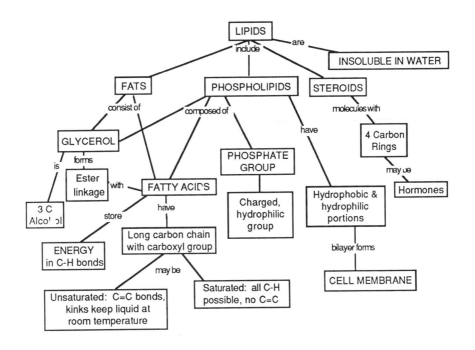

3.

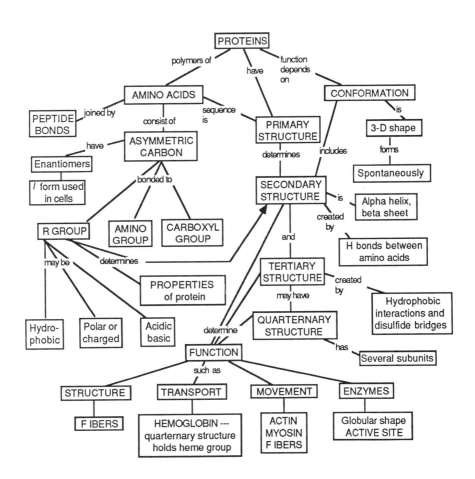

4.

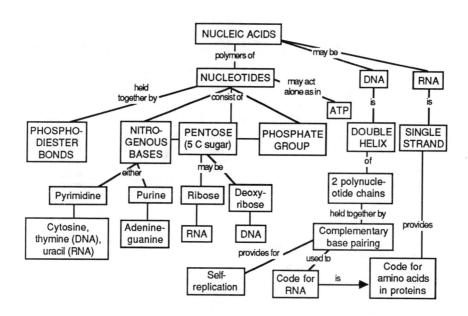

Answers to Test Your Knowledge

Matching:

1. A	5. C	9. A
2. B	6. D	10. A
3. D	7. B	
4. C	8. C	

Matching of Formulas:

1. D,B	5. C
2. A	6. A
3. C,E,F	7. E,F
4. F,G	

Multiple Choice:

1. d	5. c	9. d B
2. c	6. b	10. d
3. a	7. c	11. a
4. c	8. a	12. c

Fill in the Blanks:

1. thymine	6. quarternary
2. guanine	7. primary structure
3. purines	8. glycogen
4. nucleoside	9. phospholipids
5. nucleotide	10. peptide bonds

CHAPTER 6: INTRODUCTION TO METABOLISM

Suggested Answers To Structure Your Knowledge

1.

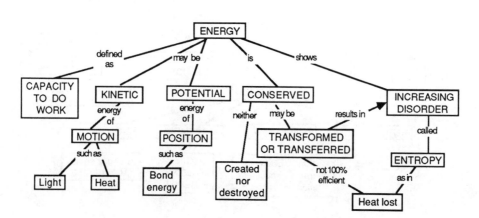

2.

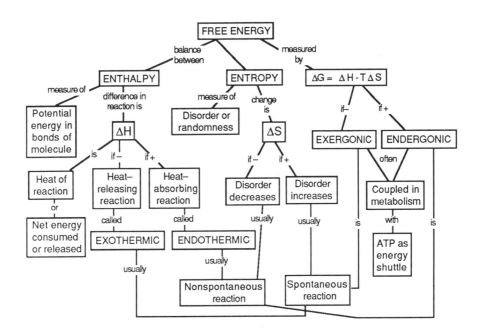

3. a. Enzymes are essential for the "cold chemistry" of life because they lower the activation energy of the reactions they catalyze and allow those reactions to occur extremely rapidly at a temperature conducive to life. They are also highly selective, so that by regulating the enzymes it produces, the cell can regulate which of the myriad of possible chemical reactions take place at any given time.

b. The three-dimensional conformation of a protein permits it to have a characteristically shaped active site with specific chemical attributes. The shape and chemistry of the active site result in the specificity of an enzyme for a particular substrate.

c. A particular substrate binds to the active site on its enzyme with hydrogen and ionic bonds. The enzyme may change shape slightly, closing around the substrate. This induced fit strains the substrate's bonds or brings reactive groups in close proximity so that the transition state of the molecule(s) is reached and the reaction may proceed more easily. Once the reaction takes place, the substrate leaves the active site and another one enters. The speed of the reaction will increase with increasing substrate concentration until all enzyme molecules are saturated.

d. One way a cell can control enzyme action is through feedback inhibition, in which the product of a pathway then acts as an inhibitor of an enzyme early in the pathway. The cell can regulate which enzymes are produced and in what quantity through control of gene expression. The compartmentalization of the cell can order metabolic pathways and their substrates in space. Multienzyme complexes, where enzymes are sequentially arranged, may be located within cellular membranes.

Answers to Test Your Knowledge

Multiple Choice:

1. b	6. c	11. d
2. c	7. b	12. a
3. a	8. d	13. c
4. a	9. b	14. a
5. d	10. a	15. c

Fill in the Blanks:

1. metabolism	6. activation energy
2. anabolic	7. noncompetitive
3. kinetic	8. coenzymes
4. enthalpy	9. feedback inhibition
5. entropy	10. allosteric

Unit II

The Cell

CHAPTER 7

A TOUR OF THE CELL

FRAMEWORK

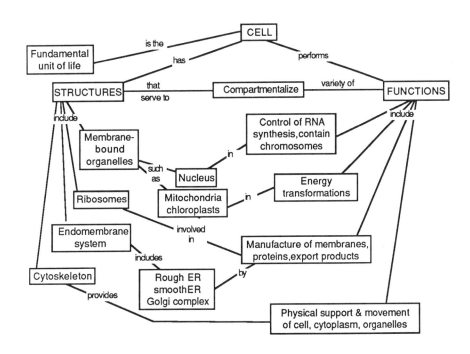

This chapter deals with the fundamental unit of life—the cell. The complexities in the processes of life are reflected in the complexities of the structure of the cell. It is easy to become overwhelmed by the number of new vocabulary terms and definitions for this array of cell organelles and membranes. Do not lose sight of the fascinating and intricate relationship between structure and function seen in the cell's components. The concept map attempts to place some structure on the wealth of detail found in a tour of the cell.

CHAPTER SUMMARY

The cell is the basic structural and functional unit of all living organisms. In the hierarchy of biological organization, the capacity for life emerges at the level of the cell. The cell contains an intricately integrated collection of subunits that perform and mediate the functions associated with life.

The Geography of the Cell

The kingdom Monera, which includes bacteria and blue-green algae, is characterized by having *prokaryotic cells,* cells with no nuclear membrane or membrane-bound organelles. The genetic material of prokaryotic cells consists of a single, circular DNA molecule concentrated in a region called the nucleoid. *Eukaryotic cells* are found in the other four kingdoms of life: protoctists, plants, fungi, and animals. A eukaryotic cell is much more structurally complex. It has a true nucleus bounded by a nuclear envelope or membrane and numerous organelles suspended in a semifluid medium called *cytosol.* The *cytoplasm* is the region between the nucleus and the membrane enclosing the cell.

Almost all cells are microscopic; their sizes are measured in micrometers (μm; 1 mm = 1000 μm). The smallest known cells are certain bacteria with diameters of 0.1 μm— just large enough to pack in the DNA—ribosomes and enzymes necessary to reproduce and sustain life. Most bacterial cells range from 1 to 10 μm in diameter, whereas eukaryotic cells are ten times larger, ranging from 10 to 100 μm. Plant cells are typically larger than animal cells.

The small size of cells is dictated by geometry and the requirements of metabolism. Area is proportional to the square of the linear dimension and volume to its cube. Eukaryotic cells, with 10 times the diameter of bacterial cells, have 100 times as much surface area but 1000 times greater volume. The *plasma membrane* must regulate sufficient exchange of oxygen, nutrients, and wastes to provide for the metabolism of the cell, and the nucleus must control the metabolic processes within the cell. The microscopic size of cells provides more surface area for exchange relative to the volume of the cell and a smaller volume of cytoplasm to be regulated by the nucleus.

The difference in the volume–to–surface-area ratio in eukaryotic compared to prokaryotic cells is partly compensated for by an extensive internal membrane system. These membranes compartmentalize the eukaryotic cell, providing local environments for specific metabolic functions, and participate in many metabolic processes through membrane-bound enzymes. The development of the electron microscope and the ultracentrifuge has allowed scientists both to view and to isolate for study these cellular compartments or structures. The understanding of cell structure and functions parallels the technological advances in these instruments.

Membranes are composed of phospholipid molecules, in a two-layer arrangement called a bilayer, and proteins. These lipids and proteins are held together by hydrophobic interactions and other noncovalent bonds. Membrane proteins serve various functions: enzymes with active sites exposed to the inside or outside of the membrane, solute pumps, receptor sites for chemical signals, and channels for passage of specific molecules. Carbohydrate molecules are often attached to the outer plasma membrane and serve as identification or recognition signals. The specific molecular composition of a membrane varies according to its functions. These structures play a key role in the organization and functioning of a cell's metabolism.

The Nucleus

The nucleus is surrounded by the *nuclear envelope,* a double membrane perforated by pores that may regulate the movement of specific molecules between the nucleus and the cytoplasm. The inner membrane is associated with a layer of protein that helps to maintain the shape of the nucleus and may contribute to the organization of the genetic material.

Most of the cell's genetic material is located in the nucleus. DNA, along with associated proteins, is organized into a characteristic number of *chromosomes* for each eukaryotic species. Chromosomes are visible only when coiled and condensed in a dividing cell; otherwise they appear as a mass of stained material called *chromatin.*

The *nucleolus,* a round structure visible in the nondividing nucleus, contains specialized regions of chromosomes, called nuclear organizers, that have multiple copies of genes for ribosome synthesis. *Ribosomes* are constructed in the nucleolus from RNA, produced in the nucleolus and elsewhere in the nucleus, and from proteins transported in from the cytoplasm. Ribosomes are transported into the cytoplasm, where protein synthesis occurs. The genetic instructions for specific proteins are transcribed from DNA into *messenger RNA* in the nucleus. The messenger RNA passes into the cytoplasm, where it complexes with ribosomes that translate the message into the primary sequence of proteins. The number of ribosomes is a measure of the rate of protein synthesis within a cell.

Ribosomes

Ribosomes are composed of a large and a small subunit that join together when they attach to a messenger RNA and begin protein synthesis. The difference in molecular composition of the smaller ribosomes of prokaryotes has medical consequences; some antibiotics specifically inhibit the prokaryotic ribosomes of bacteria without harming the activity of the patient's ribosomes.

Ribosomes usually occur in clusters, called polyribosomes, or *polysomes,* attached to a single messenger RNA. Most of the proteins produced by *free ribosomes* are used within the cytosol. *Bound ribosomes,* attached to the endoplasmic reticulum, usually make proteins that will be included within membranes or exported from the cell.

The Endomembrane System

The *endomembrane system* of a cell consists of the nuclear envelope, endoplasmic reticulum, Golgi complex, lysosomes, microbodies, vacuoles, and the plasma membrane. These membranes are all related either through direct contact or by the transfer of membrane segments by membrane-bound sacs called *vesicles*.

The *endoplasmic reticulum (ER)* is the most extensive portion of the endomembrane system. It is continuous with the outer membrane of the nuclear membrane and encloses a lumen, or network of interconnected compartments. Ribosomes are attached to the cytoplasmic surface of rough ER. *Rough ER* manufactures membrane and secretory proteins. *Smooth ER* lacks ribosomes. One of its functions is to transfer products formed by the rough ER by budding off tiny membrane-bound transport vesicles.

Proteins intended for secretion are manufactured by membrane-bound ribosomes and then threaded into the lumen of the ER, where they fold into their native configuration. Most secretory proteins are glycoproteins, and the carbohydrate is covalently bonded to the protein by enzymes built into the ER membrane.

Rough ER manufactures membranes by inserting proteins formed by the ribosomes into the membrane and assembling phospholipids with the aid of enzymes built into the membrane and precursors obtained from the cytosol.

Most smooth ER functions in the discharge of vesicles that transport membranes and products from the ER lumen to various destinations. Muscle cells have specialized smooth ER responsible for pumping calcium ions from the cytosol into the ER lumen. The smooth ER of liver cells has detoxifying enzymes built into the membrane. Some smooth ER functions in the manufacture of hormones and steroids.

Free ribosomes and ER-bound ribosomes are identical. It appears that secretory proteins contain a *signal sequence* of amino acids at the beginning of the polypeptide that enables the ribosome to attach to a receptor site on the ER membrane. The growing polypeptide moves into the ER lumen, where the signal sequence is removed by an enzyme. Signal sequences, coded for by mRNA and attached to polypeptide chains, are thought to direct various proteins to specific sites within the cell.

Transport vesicles, pinched off from the ER, travel first to the *Golgi complex*, where products of the ER are modified, stored, and directed to other locations. *Dictyosomes* are individual Golgi complexes in plant cells.

The Golgi complex consists of a stack of flattened sacs. Sacs at one end of the complex are the forming face, closely associated with smooth ER. The other end of the complex is called the maturing face, from which vesicles pinch off to travel to other sites.

Lysosomes are membrane-bound sacs of hydrolytic enzymes used by the cell to digest macromolecules. An acidic pH value is maintained within the lysosome by the membrane's inward pumping of hydrogen ions from the cytosol. The cell can protect itself from unwanted digestion and also provide an optimal pH value by containing hydrolytic enzymes within lysosomes.

Lysosomes function both to digest food particles ingested by cells that use *phagocytosis* and to recycle the cell's own macromolecules by engulfing organelles or small bits of cytosol. During development or metamorphosis, lysosomes may be programmed to destroy their cells in order to create a specific body form. Storage diseases are inherited defects in which a lysosomal hydrolytic enzyme may be missing and lysosomes become packed with indigestible substances.

Microbodies are small, single-membrane-bound compartments filled with particular enzymes for a specific metabolic pathway. *Peroxisomes* are microbodies that produce hydrogen peroxide, H_2O_2, by transferring hydrogen from various substrates to oxygen. Some peroxisomes may use oxygen to break down fats to smaller molecules; others may detoxify alcohol or other poisons. Hydrogen peroxide is itself toxic, but peroxisomes have an enzyme to convert it to water. Compartmentalizing these essential enzymes together is important to cellular function.

Glyoxysomes, commonly found in the tissues of germinating seeds, contain enzymes that convert the oils stored in the seed to sugar for the developing seedling.

Vacuoles are also membrane-bound sacs within the cell, distinguished by being larger than vesicles. Food vacuoles are formed as a result of phagocytosis. Contractile vacuoles pump excess water out of freshwater protoctists. A large *central vacuole* is found in mature plant cells, surrounded by a membrane called the *tonoplast*. This vacuole stores organic compounds and inorganic ions for the cell. Hydrolytic enzymes are compartmentalized in the central vacuole and digest stored macromolecules and recycle organic compounds from worn-out organelles. Pigments, poisonous or unpalatable compounds that may protect the plant from predators, and dangerous metabolic byproducts also may be contained in the vacuole. Plant cells can increase in size with a minimal addition of new cytoplasm as their vacuoles absorb water and expand. The cytoplasm forms a thin band between the plasma membrane and the tonoplast, maintaining a good membrane surface-area-to-volume ratio even in large plant cells. The large central vacuole forms from the coalescence of smaller vacuoles derived from the endoplasmic reticulum and Golgi complex.

The ER and Golgi complex are capable of producing a unique molecular composition for each membrane within the endomembrane system. It appears that vesicles transferring these membranes have exterior recognition molecules that enable them to fuse with the appropriate membrane. Thus, each component of the endomembrane system is related but unique, with structural modifications designed for its specific function.

Energy Transducers: Mitochondria and Chloroplasts

Cellular respiration, the catabolic processing of fuels to produce ATP with the help of oxygen, occurs within the *mitochondria* of most eukaryotic cells. *Chloroplasts*, found only in plants and photosynthetic protoctists, produce organic compounds from carbon dioxide and water by absorbing solar energy. Mitochondria and chloroplasts are not considered to be part of the endomembrane system because their membrane proteins are made, not by the ER, but by ribosomes either free in the cytosol or contained within these organelles. They also contain DNA that directs the synthesis of some of their proteins.

Mitochondria are found in nearly all eukaryotic cells. Their number can range from hundreds to thousands, depending on the metabolic needs of the cell. Two membranes, each a phospholipid bilayer with unique embedded proteins, surround a mitochondrion. The smooth outer membrane allows the passage of small molecules but blocks the passage of macromolecules from the cytosol into the intermembrane space between the membranes. The inner membrane is thrown into folds called *cristae*, creating a large membrane surface area and surrounding the *mitochondrial matrix*. Within this matrix are several different enzymes that control many of the metabolic steps of cellular respiration. Other respiration enzymes are built into the extensive inner membrane.

Plastids are plant organelles that include *leucoplasts*, which store starch; *chromoplasts*, which contain pigments; and *chloroplasts*, which contain the green pigment chlorophyll and function in photosynthesis. All three types of plastids develop in unspecialized cells from *proplastids*, the fates of which depend both on the location and the environment of the cell. Thus, chloroplasts develop only when already existing chloroplasts divide or when proplastids are exposed to light.

Chloroplasts are composed of two membranes enclosing the *stroma*; in which are found the *thylakoids*, a system of flattened sacs that may be stacked together to form structures called *grana*. Chlorophyll and other pigment molecules that absorb light energy, along with other enzymes and molecules involved in photosynthesis, are embedded in the thylakoid membrane. The ATP produced by enzymes within the thylakoid membrane is released to the stroma; where the enzymes of the metabolic pathway for the synthesis of sugar are located.

The Cytoskeleton

The *cytoskeleton* is a network of fibers that function to give mechanical support, maintain or change cell shape, anchor or direct the movement of organelles, and control movement of cilia, flagella, pseudopods, and even contraction of the whole cell. At least three types of fibers are involved in the cytoskeleton: microtubules, microfilaments, and intermediate filaments.

Current research is focusing on how the components of the cytoskeleton are integrated. High-voltage electron microscopy seems to reveal that microtubules radiate from an organizing region called the *cell center* to provide the forms that organize the cytoskeleton. One current theory suggests the existence of short protein fibers that cross-link all the elements of the cytoskeleton into one great network known as a "microtrabecular lattice." Other researchers argue that the cross-links are an artifact of sample preparation.

Microtubules are hollow rods constructed of two kinds of globular proteins called tubulins. In addition to providing the major framework of the cytoskeleton, microtubules serve as tracks to guide the movement of organelles, such as vesicles moving from the Golgi complex to the plasma membrane. Bundles of microtubules near the plasma membrane contribute to cell shape.

The separation of chromosomes in a dividing cell is associated with microtubules assembled from the cell center. In animal cells, *centrioles*, composed of nine sets of triplet microtubules arranged in a ring, may help to organize microtubule assembly for cell division.

Cilia and *flagella* are extensions of eukaryotic cells, composed of and moved by microtubules. Cilia are numerous and short; flagella occur one or two to a cell and are longer. Protoctists use cilia or flagella to swim (or crawl; as Peter Satir explained in the interview) through aqueous media. Sperm of animals and lower plants are propelled by flagella. Cilia or flagella attached to stationary cells of a tissue serve to sweep fluid past the cell.

Cilia and flagella are composed of a core of nine doublets of microtubules arranged in a ring and surrounding two single microtubules (a nearly universal 9 + 2 arrangement). The sliding of the microtubule doublets past each other occurs as arms, composed of the protein dynein, alternately attach to adjacent doublets, bend down, and then release and bend back. This action, driven by ATP, causes the bending of the flagella or cilia. A *basal body*, structurally identical to a centriole, anchors the nine doublets of the cilium or flagellum to the cell.

Microfilaments are solid rods consisting of a helix of two chains of molecules of the globular protein *actin*. In muscle cells, thousands of microfilaments of actin interdigitate with thicker filaments of *myosin*. The sliding of actin and myosin filaments past each other, driven by ATP-powered arms extending between the filaments, causes the shortening of the cell and thus the contraction of muscles.

Microfilaments seem to be present in nearly all eukaryotic cells. They function in support, such as in the core of microvilli that project from the surface of cells specialized for absorption of material across the plasma membrane. Miniature versions of the actin and myosin muscle arrangement are found in many cells in areas responsible for localized contractions. The pinching apart of animal cells when they divide and the extension and retraction of pseudopods involve actin–myosin aggregates. Cyclosis, or *cytoplasmic streaming*, in plant cells probably involves microfilaments.

Intermediate filaments are intermediate in size between microtubules and microfilaments. They are abundant in many cells and contribute to the cytoskeleton, but little is known about their exact function.

The Cell Surface

Most cells synthesize and secrete some sort of covering over their plasma membrane. A *cell wall* is a diagnostic feature of plant cells, and is also found in prokaryotes, fungi, and some protoctists. Plant cell walls are composed of fibers of *cellulose* embedded in a matrix of other polysaccharides and some protein. The exact composition of the wall varies between species and even between cell types in the same plant.

The primary cell wall, secreted by a young plant cell, is relatively thin and flexible and can stretch as the cell grows. Adjacent cells are connected by the *middle lamella*, a thin layer of polysaccharides called pectins that glue the cells together. When they stop growing, some cells secrete a thicker and stronger secondary cell wall between the plasma membrane and primary cell wall.

Animal cells often secrete a *glycocalyx* made of sticky oligosaccharides (molecules of a "few" saccharides) that serves to strengthen the cell surface and glue cells together. The glycocalyx probably functions in cell–cell recognition.

Intercellular junctions between membranes of neighboring cells improve communication and interaction between cells. The three main types of intercellular junctions between animal cells are desmosomes, tight junctions, and gap junctions. *Desmosomes* are clumps of intercellular filaments that connect two cells and terminate inside each cell as a disk of dense material reinforced by intermediate filaments. Desmosomes connect cells into strong epithelial sheets. *Tight junctions* are connections between cells that create an impermeable layer within the intercellular space. Specialized proteins built into one plasma membrane bond to similar proteins in the other membrane. Tight junctions are found in epithelial layers that separate two solutions and must prevent intercellular transport across the layer of cells. *Gap junctions* allow for the exchange of sugars, amino acids, and small molecules between cells. They consist of special doughnut-shaped proteins embedded in bordering plasma membranes that connect across the intercellular space.

Plasmodesmata are channels in plant cell walls through which strands of cytoplasm connect bordering cells, and water and small solutes can move. The plasma membranes of adjacent cells are continuous through the channel.

STRUCTURE YOUR KNOWLEDGE

1. This table lists the general functions performed by a cell. List the structures associated with each of these functions.

FUNCTIONS OF CELL	ASSOCIATED ORGANELLES AND STRUCTURES
Reproduction of cell	
Energy conversions	
Directions and control	
Manufacturing	
Digestion, recycling	
Specific pathways	
Cell–cell interaction	
Movement, and structural integrity	

2. Create a modified concept map or flow chart to trace the development of a secretory product (such as a digestive enzyme in a pancreatic cell) from the DNA code to its export from the cell.

TEST YOUR KNOWLEDGE

MULTIPLE CHOICE: *Choose the one best answer.*

1. Which of the following is/are not found in a prokaryotic cell?
   a. ribosomes
   b. plasma membrane
   c. mitochondria
   d. both a and c

2. Eukaryotic cells can be larger than prokaryotic cells because
   a. their plasma membrane is more permeable.
   b. their internal membrane system allows compartmentalization of functions and extra surface area for exchange and enzyme location.
   c. their DNA is localized in the nucleus.
   d. they have a better surface-area–to–volume ratio.

3. Which of the following is not a similarity among the nucleus, chloroplasts, and mitochondria?
   a. They all contain DNA.
   b. They all are bounded by a double phospholipid bilayer membrane.
   c. They can all divide to reproduce themselves.
   d. They are all derived from the endoplasmic reticulum system.

4. Membrane proteins can perform all the following functions except
   a. serving as pumps and channels for transport of molecules.
   b. serving as enzymes in metabolic pathways.
   c. reading of mRNA and synthesis of polypeptide chains.
   d. serving as receptor sites for chemical signals on developing polypeptide chains.

5. The nucleolus functions in
   a. synthesis and storage of components of ribosomes.
   b. formation of spindle fiber.
   c. condensation of chromosomes.
   d. direction of RNA synthesis.

6. The largest number of ribosomes
   most likely would be found in a cell
   a. with a high metabolic rate.
   b. that produces secretory products.
   c. with many cilia.
   d. that is actively dividing.

7. Which structure is not considered to be
   part of the endomembrane system?
   a. mitochondrion
   b. smooth ER
   c. nuclear envelope
   d. lysosome

8. A plant cell grows larger primarily by
   a. increasing the number of vacuoles.
   b. synthesizing more cytoplasm.
   c. taking up water into its central vacuole.
   d. synthesizing more cellulose.

9. The innermost portion of a plant cell wall is the
   a. primary wall.
   b. secondary wall.
   c. middle lamella.
   d. glycocalyx.

10. The contractile elements of muscle cells are
    a. smooth ER.
    b. centrioles.
    c. microtubules.
    d. microfilaments.

11. Microtubules are components
    of all of the following except
    a. centrioles.
    b. the spindle apparatus for separating
       chromosomes in cell division.
    c. the pinching apart of the cytoplasm in
       animal cell division.
    d. flagella and cilia.

12. Of the following, which is probably the
    most common route for membrane flow in
    the endomembrane system?
    a. rough ER—>Golgi—>lysosomes—>vesicles—>
       plasma membrane
    b. rough ER—>smooth ER—>Golgi—>vesicles—>
       plasma membrane
    c. nuclear envelope—>rough ER—>Golgi—>
       smooth ER—>lysosomes
    d. rough ER—>vesicles—>Golgi—>smooth ER—>
       plasma membrane

13. Proteins to be secreted by the
    cell are generally synthesized
    a. with signals that result in the ribosomes
       binding to ER.
    b. by free polysomes.
    c. by the nucleolus.
    d. within the Golgi complex.

14. Many of the proteins made
    by rough ER become part of
    a. the ribosomes.
    b. the cytosol.
    c. membranes.
    d. the cytoskeleton.

15. Plasmodesmata in plant
    cells are similar in function to
    a. desmosomes.
    b. tight junctions.
    c. gap junctions.
    d. glycocalyx.

FILL IN THE BLANKS *with the appropriate cellular organelle or structure.*

1. _____ transports membranes and
   products to various locations

2. _____ infolding of mitochondrial
   membrane with attached enzymes

3. _____ sticky, supportive coat on
   animal cells

4. _____ small sacs with enzymes for
   specific metabolic pathway

5. _____ system of flattened sacs inside
   chloroplasts

6. _____ anchoring structure for cilia and
   flagella

7. _____ semifluid medium between
   nucleus and plasma membrane

8. _____ system of fibers that maintains
   cell shape, anchors organelles

9. _____ connection between cells that
   creates an impermeable layer

10. _____ membrane surrounding central
    vacuole of plant cells

MATCHING: *Match the function and structure with the proper cell organelle.*

FUNCTION	STRUCTURE
A. produce rRNA, assemble ribosomes	a. network of membranes with attached ribosomes
B. manufacture proteins	b. double-membrane sac containing stacks of membrane sacs
C. contain chromosomes, control center	c. large and small subunits, attach to mRNA
D. site of cellular respiration	d. sac with hydrolytic enzymes, acid pH
E. site of photosynthesis	e. round structure with multiple copies of genes for rRNA
F. store food, pump water, form center of plant cells	f. single-membrane-bound sac
G. maintain cell shape	g. surrounded by double membrane, inner layer of protein, with pores
H. form vesicles to transport membranes, rough ER products	h. large sac formed from small vesicles
I. digest macromolecules	i. double-membrane sac with cristae
J. manufacture membranes and secretory products	j. stack of flattened sacs
K. process products of rough ER	k. membrane network without attached ribosomes

FUNCTION	STRUCTURE	ORGANELLE
1. _____	_____	chloroplast
2. _____	_____	Golgi complex
3. _____	_____	lysosome
4. _____	_____	mitochondria
5. _____	_____	nucleolus
6. _____	_____	nucleus
7. _____	_____	ribosome
8. _____	_____	rough ER
9. _____	_____	smooth ER
10. _____	_____	vacuole

CHAPTER 8

TRAFFIC ACROSS MEMBRANES

FRAMEWORK

This chapter presents the fluid–mosaic model of membrane structure and relates the molecular arrangement of biological membranes to their function of regulating the passage of substances into and out of the cell and between intracellular compartments.

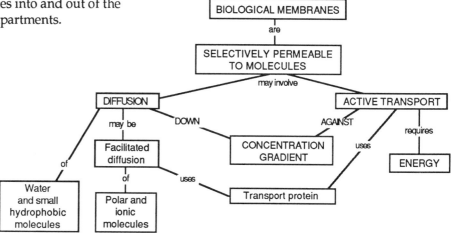

CHAPTER SUMMARY

The plasma membrane is the boundary of life, providing the cell with the potential to maintain a unique internal environment and to control the movement of materials into and out of the cell. Biological membranes are *selectively permeable*; their ability to discriminate in the chemical exchanges they allow is based on their unique structure.

Models of Membrane Structure

Scientists have created a series of models to explain membrane structure and function. The observation that lipid-soluble molecules rapidly entered cells led to the conclusion that membranes were composed of lipids. In the early 1900s, membranes isolated from red blood cells (called ghosts) were found to consist of proteins as well as lipids.

The *phospholipid* components of membranes are am-

phipathic, having both a hydrophobic region consisting of two long, nonpolar hydrocarbon tails and a hydrophilic head created by the phosphate group. A bilayer of phospholipids, with the hydrophobic tails in the center and the hydrophilic heads facing the aqueous solution on both sides and covered with a coat of globular proteins, was proposed as the molecular model of membranes by Davson and Danielli in 1935. This molecular-sandwich model, later applied to all membranes in the cell, was consistent with the observed thickness of plasma membranes (adjusting the globular proteins to layers of protein in the pleated-sheet configuration) and the apparent triple-layer staining results seen with the electron microscope.

The molecular variability among cellular membranes and the amphipathic nature of membrane proteins created problems in the Davson–Danielli model (where would the hydrophobic regions fit between the aqueous environment and the hydrophilic heads of the phospholipids?). In 1972, Singer and Nicolson proposed their *fluid–mosaic model* in which the amphipathic membrane proteins are embedded in the phospholipid bilayer with their hydrophilic regions extending out into the aqueous environment. The phospholipid bilayer is envisioned as a fluid with a mosaic of individually inserted protein molecules shifting laterally within it. This model is consistent with the known properties of membranes and is supported by evidence from *freeze-fracture* electron microscopy that shows the interior of the bilayer with proteins dispersed in a smooth matrix. The fluid–mosaic model is currently the most accepted and useful model for organizing the existing knowledge of membrane structure.

Membranes are held together primarily by hydrophobic attractions that allow the lipids and proteins to drift about laterally, keeping their hydrophobic regions in contact. The intermingling of membrane proteins of a hybrid mouse–human cell illustrates the fluid nature of membranes. This fluidity is affected by temperature; an abundance of phospholipids with unsaturated hydrocarbon tails maintains membrane fluidity at lower temperatures. The steroid cholesterol, common in animal cell membranes, also prevents the close packing of phospholipids and thus enhances fluidity. A cell may change the molecular composition of its membrane in response to changing temperature.

Each membrane has its own unique complement of membrane proteins, which determine the specific function of that membrane. *Integral proteins* extend into the hydrocarbon regions of the lipids. They may be unilateral and reach partway through the membrane or transmembrane, having two hydrophilic ends and a hydrophobic midsection. *Peripheral proteins* are attached to the surface of the membrane. Membranes are asymmetric; they have distinct inner and outer faces related to the directional orientation of their proteins, the composition of the lipid bilayers, and, in the plasma membrane, the attachment of carbohydrates to the exterior surface.

Cell–cell recognition, important in development and the immune response to foreign cells, is most likely accomplished by the ability of cells to recognize membrane carbohydrates. These branched oligosaccharides (fewer than 15 sugar units) are covalently bonded either to lipids or, most often, to proteins. The identity and location of these glycolipids and glycoproteins vary from species to species, individual to individual, and even among cell types.

The critical ability of a membrane to regulate the passage of substances between cellular compartments is determined by its supramolecular structure, the unique architecture determined by its many molecules arranged into a higher level of organization. The plasma membrane permits a regular exchange of nutrients, waste products, oxygen, and inorganic ions. This steady traffic of small molecules is regulated in a variety of ways.

Traffic of Small Molecules

Biological membranes are selectively permeable; there is a difference in the ease and rate with which small molecules pass through them. The hydrophobic center of the lipid bilayer impedes passage of ions and polar molecules. The rates at which hydrophobic molecules cross an artificial membrane are related to their lipid solubilities and size. Small polar, but uncharged, molecules, such as H_2O and CO_2, can cross the synthetic membrane rapidly; however, large polar molecules, such as glucose, and all ions cannot. These molecules can enter a living cell because of the membrane proteins that regulate permeability.

Transport proteins span the plasma membrane and provide the mechanism for movement of ions and moderately sized polar molecules (such as sugars). These highly specific proteins provide either a channel or physical binding and transport for the molecules they move.

Diffusion is the movement of a substance down its concentration gradient due to random molecular motion. This spontaneous process decreases free energy and increases entropy (disorder, randomness) in a system. Much of the exchange across membranes occurs by diffusion, including the important exchange of O_2 and CO_2 in respiration. The cell does not expend energy when substances diffuse across membranes down their concentration gradient; the process is called *passive transport*.

The diffusion of water across a selectively permeable membrane is a special case of passive transport called *osmosis*. Water will move from a *hypotonic* solution across a membrane into a *hypertonic* solution—in other words, from a region with a lesser concentration of solutes into a region with a greater concentration of solutes. *Isotonic* solutions have equal solute concentrations, and there is no net movement of water between them. Water diffuses down its own concentration gradient, which is affected by the binding of water molecules to solute particles. The association between water and solute molecules lowers the proportion of unbound water that is free to cross the membrane. The direction of osmosis is determined by the difference in *total* solute concentration, regardless of the kinds of solute molecules involved.

Osmotic pressure, which can be determined by an instrument called an osmometer, is a measure of the tendency for a solution to take up water when separated from pure water by a selectively permeable membrane. This pressure is proportional to the solute concentration; it increases as the concentration of solute particles increases. Water moves from a solution with a lesser osmotic pressure (fewer solute particles—hypotonic) to a solution with a greater osmotic pressure (hypertonic).

An animal cell placed in a hypertonic environment will lose water and shrivel. If placed in a hypotonic environment, the cell will gain water, swell, and possibly lyse (burst). Cells without rigid walls must either live in an isotonic environment (salt water or isotonic body fluids) or have adaptations for *osmoregulation*, the control of water balance. Cells in fresh water environments may have membranes that are less permeable to water and contractile vacuoles that expel excess water.

The cell walls of plants, fungi, prokaryotes, and some protoctists play a role in water balance within hypotonic environments. The inward movement of water is offset by pressure within the cell against its cell wall, resulting in an equilibrium movement of water and creating a *turgid* cell. Turgid cells provide mechanical support for nonwoody plants. Plant cells in an isotonic surrounding are *flaccid*. In a hypertonic medium, plant cells undergo *plasmolysis*, the pulling away of the plasma membrane from the cell wall as the cell shrivels.

Facilitated diffusion is the diffusion of polar molecules and ions across a membrane with the aid of a *carrier protein*. Carrier proteins are similar to enzymes in several ways: they are specialized for the solute they transport, presumably with a specific binding site for that solute; they speed up the transport of solute; they can reach a maximum rate of transport when all carrier proteins are saturated by a high concentration of the solute; and they can be inhibited by molecules that resemble their normal solute. A proposed model for the mechanism of carrier-protein function involves a change in conformation of the protein caused by the binding of the solute that serves to translocate the binding site from one side of the membrane to the other. Facilitated diffusion is passive transport; it speeds diffusion but cannot change the direction of movement of a solute down its concentration gradient.

Active transport, requiring the expenditure of energy by the cell, can move solutes against their concentration gradients. Active transport is essential for a cell to be able to maintain internal concentrations of small molecules that are different from the environment. Transport proteins that engage in active transport use cellular energy, usually in the form of ATP, to pump molecules against their concentration gradient. The terminal phosphate group of ATP may be transferred to the protein, inducing it to change its conformation and translocate the bound solute across the membrane. The *sodium–potassium pump*, also called sodium–potassium ATPase because it acts as an enzyme to hydrolyze ATP, allows the cell to exchange Na^+ and K^+ across the membrane in animal cells, creating a greater concentration of potassium ions and a lesser concentration of sodium ions within the cell.

Cells have a *membrane potential*, a voltage across their plasma membrane created by the electrical potential energy resulting from the separation of opposite charges. Due to several factors, the cytoplasm is negatively charged compared to the extracellular fluid. At cellular pH, most proteins and other macromolecules are negatively charged. Ions diffuse through the selectively permeable plasma membrane at different rates. For instance, K^+ leaks out of the cell faster than Na^+ leaks back in. There is also evidence that the sodium–potassium pump actually pumps three Na^+ ions out of the cell for every two K^+ ions it pumps in. An electrogenic pump is a transport protein that generates voltage across a membrane by active transport of ions. A proton pump that transports H^+ out of the cell helps to generate membrane potential in plants, fungi, and bacteria.

The membrane potential favors the diffusion of cations into the cell and anions out of the cell. Diffusion of an ion is affected by both the membrane potential and its own concentration gradient; thus an ion diffuses down its *electrochemical gradient*.

Co-transport is a mechanism through which the active transport of one solute is indirectly driven by an ATP-powered pump that transports another solute against its gradient. Energy is stored by concentrating this substance on one side of the membrane. As the solute diffuses back across the membrane through specific transport proteins, other solutes may be co-transported against their own concentration gradients. The proton pump of bacterial cells may drive the active transport of amino acids and sugars by coupling them with hydrogen ions that diffuse through specific symport proteins. (A symport translocates two different molecules simultaneously in the same direction. An antiport exchanges two solutes by transporting one into the cell and the other out of the cell.)

Traffic of Large Molecules

Large molecules, such as proteins and polysaccharides, are transported across the plasma membrane through the processes of exocytosis and endocytosis. In *exocytosis*, the cell secretes macromolecules by the fusion of vesicles containing these molecules with the plasma membrane. In *endocytosis*, a region of the plasma membrane sinks inward and pinches off to form a vesicle containing material that had been outside the cell. *Phagocytosis* is a form of endocytosis in which pseudopodia wrap around a food particle and create a vacuole. The vacuole fuses with a lysosome containing hydrolytic enzymes. In *pinocytosis*, droplets of extracellular fluid are taken into the cell in small vesicles. *Receptor-mediated endocytosis* allows for the acquisition of specific substances that bind with receptor sites on the cell surface.

Membranes and ATP Synthesis

The existence of concentration gradients across a membrane can serve as an energy source for the generation of ATP. The diffusion of H^+ down its concentration gradient through a transport protein can cause the protein to add a phosphate group to ADP. The membranes of mitochondria and chloroplasts make ATP by using energy from food and light, respectively, to generate concentration gradients of H^+.

STRUCTURE YOUR KNOWLEDGE

1. Relate the structure of membranes as depicted by the fluid–mosaic model to their permeability to nonpolar hydrophobic molecules and polar or ionic molecules.
2. Compare facilitated diffusion and active transport in relation to the type of molecules carried, the membrane components involved, and the energy required.
3. Create a concept map to illustrate your understanding of the important process of osmosis.

TEST YOUR KNOWLEDGE

MULTIPLE CHOICE: *Choose the one best answer.*

1. Glycoproteins and glycolipids are important for
   a. facilitated diffusion.
   b. active transport.
   c. cell–cell recognition.
   d. co-transport.

2. Amphipathic molecules
   a. have hydrophilic and hydrophobic regions.
   b. include phospholipids and integral membrane proteins.
   c. may be held together by hydrophobic interactions.
   d. include all of the above.

3. Which of the following is not true about osmosis?
   a. It increases free energy in a system.
   b. Water moves from a hypotonic to a hypertonic solution.
   c. Osmotic pressure increases with increasing solute concentration.
   d. It increases the entropy in a system.

4. One of the problems with the Davson–Danielli molecular-sandwich model for membranes was that
   a. it did not explain the easy permeability by hydrophobic molecules.
   b. it did not account for the actual thickness of the plasma membrane.
   c. it could not explain where the hydrophobic regions of the membrane proteins would align.
   d. it could not account for the membrane potential.

5. Evidence for the fluid–mosaic model of membrane structure came from
   a. the freeze-fracture technique of electron microscopy.
   b. the movement of proteins in hybrid cells.
   c. the amphipathic nature of membrane proteins.
   d. all of the above.

6. Membrane proteins that function in active transport are likely to be
   a. peripheral proteins.
   b. integral proteins.
   c. ATPases.
   d. both b and c.

7. Ions diffuse across membranes down their
   a. electrochemical gradients.
   b. chemical gradients.
   c. electrical gradients.
   d. concentration gradients.

8. The fluidity of membranes in cold weather may be maintained by
   a. increasing the number of phospholipids with saturated hydrocarbon tails.
   b. activating an H^+ pump.
   c. increasing the concentration of cholesterol in the membrane.
   d. thickening the phospholipid layer.

9. Water, a polar molecule, can easily pass through membranes because
   a. it is small enough to pass between the lipids of the membrane.
   b. it always has a strong concentration gradient.
   c. it creates enough pressure to force its way through the membrane.
   d. it has specialized carrier molecules that facilitate its diffusion.

10. A plant cell placed in a hypotonic environment will
    a. plasmolyze.
    b. lyse.
    c. become turgid.
    d. become flaccid.

11. Which of the following is not true of the carrier molecules involved in facilitated diffusion?
    a. They increase the speed of transport across a membrane.
    b. They can concentrate solute molecules on one side of the membrane.

    c. They have specific binding sites for the molecules they transport.

    d. They may undergo a conformational change upon binding of solute.

12. The membrane potential of a cell may be created
    a. by the pumping of ions across the cell membrane.
    b. the selective permeability of membranes to various ions.
    c. an electrogenic pump.
    d. all of the above.

13. Co-transport may involve
    a. passive transport of two solutes.
    b. a symport that actively transports one solute in tandem with another that is diffusing down its concentration gradient.
    c. ion diffusion against the electrochemical gradient.
    d. receptor-mediated endocytosis.

14. Exocytosis involves all of the following except
    a. the pinching in of the plasma membrane to form a vesicle.
    b. the fusion of a vesicle, usually from the ER or Golgi complex, with the plasma membrane.
    c. a mechanism to transport carbohydrates to the outside of plant cells during the formation of cell walls.
    d. a mechanism to rejuvenate the plasma membrane.

15. Solute gradients can function as an energy source in
    a. co-transport.
    b. the synthesis of ATP.
    c. the membrane potential of a membrane.
    d. all of the above.

Use this U-tube setup to answer questions 16-18.
    The solutions in the two arms of this U-tube are separated by a membrane that is permeable to water and glucose but not to sucrose. Side A is filled with a solution of 2 M sucrose and 1 M glucose. Side B is filled with 1 M sucrose and 2 M glucose.

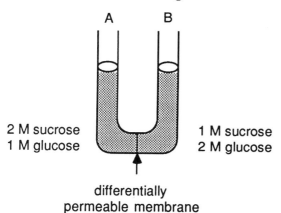

A   B

2 M sucrose
1 M glucose

1 M sucrose
2 M glucose

differentially
permeable membrane

16. Initially, the solution in side A, with respect to that in side B, is
    a. hypotonic.
    b. hypertonic.
    c. isotonic.
    d. saturated.

17. After the system reaches equilibrium, what changes are observed?
    a. The water level is higher in side A than in side B.
    b. The water level is higher in side B than in side A.
    c. The molarity of glucose is higher in side A than in side B.
    d. The molarity of sucrose has increased in side A.

18. During the period before equilibrium is reached, which molecule(s) will show net movement through the membrane?
    a. water
    b. glucose
    c. water and glucose
    d. water and sucrose

19. Most marine organisms are
    a. hypertonic to their environment.
    b. isotonic to their environment.
    c. hypotonic to their environment.
    d. turgid in their environment.

20. Pinocytosis involves
    a. the fusion of a small vacuole with a lysosome.
    b. receptor-mediated endocytosis.
    c. the pinching in of the plasma membrane around droplets of external fluid.
    d. all of the above.

CHAPTER 9

RESPIRATION:
HOW CELLS HARVEST CHEMICAL ENERGY

FRAMEWORK

This chapter details the intricate processes through which cells harvest energy from organic molecules. The catabolic pathways of glycolysis and respiration capture the chemical energy in glucose and other fuels and store it in ATP. *Glycolysis*, occurring in the cytosol, produces a small amount of ATP, and pyruvic acid and NADH, which then enter the mitochondrion for *respiration*. The mitochondrion consists of a matrix in which the enzymes of the *citric acid cycle* are localized, a highly folded inner membrane in which enzymes and the molecules of the electron transport chain are embedded, and an intermembrane space between the double membranes to temporarily house H^+ pumped across the inner membrane during the *redox reactions* of the *electron transport chain*. This *proton motive force* drives ATP formation by ATPases located in the membrane through a process known as *chemiosmosis*.

CHAPTER SUMMARY

Cells require energy to perform the work associated with living: synthesis of macromolecules, pumping of substances across membranes, movement, reproduction, and maintenance of their complex, ordered structure. For most cells, this energy ultimately comes from the sun through the organic molecules produced by photosynthetic organisms. Cells obtain the chemical energy stored in their food by systematically degrading, with the help of enzymes, the energy-rich molecules into simpler, energy-poor waste products.

Cellular respiration, the most common catabolic pathway, breaks down glucose (or other energy-rich organic compounds) with the use of oxygen to obtain energy in the usable form of ATP. This exergonic process has a free energy change of –686 kcal/mol glucose. The waste products of respiration, CO_2 and H_2O, are the raw materials chloroplasts use for photosynthesis to produce glucose and return oxygen

to the air. The chemical elements essential to life are recycled. Energy, however, is not. It is converted from the ordered form of light, stored temporarily in chemical bonds, to the disordered form of heat.

How Cells Make ATP: An Introduction

ATP is very reactive because of the instability of the bonds between its three phosphate groups. Hydrolysis of the terminal high-energy phosphate bond of one mole of ATP releases 7.3 kcal. Generally, the phosphate group with its unstable bond is shifted by enzymes to another molecule. This phosphorylation primes the molecule to undergo a change that performs work. In doing this work, the substance loses the phosphate group by hydrolysis. The cell must regenerate its supply of ATP by phosphorylating ADP—an endergonic reaction requiring 7.3 kcal/mol. Respiration supplies the energy for ATP synthesis.

Respiration occurs in three stages: *glycolysis*, the *citric acid cycle*, and *electron transport*. Glycolysis and the citric acid cycle generate some ATP, but they primarily provide electrons for the electron transport chain that powers the enzymes that make ATP. For each molecule of glucose respired to carbon dioxide and water, the cell makes 36 to 38 molecules of ATP.

Substrate-level phosphorylation involves the direct transfer of phosphate to ADP from compounds with high-energy phosphate bonds. Both ADP and a high-energy intermediate compound, from glycolysis or the citric acid cycle, are bound to an enzyme and the phosphate group is shifted to ADP. This reaction is energetically possible because the $-\Delta G$ for the hydrolysis of the intermediate compound is greater than the $+\Delta G$ for the phosphorylation of ADP. A small percentage of ATP is formed this way in most cells.

The electron transport chain produces most cellular ATP through a process called *chemiosmosis*. The molecules of the electron transport chain, located in the inner membrane, the cristae, of the mitochondrion, pump H^+ from the mitochondrial matrix into the intermembrane space. The proton gradient thus formed stores energy that can be used by the enzyme *ATPase* to make ATP.

Mitchell received the Nobel prize for his formulation of the chemiosmotic theory of ATP synthesis. The energy for the generation of the proton gradient that powers the ATPases (also called ATP synthetases) comes from light in photosynthesis or from catabolism of food in respiration.

Redox Reactions in Metabolism

Oxidation-reduction or *redox reactions* involve the partial or complete transfer of one or more electrons from one reactant to another. The substance that loses electrons is *oxidized* and acts as a *reducing agent* to the substance that gains electrons and is thus *reduced*. The substance that is reduced acts as an *oxidizing agent* by accepting electrons from the substance that is oxidized. Oxygen, which is highly electronegative, attracts electrons strongly and is one of the most powerful oxidizing agents.

Organic molecules with an abundance of hydrogen are rich in high-energy electrons. As these electrons move to electronegative atoms, they give up potential energy. Spontaneous redox reactions release energy most often as heat. In respiration, the energy from the oxidation of glucose and the reduction of oxygen is captured in ATP. Respiration is a redox process that transfers the high-energy electrons of hydrogen in organic molecules to oxygen, producing water with low-energy electrons.

The electrons that are removed in the catabolic breakdown of glucose are usually transferred to a special oxidizing agent and electron acceptor called *NAD$^+$*. Dehydrogenases, the enzymes that catalyze these redox reactions, remove two hydrogen nuclei with two electrons from the substrate. The two electrons and one of the hydrogen nuclei are passed to NAD$^+$, forming the reduced compound *NADH*, and the other H$^+$ is released to the environment. NAD$^+$ functions as a coenzyme in the enzymatic process; it enters the active site along with the substrate. NADH captures most of the energy of the high-energy electrons and is used in *oxidative phosphorylation*, the redox reactions of the electron transport chain that lead to the chemiosmotic production of ATP.

Aerobic and Anaerobic Catabolism

In the three major catabolic schemes—aerobic respiration, anaerobic respiration, and fermentation—glucose or other fuels are oxidized and their high-energy electrons are passed to NAD$^+$. The ultimate acceptor of these electrons is oxygen for *aerobic respiration*. Organisms that rely on aerobic respiration for energy are called *strict aerobes*; they must have oxygen to survive. *Strict anaerobes*, found in a few groups of bacteria, are poisoned by oxygen. The final electron acceptor in anaerobic respiration is a substance such as sulfate or nitrate. Both anaerobic and aerobic respiration are considered to be cellular respiration; they involve catabolic pathways that use electron transport chains to make ATP. The physical process of breathing or respiration that provides oxygen for cells that use aerobic respiration is different from the chemical use of oxygen in cellular respiration.

Fermentation produces ATP by substrate-level phosphorylation without the use of an electron transport chain. *Facultative anaerobes* can make ATP by fermentation or respiration, depending upon whether oxygen is available.

Glycolysis

Glycolysis, the breakdown of glucose to pyruvic acid, is common to fermentation and respiration. This metabolic method, occurring in all organisms, probably evolved as the method of ATP formation by early prokaryotes before oxygen was available in the atmosphere. The cytoplasmic location of glycolysis, not using any of the membrane-bound organelles of eukaryotic cells, also implies great antiquity. In glycolysis, a glucose molecule is split into two three-carbon molecules, two molecules of NAD$^+$ are reduced, and there is a net production of two ATPs by substrate-level phosphorylation. Most of the energy of this exergonic pathway ($-\Delta G$ = 140 kcal/mol glucose) is conserved in the high-energy bonds of NADH and ATP.

When oxygen is present, glycolysis serves as the first step of respiration, with the citric acid cycle and electron transport continuing in the mitochondria. Without oxygen, glycolysis is part of fermentation—the anaerobic conversion of sugar to a waste product that regenerates NAD$^+$ and captures a small percentage (2 ATP/sugar molecule) of the energy stored in glucose.

In *alcohol fermentation*, pyruvic acid is converted into ethanol (ethyl alcohol) and CO_2 is released. The alcohol fermentation of the yeasts used in baking or brewing provides CO_2 for leavening bread or alcohol for beer and wine production. In *lactic acid fermentation*, pyruvic acid is reduced by NADH to form lactic acid and recycle NAD$^+$; no CO_2 is released. The dairy industry uses the lactic acid fermentation of certain fungi and bacteria to make cheese and yogurt. Muscle cells make ATP by lactic acid fermentation in vigorous exercise when energy demand is high and oxygen supply is low. Aerobic exercise occurs when the heart and lungs adjust to the stress and increase their output so that muscle cells can revert to aerobic respiration.

The Citric Acid Cycle

When oxygen is present, the pyruvic acid from glycolysis (containing much of the energy that was stored in the glucose molecule) enters the mitochondrion to undergo the citric acid cycle. Within a multienzyme complex, acetyl coenzyme A, or *acetyl CoA*, is formed from pyruvic acid by the removal of the carboxyl group and release of CO_2, and the attachment of a coenzyme to the remaining acetyl group. This molecule then shunts its acetyl group into the pathway known as the *Krebs cycle*, tricarboxylic acid cycle, or citric acid cycle.

For each turn of the Krebs cycle, two carbons enter in the reduced state in an acetyl group, two carbons exit completely oxidized as CO_2, three molecules of NAD^+ are reduced to NADH, one molecule of reduced *FADH$_2$* is formed, and one ATP molecule is made by substrate-level phosphorylation. The NADH and FADH$_2$ proceed to the electron transport chain, where most of the ATP formed from respiration is made by oxidative phosphorylation.

The Electron Transport Chain and Oxidative Phosphorylation

The *electron transport chain* is a series of electron carrier molecules located in the inner membrane of the mitochondrion. Through a sequence of redox reactions, the electrons from NADH are passed from molecule to molecule until they finally reach oxygen. Most components of the chain are proteins with tightly bound coenzymes (prosthetic groups) that shift between reduced and oxidized states as they accept and donate electrons. NADH passes its high-energy electrons to a flavoprotein named FMN, which becomes oxidized as it passes the electrons to the next electron carrier, ubiquinone (Q). Q passes the electrons on to a series of molecules called *cytochromes*, proteins with a *heme* prosthetic group. The last cytochrome of the chain passes its electrons to oxygen, which picks up a pair of H^+ from the aqueous medium and forms water.

Electron transfer from NADH to oxygen is very exergonic ($-\Delta G = 53$ kcal/mol). The electron transport chain releases this energy in a series of shorter steps as electrons are passed on to the next, more electronegative acceptor. FADH$_2$, the other reduced electron acceptor of the citric acid cycle, adds its electrons to the chain at a lower energy level, and thus one-third less energy for ATP synthesis is derived from FADH$_2$ than from NADH.

Some electron carriers in the chain accept and release H^+ along with the electrons, whereas others only transport electrons. The electron transport chain generates a proton gradient when some carriers accept H^+ ions from the matrix side of the inner membrane and release them to the intermembrane space. The electron carriers are collected into three complexes that span the membrane. Electrons are transferred between the complexes by carriers that are mobile in the membrane. H^+ ions are pumped across the membrane at three points along the electron transport chain.

The potential energy stored in the proton gradient is referred to as the *proton-motive force*. The force consists of an electrochemical gradient created by the concentration gradient and the voltage across the membrane due to the uneven H^+ concentrations. The electrochemical gradient drives H^+ across the membrane through the ATPase complexes that synthesize ATP. This chemiosmotic mechanism accounts for most of the ATP formed from cellular respiration.

Respiratory poisons provide evidence for the chemiosmotic model. Some poisons, such as cyanide, block the passage of electrons along the electron transport chain. No protons are pumped, and no ATP is made. A class of poisons called uncouplers make the membrane leaky to H^+. The electron chain works, oxygen consumption increases, but leakage of ions destroys the proton gradient and no ATP is made. Another group of poisons inhibits the ATPase. The proton gradient generated is of a higher magnitude than usual because the gradient is not diminished by H^+ moving through the ATPase, and no ATP is made.

The tally for ATP formed from the oxidation of a molecule of glucose to six molecules of carbon dioxide includes the substrate-level phosphorylation of ATP in glycolysis (4 ATP) and the citric acid cycle (2 ATP) and the ATP generated by chemiosmotic mechanisms in the electron transport chain (34 ATP—3 for every 2 NADH and 2 for every FADH$_2$) minus the 2 ATP required initially to phosphorylate glucose and 2 ATP for the passage of the 2 NADH formed during glycolysis into the mitochondrion. The total is 36 ATP. In prokaryotes, the yield is 38 because NADH from glycolysis does not have to pass its electrons across a mitochondrial membrane. The yield of ATP per molecule of glucose oxidized is only an estimate because electron transport and ATP synthesis are loosely linked by the proton gradient. The use of the gradient to transport solutes across the mitochondrial membrane, and differences in the permeability of the membrane to H^+, may affect the yield of ATP from respiration.

Catabolism of Other Molecules

Fats, proteins, disaccharides, starches, and glycogen can all be used by cellular respiration to make ATP. Polysaccharides are hydrolyzed to provide glucose. Digestion of disaccharides provides glucose and other monosaccharides that can be enzymatically converted to glucose for respiration. Proteins can be used for fuel after they are digested into their constituent amino acids and then deaminated by the removal of their amino groups. Depending on its structure, the amino acid can enter into respiration at several sites (pyruvic acid, acetyl CoA, or as an intermediate of the citric acid cycle). The digestion of fats yields glycerol, which is converted to an intermediate of glycolysis, and fatty acids, which are broken down to two-carbon fragments that enter the citric acid cycle as acetyl CoA. Fats are important fuels because they are rich in hydrogen, which has high-energy electrons. A gram of fat produces more than twice as much ATP as does a gram of carbohydrate.

Biosynthesis

Not all the organic molecules of food are oxidized as fuel to make ATP; many provide the carbon skeletons the cell needs for biosynthesis. Some monomers from digestion, such as amino acids, can be directly incorporated into the cell's macromolecules. Many of the intermediates of glycolysis and the citric acid cycle serve as precursors for the cell's anabolic pathways.

The molecules of carbohydrates, fats and proteins can all be interconverted through the intermediates of metabolism to provide for the cell's needs. Metabolism is a very intricate and adaptable network.

Control of Respiration

Metabolic pathways are strictly controlled under the basic principles of supply and demand. Through feedback inhibition, the end product of a pathway inhibits the enzyme that initiates the pathway. Respiration itself is controlled by the supply of ATP in the cell. The allosteric enzyme that controls the third step of glycolysis, phosphofructokinase, is inhibited by ATP and activated by ADP or AMP. Phosphofructokinase is also inhibited by citric acid transported from the mitochondria into the cytosol, thus synchronizing the rates of glycolysis and the citric acid cycle. Other allosteric enzymes located at key intersections help to maintain metabolic balance. The processes through which cells harvest energy and raw materials from food molecules are intricately connected and controlled.

STRUCTURE YOUR KNOWLEDGE

1. The following table lists the major divisions of the energy-producing metabolic pathways of the cell. Fill in the location of the pathways and the major molecules that enter and exit for each molecule of glucose processed.

PROCESS	LOCATION	MOLECULES IN	MOLECULES OUT
Glycolysis			
Fermentation			
Citric acid cycle			
Electron transport chain			

2. There are two ways in which ATP is produced by the cell during respiration. Create a concept map to organize your understanding of substrate-level phosphorylation and oxidative phosphorylation.

TEST YOUR KNOWLEDGE

MULTIPLE CHOICE: *Choose the one best answer.*

1. Which of the following is not true of a substance that is phosphorylated?
   a. It has an unstable phosphate bond.
   b. Its reactivity has been increased; it is primed to do work.
   c. It has entered the oxidative phosphorylation process.
   d. None of the above is true.

2. Chemiosmosis involves
   a. the diffusion of water down a concentration gradient.
   b. a proton-motive force that then drives ATP formation.
   c. a proton gradient that drives the redox reactions in the cristae.
   d. an ATPase that pumps H^+ across the mitochondrial membrane.

3. In a redox reaction,
   a. the substance that is reduced gains energy.
   b. the substance that is oxidized gains energy.
   c. the oxidizing agent donates electrons.
   d. the reducing agent is the most electronegative.

4. Which of the following is not true of oxidative phosphorylation?
   a. It uses oxygen as the ultimate electron donor.
   b. It involves the redox reactions of the electron transport chain.
   c. It involves an ATPase located in the inner mitochondrial membrane.
   d. It produces three ATP for every two NADH that are oxidized.

5. Substrate-level phosphorylation
   a. involves the shifting of a phosphate group from ATP to a substrate.
   b. can use NADH or $FADH_2$.
   c. accounts for all the ATP formed by anaerobic respiration.
   d. takes place both in the cytosol and in the mitochondrial matrix.

6. Facultative anaerobes
   a. are poisoned by oxygen.
   b. make ATP by fermentation or aerobic respiration, depending on availability of oxygen.
   c. must use some molecule other than oxygen as the final electron acceptor.
   d. include plants that can produce their own oxygen but undergo anaerobic respiration at night.

7. Glycolysis is not as energy-productive as respiration because
   a. NAD^+ is regenerated without the high-energy electrons passing through the electron transport chain.
   b. it is the pathway common to fermentation and respiration.
   c. it does not take place in a specialized membrane-bound organelle.
   d. pyruvic acid is more oxidized than CO_2; it still contains much of the energy from glucose.

8. The rising of yeast bread is caused by the release of
   a. alcohol vapor.
   b. oxygen.
   c. carbon dioxide.
   d. lactic acid.

9. When pyruvic acid is converted to acetyl CoA
   a. CO_2 and ATP are released.
   b. a multienzyme complex breaks off the carboxyl group and attaches a coenzyme.
   c. one turn of the citric acid cycle is completed.
   d. NAD^+ is regenerated so that glycolysis can continue.

10. How many molecules of CO_2 are generated for each turn of the Krebs cycle?
    a. 2
    b. 3
    c. 4
    d. 6

11. Support for the chemiosmotic theory comes from
    a. the production of only two ATP from the oxidation of $FADH_2$.
    b. the difference in the membrane potential between the inner mitochondrial membrane and the plasma membrane.
    c. the study of poisons called uncouplers that prevent the formation of the proton gradient and result in no synthesis of ATP.
    d. the presence of the electron transport molecules on the inside of the cristae where they can bind to H^+.

12. Which of the following reactions is incorrectly paired with its location?
    a. ATP synthesis—inner membrane of the mitochondrion
    b. fermentation—cell cytoplasm
    c. glycolysis—cell cytoplasm
    d. citric acid cycle—cristae of mitochondrion

13. In the reaction $C_6H_{12}O_6 + 6 O_2 \longrightarrow 6 CO_2 + 6 H_2O$,
    a. oxygen becomes reduced.
    b. glucose becomes reduced.
    c. oxygen becomes oxidized.
    d. water is a reducing agent.

14. What do muscle cells in oxygen deprivation gain from fermentation of pyruvic acid?
    a. ATP and lactic acid
    b. NAD^+ and lactic acid
    c. CO_2 and lactic acid
    d. ATP and $NADP^+$

15. Glucose, made from six radioactively labeled carbon atoms, is fed to yeast cells in the absence of oxygen. How many molecules of radioactive alcohol (C_2H_5OH) are formed from each molecule of glucose?
    a. 0
    b. 1
    c. 2
    d. 3

16. Which of the following produces the most ATP per gram?
    a. glucose, because it is the starting place for glycolysis
    b. glycogen or starch, because they are polymers of glucose
    c. fats, because they are richer in hydrogen with high-energy electrons
    d. proteins, because of the energy stored in their tertiary structure

17. Fats and proteins can be used as fuel in the cell because
    a. they are converted to glucose by enzymes.
    b. they are converted to intermediates of glycolysis or the citric acid cycle.
    c. they can pass through the mitochondrial membrane to enter the citric acid cycle.
    d. they contain high-energy phosphate bonds.

18. Which of the following factors would not necessarily slow down or turn off cellular respiration?
    a. lack of glucose
    b. lack of oxygen
    c. feedback inhibition
    d. lack of NAD^+

19. Which is not true of the enzyme phosphofructokinase?
    a. It is an allosteric enzyme.
    b. It is inhibited by citric acid.
    c. It is the pacemaker of glycolysis and respiration.
    d. It is inhibited by ADP or AMP.

20. We can produce and store fat in our bodies even if our diet is fat-free because
    a. carbohydrates such as breads and potatoes are fattening.
    b. a genetic predisposition may interfere with feedback inhibition.
    c. excess carbohydrates and proteins can be converted to fat through intermediates of glycolysis and the citric acid cycle.
    d. fats contain more energy than proteins or carbohydrates.

CHAPTER 10

PHOTOSYNTHESIS

FRAMEWORK

This chapter describes how the process of photosynthesis captures light energy from the sun and stores it in high-energy chemical bonds of organic molecules. In the *light reactions*, the absorption of photons of light by *chlorophyll* and other pigments, assembled into *photosystems* in the *thylakoid membranes* of *chloroplasts*, transfers some excited high-energy electrons to *NADPH* and passes other excited electrons down a redox electron transfer chain that results in *photophosphorylation* by the chemiosmotic synthesis of *ATP*. *Water* is split to replace the electrons of chlorophyll trapped in NADPH, and *oxygen* is evolved. In the *dark reactions* of the *Calvin–Benson cycle* taking place in the *stroma*, CO_2 is fixed to *Rubisco* and reduced, with the aid of ATP and NADPH from the light reactions, to form *glyceraldehyde phosphate*, an energy-rich sugar. C_4 and CAM plants have adaptations that supply a high concentration of CO_2 to *Rubisco carboxylase* in order to avoid *photorespiration*.

CHAPTER SUMMARY

In photosynthesis, the chloroplasts of plants capture the light energy of the sun and convert it into chemical energy stored in the bonds of glucose and other organic molecules made from carbon dioxide and water. *Autotrophs* "feed themselves" in the sense that they make their own organic molecules from inorganic raw materials. Plants, some protoctists, and certain prokaryotes are *photosynthetic autotrophs*, which use light as the source of energy for the manufacture of organic molecules. Some bacteria are *chemosynthetic autotrophs*, organisms that produce their organic compounds using energy obtained from oxidizing inorganic substances.

Heterotrophs eat other organisms to obtain energy and organic molecules. They may eat plants or animals or decompose organic litter, but they are ultimately dependent on photosynthetic autotrophs for food and oxygen.

Chloroplasts: Sites of Photosynthesis

Chloroplasts, found mainly in the *mesophyll* tissue of the leaf, contain *chlorophyll*, the green pigment that absorbs the light energy that drives photosynthesis. A chloroplast consists of a double membrane surrounding the dense fluid called the *stroma* and an elaborate *thylakoid membrane* system enclosing the thylakoid space. Thylakoid sacs may be layered to form *grana*. Chlorophyll is contained in the thylakoid membrane and the conversion of light energy into chemical energy occurs in the thylakoids. The production of sugar from CO_2 occurs in the stroma. CO_2 enters the leaf through *stomata*. Water for photosynthesis is supplied through vascular bundles from the root.

Photosynthetic prokaryotes lack chloroplasts but contain chlorophyll in their plasma membrane or in other membranes. The photosynthetic membranes of cyanobacteria are often arranged into parallel stacks of flattened sacs similar to thylakoids. Chloroplasts may have evolved from cyanobacteria contained in larger host cells.

How Plants Make Food: An Overview

The process of photosynthesis is represented as follows:

$$6\,CO_2 + 12\,H_2O + \text{light energy} \longrightarrow$$
$$C_6H_{12}O_6 + 6\,O_2 + 6\,H_2O$$

This equation indicates that water is both consumed and created during photosynthesis. If only the net consumption is considered, the equation for photosynthesis is the reverse of respiration. Although these two reactions are distinct, they share many of the same intermediate compounds, an electron transport chain, and a chemiosmotic mechanism of ATP formation.

Using evidence from bacteria that use H_2S rather than water for photosynthesis and create sulfur instead of oxygen as a byproduct, van Niel hypothesized that all photosynthetic organisms need a hydrogen source and that plants split water as their hydrogen source, releasing oxygen. Twenty years later, scientists confirmed this hypothesis by using a heavy isotope of oxygen (^{18}O) to label the CO_2 and water used in photosynthesis. Labeled O_2 was produced only when the water was the source of the tracer.

The most important result of photosynthesis is the incorporation of hydrogen from a low potential energy state in water to sugar, where its high-energy electrons store energy. Photosynthesis is a redox process like respiration, but differs in the direction of electron flow. The electrons increase their potential energy when they travel from water to sugar. This endergonic process ($\Delta G = +686$ kcal/mol) uses light as the energy source.

How the Light Reactions Capture Solar Energy

The *light reactions* are the steps of photosynthesis that convert solar energy into chemical energy. The light energy absorbed by chlorophyll drives the transfer of two electrons and a hydrogen proton from water (which is split and releases oxygen) to the electron acceptor *NADP*+ to reduce it to form NADPH, a source of energized electrons. ATP is also formed during the light reactions in a process called *photophosphorylation*.

In the *dark reactions*, a process known as *carbon fixation* reduces CO_2 to form carbohydrate, using NADPH and energy from ATP. NADPH and ATP come from the light reactions that occur in the thylakoids; CO_2 is reduced to sugar by the dark reactions in the stroma.

In the giant thermonuclear reactor of the sun, fusion reactions of hydrogen atoms into helium convert mass to energy. This *electromagnetic energy*, also called *radiation*, travels as rhythmic wave disturbances of electrical and magnetic fields. The distance between the crests of the electromagnetic waves, called *wavelength* (λ), ranges from less than a nanometer to more than a kilometer. Within this *electromagnetic spectrum*, the narrow band of radiation from 400 nm to 700 nm is called *visible light*, because humans can see it as various colors.

Light behaves as if it consists of discrete particles, called quanta or *photons*, delivering a fixed quantity of energy. The amount of energy in a photon of light is inversely related to its wavelength. Thus a photon of red light, with a long wavelength, has less energy than a photon of violet light.

Light may be either reflected, transmitted, or absorbed when it strikes matter. Substances that absorb visible light are called *pigments*. A *spectrophotometer* measures the ability of a pigment to absorb various wavelengths of light. The *absorption spectrum* of chlorophyll *a* shows that it absorbs blue and red light best. Chlorophyll is green because it reflects or transmits green wavelengths of light.

An *action spectrum* shows the relative rates of photosynthesis under different wavelengths of light. The action spectrum for photosynthesis does not exactly match the absorption spectrum of chlorophyll *a*. Although only chlorophyll *a* can directly channel energy into the light reactions, other pigments in the chloroplasts can absorb light of different wavelengths and transmit the energy to chlorophyll *a*. These accessory pigments include chlorophyll *b*, a green pigment structurally very similar to *a*, and the *carotenoids*, yellow and orange pigments that extend the spectrum of light that can drive photosynthesis.

When a pigment molecule absorbs the energy from a photon, one of its electrons is elevated to an orbital where it has more potential energy, and the pigment is said to be in an *excited state*. Only photons that have an energy equal to the difference in energy between the excited state and *ground state*, when the electron is in its normal orbital, are absorbed by a molecule. Thus, pigment molecules with different distributions of electrons have unique absorption spectra.

The excited state is unstable. Generally, the potential energy of the electron, created from the light energy of the photon, is released as heat as the electron drops back to its ground-state orbital. Some pigments also emit a lower-energy photon of light, called *fluorescence*, as their electrons return to ground state. Chlorophyll removed from chloroplasts will fluoresce red light upon illumination.

When chlorophyll absorbs light within a thylakoid membrane, a nearby molecule called a *primary electron acceptor* traps the high-energy electron. In a redox reaction, chlorophyll is photooxidized by the absorption of light energy, and the electron acceptor is reduced.

Chlorophyll *a* and the accessory pigments are clustered together in the thylakoid membrane in assemblies of 200 to 300 pigment molecules. The one chlorophyll *a* molecule that can donate its excited electron to the primary electron acceptor is located in the *reaction center* of the pigment assembly. Other pigment molecules in the assembly act as light-gathering antennae to absorb photons and pass the energy from molecule to molecule until it reaches the reaction center. This light-harvesting unit of the thylakoid membrane, consisting of the antenna complex, reaction-center chlorophyll *a*, and the primary electron acceptor is called a *photosystem*.

There are two types of photosystems in the thylakoid membrane. The chlorophyll molecule at the reaction center of *photosystem I* is called *P700*, after the wavelength of light (700 nm) it absorbs best. At the reaction center of *photosystem II* is a chlorophyll *a* molecule called *P680*. The differences between these two molecules, and between these molecules and the rest of the chlorophyll *a* molecules in the photosystem, are the specific proteins they are bound to in the thylakoid membrane and the close proximity to their respective primary electron acceptors.

Of the two possible routes of electron flow in the light reactions, *cyclic electron flow* is the simplest, involving only photosystem I and generating only ATP, no NADPH or O_2. Electrons, often traveling in pairs, are elevated by the absorption of two photons of light and are passed by a series of redox reactions along an electron transport chain until they return to P700. Certain electron carriers in the chain can only transport electrons in the company of hydrogen ions. These molecules pick up H^+ from the stroma and then deposit them in the thylakoid compartment, storing energy in the form of the proton-motive force of a H^+ gradient. This energy is tapped by ATPase enzymes built into the thylakoid membrane, apparently through a chemiosmotic mechanism similar to that found in mitochondria. The synthesis of ATP in chloroplasts is called photophosphorylation because it is driven by light energy. *Cyclic photophosphorylation* refers to ATP formed during cyclic electron flow.

In *noncyclic electron flow*, electrons pass continuously from water to $NADP^+$. Both photosystems are involved in this process. Two electrons excited from P700 of photosystem I are trapped in their high-energy state in NADPH. The oxidized P700 now becomes a potent oxidizing agent, attracting electrons supplied from photosystem II. A pair of electrons, elevated when P680 absorbs light energy, is trapped by the primary electron acceptor of photosystem II and passed through the electron transport chain used in cyclic electron flow to P700. As the electrons lose their potential energy, the transport chain pumps H^+ across the thylakoid membrane, creating the proton-motive force to drive ATP synthesis in the process called *noncyclic photophosphorylation*.

Oxidized P680 has such a great affinity for electrons that it can remove them from water, splitting it into two hydrogen ions and an oxygen atom. The oxygen atom immediately combines with another oxygen to form O_2, the byproduct of photosynthesis. The net effect of noncyclic electron flow is to move electrons from their low–potential-energy state in water to their high–potential-energy state in NADPH. For each pair of electrons transferred, two photons of light must be absorbed by each photosystem, an NADPH molecule is formed, and at least one molecule of ATP is created.

It is not certain whether cyclic phosphorylation is important in the light reactions of plants. It may provide ATP for the cell when there is no need for additional NADPH. The primary source of ATP comes from chemiosmosis driven by noncyclic electron flow.

Chemiosmosis in mitochondria and chloroplasts is very similar. The key difference is that in respiration chemical energy from food is being converted into ATP, whereas in chloroplasts light energy is the source of the chemical energy stored in ATP.

In the light, the proton gradient across the thylakoid membrane is as great as 3 pH units. The gradient is eliminated when the lights are turned off. Additional evidence for the chemiosmotic theory came from the work of Jagendorf, in which he got thylakoids to make ATP in the dark by creating a pH gradient across the membrane.

How the Dark Reactions Fix Carbon

The dark reactions take the ATP and NADPH formed by the light reactions to drive the reduction of CO_2 to sugar in a metabolic cycle known as the *Calvin–Benson cycle*. *Glyceraldehyde 3-phosphate* is the three-carbon sugar formed by three turns of the cycle, fixing three molecules of CO_2. First, CO_2 is added to *Rubisco*, a five-carbon sugar with phosphate groups on both ends, by the enzyme *Rubisco carboxylase*, perhaps the most abundant protein on earth. The unstable six-carbon intermediate formed splits immediately into two molecules of 3-phosphoglyceric acid, which are then phosphorylated by ATP to form two molecules of 1,3-diphosphoglyceric acid. Hydrolysis of this phosphate bond helps to transfer a pair of high-energy electrons from NADPH to create glyceraldehyde phosphate, the reduced, energy-rich sugar formed by the Calvin–Benson cycle (and also created when glucose is split in glycolysis).

The cycle must turn three times, reducing three CO_2 molecules, to create one molecule of glyceraldehyde phosphate and regenerate the three molecules of Rubisco used in the cycle. The rearrangement of the five molecules of glyceraldehyde phosphate into three molecules of Rubisco requires three more ATPs. A total of nine ATPs and six molecules of NADPH are consumed in the formation of one glyceraldehyde phosphate. Glucose, usually thought of as the product of photosynthesis, is formed from two molecules of glyceraldehyde phosphate.

Rubisco carboxylase, the enzyme that initiates the Calvin–Benson cycle, accepts O_2 in place of CO_2 when the air spaces in the leaf have a high concentration of O_2. When oxygen is added to Rubisco, the five-carbon product splits into a three-carbon molecule that remains in the cycle, and a two-carbon compound, glycolic acid, that leaves the chloroplasts, enters a peroxisome, and is broken down to release CO_2. This seemingly wasteful process that removes fixed carbon from the Calvin–Benson cycle and produces no ATP is called *photorespiration*. It is most common on hot, dry, bright days when the leaves close their stomata to conserve water and oxygen formed from photosynthesis collects in the leaf.

C_3 *plants* incorporate CO_2 directly into the Calvin–Benson cycle and are susceptible to photorespiration. C_4 *plants* have evolved a different mode of carbon fixation that serves to minimize photorespiration. CO_2 is first added to phosphoenol pyruvic acid (PEP) with the aid of an enzyme (PEP carboxylase) that has a greater affinity for CO_2 than does Rubisco carboxylase. The four-carbon compound formed by this carbon fixation in the *mesophyll cells* of the leaf is transported through plasmodesmata into the *bundle-sheath cells* tightly packed around the veins of the leaf. The four-carbon compound is broken down to release CO_2, which is then in high enough concentrations that Rubisco carboxylase will accept CO_2 and initiate the Calvin–Benson cycle. C_4 plants illustrate the important connection between structure and function in their adaptation to photosynthesize in hot, arid areas.

CAM plants illustrate another adaptation to very dry climates. These plants, mostly cacti, close their stomata during the day to prevent water loss, but open them at night to take up CO_2 and incorporate it into a variety of organic acids. These carbon-containing compounds are stored in the vacuoles of the mesophyll cells during the night and then broken down to release CO_2 for photosynthesis during the daylight. Unlike C_4 plants, CAM plants (named for Crassulacean Acid Metabolism after the plant family in which it was first discovered) do not structurally separate carbon fixation from the Calvin–Benson cycle; instead; the two processes are separated in time.

Fate of Photosynthetic Products

Carbohydrate is transported through the plant in the form of sucrose, a disaccharide. About 50% of the organic material produced by photosynthesis is used as fuel for cellular respiration in the mitochondria of plant cells; the rest is used for carbon skeletons for synthesis of organic molecules and cellulose, stored as starch, or lost through photorespiration. The incredible productivity of photosynthesis, producing about 160 billion metric tons of carbohydrate per year, provides food, shelter, fuel, and oxygen for heterotrophs.

STRUCTURE YOUR KNOWLEDGE

1. In a diagrammatic form, outline the key events of photosynthesis.
2. Describe the difference between an absorption and an action spectrum. What is the significance of the fact that the absorption spectrum of chlorophyll *a* does not completely correspond to the action spectrum of photosynthesis?
3. Create a concept map to help you develop your understanding of the chemiosmotic synthesis of ATP in photophosphorylation.

TEST YOUR KNOWLEDGE

MULTIPLE CHOICE: *Choose the one best answer.*

1. Chemosynthetic autotrophs
   a. use hydrogen sulfide as their hydrogen source for the photosynthesis of their organic compounds.
   b. "feed themselves" by obtaining energy in the chemical bonds of organic molecules.
   c. oxidize inorganic compounds to obtain energy to drive the synthesis of their organic compounds.
   d. live as decomposers of inorganic chemicals in organic litter.

2. Which of the following processes and locations is mismatched?
   a. light reactions—grana
   b. electron transport chain—thylakoid membrane
   c. Calvin–Benson cycle—stroma
   d. ATPase—double membrane surrounding chloroplast

3. Photosynthesis is a redox process in which
   a. CO_2 is reduced and water is oxidized.
   b. $NADP^+$ is reduced and Rubisco is oxidized.
   c. CO_2, $NADP^+$, and water are reduced.
   d. O_2 acts as an oxidizing agent, and water acts as a reducing agent.

4. Which of the following is not true of a photon?
   a. The energy it contains is inversely related to its wavelength.
   b. The energy it contains is equal to the difference in energy between the ground state and excited state of all electrons.
   c. Its energy may be converted to heat or light of a longer wavelength when the electron it excites drops back to its ground-state orbital.
   d. It is thought of as a discrete package or unit of radiant energy.

5. A spectrophotometer measures
   a. the absorption spectrum of a substance.
   b. the action spectrum of a substance.
   c. the amount of energy in a photon.
   d. fluorescence.

6. Accessory pigments within chloroplasts are responsible for
   a. driving the splitting of water molecules.
   b. absorbing photons of different wavelengths of light and passing that energy to reaction center molecules.

c. reducing the primary electron acceptor.
d. extending the absorption spectrum of chlorophyll *a*.

7. Which of the following is not true of the reaction center molecule?
    a. It is always a chlorophyll *a* molecule.
    b. It is named for the wavelength of light it absorbs best.
    c. It is bound to specific proteins in the thylakoid membrane and located close to its primary electron acceptor.
    d. It accepts energized electrons from the antenna complex.

8. Cyclic electron flow in the chloroplast produces
    a. ATP.
    b. ATP and NADPH.
    c. ATP, NADPH, and O_2.
    d. Glyceraldehyde 3-phosphate.

9. The chlorophyll known as P680 is reduced by electrons from
    a. photosystem I.
    b. photosystem II.
    c. water.
    d. NADPH.

10. CAM plants avoid photorespiration by
    a. fixing CO_2 into organic acids during the night.
    b. fixing CO_2 into four-carbon compounds in the bundle-sheath cells.
    c. fixing CO_2 into four-carbon compounds in the mesophyll.
    d. using an enzyme, PEP carboxylase, that outcompetes Rubisco carboxylase for CO_2.

11. Electrons that flow through the two photosystems have their lowest potential energy in
    a. P700
    b. P680.
    c. the primary electron acceptor of photosystem II.
    d. water.

12. Chloroplasts can make carbohydrate in the dark if provided with
    a. ATP and NADPH and CO_2.
    b. an artificially induced proton gradient.
    c. organic acids or four-carbon compounds such as malic acid.
    d. a source of hydrogen.

13. In the chemiosmotic synthesis of ATP, H^+ diffuses through the ATPase
    a. from the stroma into the thylakoid compartment.
    b. from the thylakoid compartment into the stroma.

c. from the cytoplasm into the matrix.
d. from the cytoplasm into the stroma.

14. How many "turns" of the Calvin–Benson cycle are required to produce one molecule of glyceraldehyde phosphate?
    a. 1
    b. 2
    c. 3
    d. 6

15. In green plants, most of the ATP for synthesis of proteins, cytoplasmic streaming, and other cellular activities comes from
    a. photosystems I and II.
    b. the dark reactions.
    c. photophosphorylation.
    d. oxidative phosphorylation.

16. The six molecules of glyceraldehyde phosphate formed from three turns of the Calvin–Benson cycle are converted into
    a. three molecules of glucose.
    b. one glyceraldehyde phosphate and three molecules of Rubisco.
    c. one molecule of glucose and two molecules of Rubisco.
    d. one glyceraldehyde phosphate and three four-carbon intermediates.

17. NADPH is used in the dark reactions to reduce
    a. an unstable six-carbon intermediate.
    b. Rubisco.
    c. 1,3-diphosphoglyceric acid.
    d. glyceraldehyde phosphate.

18 – 25. Indicate whether the following events occur during
    a. respiration.
    b. photosynthesis.
    c. both a and b.
    d. neither a nor b.

18. ____ Chemiosmotic synthesis of ATP

19. ____ Reduction of oxygen

20. ____ Reduction of CO_2

21. ____ Reduction of NAD^+

22. ____ Oxidation of water

23. ____ Oxidation of $NADP^+$

24. ____ Oxidative phosphorylation

25. ____ Electron flow along a cytochrome chain

CHAPTER 11

REPRODUCTION OF CELLS

FRAMEWORK

This chapter details the process through which cells create new cells, with identical genetic material. There are two critical stages of the process—the replication of the cell's DNA, and the parceling out of the duplicate sister chromatids such that the new daughter cells get one of each of the cell's chromosomes. Key concepts that you will encounter in this chapter include:

binary fission	polar fibers
mitosis, interphase	asters
cell cycle	cell center
G1, S, G2, M phases	cytokinesis
sister chromatids	cleavage furrow
centromeres	cell plate
mitotic spindle	cell culture
centrioles	contact inhibition
kinetochore fibers	restriction point

CHAPTER SUMMARY

Cell division, the process through which a cell reproduces itself, is the basis for the perpetuation of life. Cell division creates duplicate offspring in unicellular organisms and provides for growth, development and repair in multicellular organisms. The process of recreating a structure as intricate as a cell necessitates the exact duplication and equal division of the DNA and genes that contain the cell's genetic program.

Bacterial Reproduction

Prokaryotes reproduce themselves by a process known as *binary fission*. First, two copies of the single circular DNA molecule in bacteria are attached to the plasma membrane. Growth of the membrane separates the two copies of the chromosome. After the cell has reached about twice its original size, the membrane pinches in and a cell wall develops between the chromosomes, creating two daughter cells.

Eukaryotic Chromosomes and Their Duplication

Dividing the tens of thousands of genes of a eukaryotic cell involves a more elaborate process. Through DNA replication, the cell duplicates the genes on its chromosomes prior to cell division. Grouping genes on chromosomes simplifies the management and distribution of genes in cell division. The duplicated chromosomes separate during *mitosis* and the cytoplasm divides (*cytokinesis*) so that two separate daughter cells are formed.

Somatic cells (cells that are not sperm or egg) all have the same number of chromosomes in their nuclei. Sperm and egg cells have one-half the chromosome complement of somatic cells. Each species has a characteristic number of chromosomes.

Each chromosome is a very long DNA molecule with associated proteins that help structure the chromosome and control the activity of the genes. This DNA–protein complex, called chromatin, is organized into a coiled and folded fiber. When the cell is not dividing, the fibers are dispersed, and individual chromosomes are not visible with the light microscope.

Replicated chromosomes result in two identical *sister chromatids*, joined together at a specialized region called a *centromere*. Toward the end of mitosis, the two sister chromatids separate from each other, and daughter cells that have one copy of each chromosome are formed.

Reproduction of Cellular Organelles

In addition to replicating its genetic material, a cell usually increases its mass and supply of organelles (especially ribosomes) and membranes before division. Mitochondria and chloroplasts both contain small, circular DNA molecules that replicate and divide in a manner similar to bacterial DNA. Lipids for these organelles are supplied by the cell's ER, and some proteins essential for their functioning are coded for by the cell's chromosomal DNA. These organelles reproduce on their own and are roughly divided between the two daughter cells during cytokinesis. Centrioles, associated with the microtubule apparatus involved in mitosis, duplicate on their own before cell division.

The Cell Cycle

The *cell cycle* involves the *M phase* (mitotic phase), in which the chromosomes and cytoplasm are divided, and *interphase*, during which most of the cell's growth, metabolic activities, and chromosome replication occur. The length of the cell cycle varies greatly depending on the cell type and its physiological state.

Interphase, usually lasting 90% of the cell cycle, involves the *G1 phase* (the "first gap" period), the *S phase* (synthesis phase), and the *G2 phase* ("second gap" period). Most organelles and cell components are produced continuously throughout the G1, S, and G2 phases; DNA synthesis occurs only during the S phase.

Although mitosis is a dynamic process of chromosome separation and cytoplasm division, it is conventionally described in four stages: *prophase, metaphase, anaphase,* and *telophase.*

During prophase, the nucleoli disappear and the chromatin fibers coil and fold into visible chromosomes consisting of sister chromatids joined at the centromere. The centriole pairs (present in animal cells) move to opposite poles of the cell and the microtubules of the spindle (and aster in animal cells) become arranged. The nuclear membrane disintegrates, and microtubules attach to the chromosomes.

In metaphase, the centromeres of all the chromosomes become aligned on the *metaphase plate,* a plane equidistant between the spindle's two poles.

Anaphase involves the physical separation of the sister chromatids; these split at the centromeres and move along the spindle apparatus toward opposite poles. The poles of the cell move farther apart.

By telophase, equivalent sets of chromosomes have gathered at the two poles of the cell. Nuclear envelopes form from fragments of the parent nuclear envelope and contributions from the endomembrane system. Nucleoli reappear, the chromatin fibers unravel, and cytokinesis occurs.

The *mitotic spindle* consists of fibers made of microtubules and proteins. In animal cells, the spindle begins as two asters, radial arrays around each centriole pair. Both plant and animal cells develop *polar fibers* extending from the poles to the equator of the cell. The organizing center of the spindle is called the cell center or mitotic center. This amorphous cloud of material, visible only with electron microscopy, is present in both plants and animal cells and is associated with the centrioles in animal cells. Little is known of the details of its action.

The chromosomes interact with the spindle by special links at the centromere called *kinetochores,* one for each sister chromatid. Special microtubules called *kinetochore fibers* attach to each kinetochore. Interaction between the kinetochore fibers and the polar fibers sets the chromosomes into agitated motion until they line up on the metaphase plate.

The extension of the spindle poles away from each other as the cell elongates probably is due to the addition of subunits to the polar fibers and the sliding of fibers past each other where they overlap at the equator. This movement is thought to involve the hydrolysis of ATP by an enzyme similar to dynein, the protein responsible for movement of cilia and flagella.

The mechanism for the separation and movement of sister chromatids along the spindle is not known. It appears to be different from the force generated for pole separation, because blocking dynein's action does not prevent chromosome movement. A low-energy process such as the dissociation of the protein subunits of kinetochore microtubules may be involved.

During late anaphase or early telophase, cytokinesis begins. *Cleavage* separates the two daughters in animal cells. The shallow groove in the cell surface along the metaphase plate, known as the *cleavage furrow,* forms as the *contractile ring* of microfilaments inside the cell begins to contract. The cleavage furrow deepens until the dividing cell is pinched in two.

In plant cells, a *cell plate* forms at the site of the old metaphase plate from the fusion of membrane vesicles derived from the dictyosomes (Golgi complex). The double membrane, formed from the fusion of these vesicles, eventually joins the plasma membrane, creating two daughter cells with their own continuous plasma membrane. A new cell wall develops between these membranes.

Mitosis may not always be followed by cytokinesis. Multinucleated cells can thus be formed, as in the *plasmodia* of slime molds.

Control of Cell Division

Control of the timing and rate of cell division is critical to normal growth, development, and maintenance. Control mechanisms may be studied through the use of *tissue culture,* or *cell culture.* This technique involves growing cells in the laboratory in glass or plastic containers supplied with a nutrient solution called a *growth medium.*

Research using cell culture has identified factors that stimulate or inhibit cell division. Certain hormones and other chemicals are essential for normal cell division. Inhibiting protein synthesis, depleting essential nutrients, or overcrowding serves to block division. Most normal vertebrate cells will stop dividing when they come in contact with other cells, a phenomenon known as *contact inhibition.* Cancer cells do not exhibit contact inhibition when grown in cell culture.

An important control point seems to occur during the G1 phase of the cell cycle. Some mechanism determines whether the cell will switch into a nondividing state (called G0) or continue past a so-called *restriction point* into the S, G2, and M phases. Normal cells that stop dividing usually do so before the restriction point in the G1 phase. Cancer cells, if they stop dividing, seem to do so at random points in the cell cycle. Moreover, most cancer cells in cell culture appear to be "immortal"; they continue to divide indefinitely instead of stopping after the typical 20 to 50 divisions of normal mammalian cells. Much remains to be learned about the cell division processes of both normal and cancerous cells.

STRUCTURE YOUR KNOWLEDGE

1. Describe the life of one chromosome as it proceeds through an entire cell cycle.
2. Draw a sketch of a mitotic spindle that includes the components of the spindle and lists the functions of each part.

TEST YOUR KNOWLEDGE

MATCHING: *Match the event described with its place in the cell cycle.*

A. anaphase
B. G1 phase
C. G2 phase
D. metaphase
E. prophase
F. S phase
G. telophase

1. ____ most cells that have finished dividing stop here
2. ____ sister chromatids separate and chromosomes move apart
3. ____ mitotic spindle forms
4. ____ cell plate forms or cleavage furrow pinches cells apart
5. ____ chromosomes duplicate
6. ____ chromosomes line up at equatorial plane
7. ____ phase after DNA replication
8. ____ chromosomes become visible
9. ____ nuclear envelope reforms
10. ____ poles of cell move farther apart

MULTIPLE CHOICE: *Choose the one best answer.*

1. One of the major differences in the cell division of prokaryotic cells compared to eukaryotic cells is that
   a. cytokinesis does not occur.
   b. genes are not duplicated on chromosomes.
   c. no spindle apparatus forms.
   d. DNA replication does not occur.

2. Which of the following is not true of chromatin?
   a. It was named for its affinity for certain stains.
   b. It consists of fibrous proteins.
   c. The coiling and folding of its fibers may influence gene activity.
   d. It consists of the DNA–protein complex that makes up chromosomes.

3. The longest part of the cell cycle is
   a. interphase.
   b. G1 phase.
   c. G2 phase.
   d. mitosis.

4. Cytokinesis may involve
   a. the separation of sister chromatids.
   b. the contraction of the contractile ring of microfilaments.
   c. interactions between kinetochore and polar fibers.
   d. the formation of the cell center.

5. Humans have 46 chromosomes. That number of chromosomes will be found in
   a. cells in anaphase.
   b. the egg and sperm cells.
   c. the somatic cells.
   d. all the cells of the body.

6. Polar fibers
   a. are found only in animal cells.
   b. attach to the kinetochore region of the centromere.

   c. seem to be involved in moving
       the chromosomes.
   d. seem to be involved in pushing the poles
       of the cell apart.

7. Contact inhibition is seen in
    a. cancer cells.
    b. normal cells growing in cell culture.
    c. cells in tissue culture that contact the
        surface of the culture vessel.
    d. cells that stop growing in the G2 phase.

8. A cell that passes the restriction
    point most likely will
    a. continue to divide.
    b. stop dividing.
    c. continue to divide only if it is a cancer cell.
    d. experience contact inhibition.

9. Which of the following is not true of a cell plate?
    a. It forms at the site of the metaphase plate.
    b. It results from the fusion of microtubules.
    c. It fuses with the plasma membrane.
    d. A cell wall is laid down
        between its membranes.

10. Sister chromatids
    a. have one-half the amount of genetic material
        as does the original chromosome.
    b. move together toward one pole of the cell.
    c. are connected at a region
        called the centromere.
    d. are formed during prophase.

ANSWER SECTION

CHAPTER 7: A TOUR OF THE CELL

Suggested Answers to Structure Your Knowledge

1.

FUNCTIONS OF CELL	ASSOCIATED ORGANELLES AND STRUCTURES
Reproduction of cell	Chromosomes, centrioles, cell center, microtubules
Energy conversions	Mitochondria, chloroplasts
Directions and control	DNA--->RNA--->proteins (enzymes)
Manufacturing	Ribosomes, rough ER, smooth ER, vesicles, Golgi complex
Digestion, recycling	Lysosomes, central vacuole in plant cells
Specific pathways	Microbodies, perioxosomes, glyoxysomes
Cell--cell interaction	Desmosomes, tight and gap junctions, plasmodesmata
Movement, and structural integrity	Cytoskeleton: microtubules (cilia, flagella), microfilaments, intermediate filaments

2.

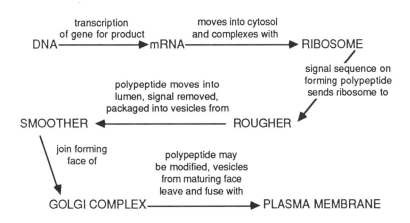

Answers to Test Your Knowledge

Multiple Choice:

1. c	6. b	11. c
2. b	7. a	12. b
3. d	8. c	13. a
4. c	9. b	14. c
5. a	10. d	15. c

Fill in the Blanks:

1. vesicles	6. basal body
2. cristae	7. cytosol
3. glycocalyx	8. cytoskeleton
4. microbodies	9. tight junction
5. grana	10. tonoplast

Matching:

1.	E, b	6.	C, g
2.	K, j	7.	B, c
3.	I, d	8.	J, a
4.	D, i	9.	H, k
5.	A, e	10.	F, h

CHAPTER 8: TRAFFIC ACROSS MEMBRANES

Suggested Answers to Structure Your Knowledge

1. The fluid–mosaic model depicts a membrane composed of a phospholipid bilayer with embedded and surface proteins. Small, nonpolar, hydrophobic molecules easily dissolve in the hydrophobic interior of the membrane and pass through by diffusion. The rate of permeability is determined by the size and relative lipid solubility of the molecule. Small polar or ionic compounds cannot pass through the hydrophobic center of the membrane. They are transported by specific carrier proteins, either through channels in the protein or by a conformational change in the protein caused by the binding of the solute, that shifts the solute from one side of the membrane to the other.

2. Facilitated diffusion and active transport both involve transport proteins in the membrane that are specific for solute molecules. In facilitated diffusion, polar or ionic molecules that could not easily permeate the membrane are moved *down* their concentration gradient. This passive transport does not require energy from the cell. Active transport is used to move any type of small molecule *against* its concentration gradient. The transport protein usually acts as an enzyme in hydrolyzing ATP to provide the energy for this transport.

3.

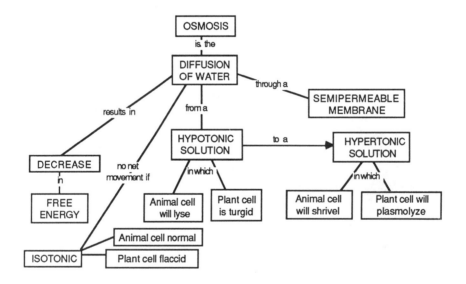

Answers to Test Your Knowledge

Multiple Choice:

1.	c	6.	d	11.	b	16.	c
2.	d	7.	a	12.	d	17.	a
3.	a	8.	c	13.	b	18.	c
4.	c	9.	a	14.	a	19.	b
5.	d	10.	c	15.	d	20.	c

CHAPTER 9: RESPIRATION: HOW CELLS HARVEST ENERGY

Suggested Answers to Structure Your Knowledge

1.

PROCESS	LOCATION	MOLECULES IN	MOLECULES OUT
Glycolysis	Cytosol	Glucose, 2 ATP (NAD$^+$, ADP, $\textcircled{P}$)	2 pyruvic acid, 2 NADH, 4 ATP (2 net ATP)
Fermentation	Cytosol	Glucose, 2 ATP (NAD$^+$, ADP, $\textcircled{P}$)	2 alcohol, 2 CO_2, or 2 lactic acid, 4 ATP (2 net ATP)
Citric acid cycle	Matrix of mitochondria	2 acetyl CoA (from pyruvic acid), (NAD$^+$, FAD, ADP)	4 CO_2, 6 NADH, 2 FADH$^+$, 2 ATP
Electron transport chain	Mitochondrial inner membrane	10 NADH (2 from glycolysis use 2 ATP to enter mito-chondria), 2 FADH$^+$, 6 CO_2	12 H_2O, 34 ATP (32 net)

2.

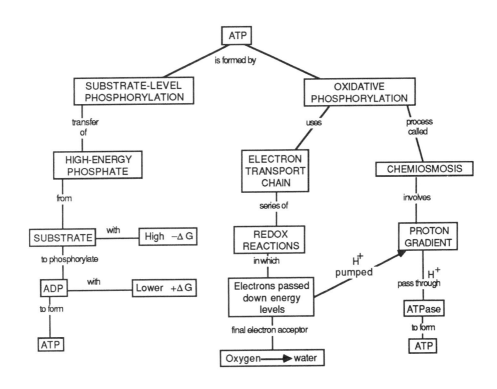

Answers to Test Your Knowledge

Multiple Choice:

1.	c	6.	b	11.	c	16.	c
2.	b	7.	a	12.	d	17.	b
3.	a	8.	c	13.	a	18.	a
4.	a	9.	b	14.	b	19.	d
5.	d	10.	a	15.	c	20.	c

CHAPTER 10: PHOTOSYNTHESIS

Suggested Answers to Structure Your Knowledge

1.

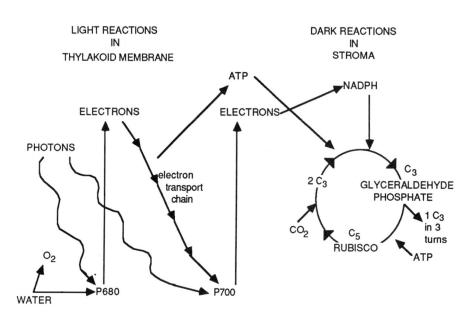

2. An absorption spectrum shows the wavelengths of light that are best absorbed by a particular pigment. An action spectrum presents the effectiveness or rate of a reaction at various wavelengths. The absorption spectrum of chlorophyll *a*, the key photosynthetic pigment, and the action spectrum for photosynthesis are not identical. Some wavelengths of light, particularly in the yellow and orange range, result in a higher rate of photosynthesis than would be indicated by the absorption of those wavelengths by chlorophyll *a*. These differences are accounted for by the existence of accessory pigments that absorb light energy from different wavelengths and pass that energy on to the reaction center molecules.

3.

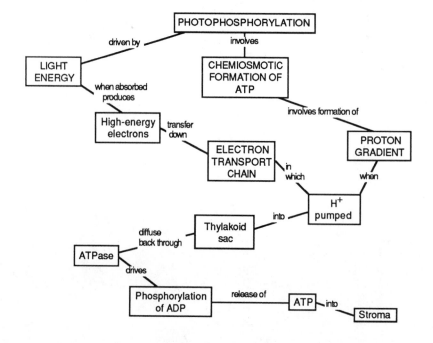

Answers to Test Your Knowledge

Multiple Choice:

1. c	6. b	11. d	16. b	21. a
2. d	7. d	12. a	17. c	22. b
3. a	8. a	13. a	18. c	23. d
4. b	9. c	14. c	19. a	24. a
5. a	10. a	15. d	20. b	25. c

CHAPTER 11: REPRODUCTION OF CELLS

Suggested Answers to Structure Your Knowledge

1. Journey of one chromosome through cell cycle:
Interphase: 90% of cell cycle; growth, metabolism, DNA replication.

G1 phase—first gap phase: the chromosome, consisting of chromatin fiber of DNA and associated proteins, is diffuse throughout the nucleus. Proteins, stabilizing the coiling and folding of the fiber, may contribute to the control of gene activity. RNA molecules are being transcribed from genes that are switched on.

S phase—synthesis of DNA: the chromosome is replicated; two exact copies, called sister chromatids, are produced and held together at a region called the centromere.

G2 phase—second gap phase: growth and metabolic activities of cell continue; RNA transcription continues.

Mitosis: cell division

Prophase: the chromosome, consisting of two sister chromatids, becomes tightly coiled and folded. Kinetochore fibers of the mitotic spindle attach to kinetochore on each chromatid. When kinetochore fibers interact with polar fibers, the chromosome begins to move.

Metaphase: the centromere of the chromosome is aligned at the metaphase plate along with the centromeres of the other chromosomes.

Anaphase: the sister chromatids separate (now considered to be individual chromosomes) and move to opposite poles.

Telophase: chromatin fiber of chromosome uncoils and is surrounded by reforming nuclear membrane.

2.

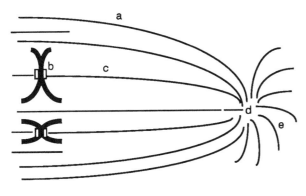

a. Polar Fibers — microtubules from pole to center; push poles apart, by sliding at region of overlap at equator and addition of tubulin molecules
b. Kinetochore— region of centromere where kinetochore fiber attaches
c. Kinetochore fibers— interact with polar fibers to move chromosomes, pull chromosomes apart
d. Cell center— region of organization of mitotic spindle (associated with paired centrioles in animal cell)
e. Aster— first part of spindle formed (only in animal cells)

Answers to Test Your Knowledge

Matching:

1.	B	6.	D
2.	A	7.	C
3.	E	8.	E
4.	G	9.	G
5.	F	10.	A

Multiple Choice:

1.	c	6.	d
2.	b	7.	b
3.	a	8.	a
4.	b	9.	b
5.	c	10.	c

CHAPTER 12

MEIOSIS AND SEXUAL LIFE CYCLES

FRAMEWORK

This chapter introduces the process of *meiosis* and the concept of *genetic variation* that is a product of *sexual life cycles* due to *crossing over, independent assortment,* and *random fertilization*. The sexual life cycles of animals, plants, algae, fungi, and protoctists all include meiosis and fertilization but vary in the timing of these events.

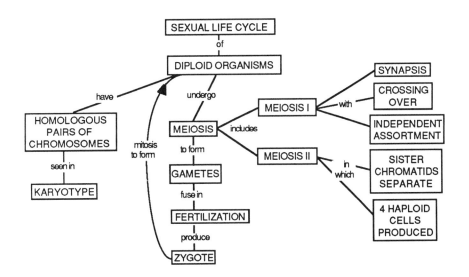

CHAPTER SUMMARY

Genetics is the study of the inheritance of traits from parents to offspring by the transmission of discrete units of information, known as *genes* and coded in the genetic material DNA. The possession of a genetic program, which is transmitted from one generation to the next, expressed by translation of genes into specific traits, and subject to modification through mutation and selection, is a characteristic unique to life.

Sexual Vs. Asexual Reproduction

In *asexual reproduction*, an individual inherits all its genes from a single parent. The offspring is usually an exact genetic copy of the parent. In *sexual reproduction*, an individual receives a unique combination of genes inherited from two parents. The genetic variation created by sexual reproduction is a result of the management of chromosomes during the sexual life cycle.

Sexual Life Cycles: The Human Example

In somatic cells, chromosomes are present in pairs known as *homologous chromosomes* or homologues. Homologues have the same length, centromere position, and staining pattern. Genes controlling the same traits are found at the same *loci*, or locations, on homologous chromosomes. A *karyotype* is an ordered display of an individual's chromosomes, made by photographing white blood cells or cheek cells that have been arrested in metaphase, cutting the individual chromosomes out of the photograph, and arranging them into homologous pairs by size and shape.

The exception to the rule of homologous chromosomes is found in the *sex chromosomes*, the X and Y chromosomes that determine the sex of a person. Females have two homologous X chromosomes; males have one X and one Y chromosome. The remaining homologous pairs of chromosomes are called *autosomes*.

Diploid cells contain a set of chromosomes from each parent, the result of the fusion of *gametes*, *haploid* egg and sperm, in the process of *fertilization* or *syngamy*. The haploid number ("N") of chromosomes for humans is 23. Somatic cells contain 46 chromosomes. The combined chromosome sets present in the fertilized egg or *zygote* are passed on to all somatic cells of the body with precision by mitosis. The sex organs then produce new cells of the *germ line*, the gametes, to pass on the genetic program to the next generation.

Meiotic Cell Division

Meiosis is a special type of cell division that halves the chromosome number in gametes to compensate for the doubling that occurs at fertilization. An alternation between diploid and haploid conditions, involving the processes of meiosis and fertilization, is characteristic of the life cycles of all sexually reproducing organisms.

Meiosis, like mitosis, is preceded by chromosome replication. This replication is then followed by two consecutive cell divisions, *meiosis I* and *meiosis II*, resulting in four haploid daughter cells.

In prophase I, homologous chromosomes pair up in a process called *synapsis*, resulting in four closely associated chromatids—two from each replicated homologue—called *tetrads*, or *bivalents*. *Crossing over*, the exchange of genetic material between nonsister chromatids, is visible during this stage by the appearance of X-shaped regions called *chiasmata*.

The synapsed homologous pairs line up together at the metaphase plate during metaphase I. During anaphase I, the homologous pairs separate, one homologue going to each pole. The centromeres do not divide, and sister chromatids remain together.

Meiosis II is identical to mitosis in that the sister chromatids separate. Because it is not preceded by a replication, meiosis II results in four *haploid* cells.

Sexual Sources of Genetic Variation

The sorting out and recombining of chromosomes during the processes of meiosis and fertilization are responsible for most of the genetic variation found from one generation to the next. This variation is the raw material on which natural selection acts.

The first meiotic division results in the *independent assortment of chromosomes*. Each homologous pair lines up independently at the metaphase plate, and the orientation of the maternal and paternal chromosomes is random. The gametes produced by meiosis can have any combination of maternal and paternal chromosomes. The number of combinations possible is equal to 2N, where N is the haploid number. For humans, this number of possible combinations of chromosomes is 2^{23} — 8 million distinct gametes that each individual can produce.

The random nature of fertilization adds to the genetic variability established in meiosis. A given pair of parents will produce a zygote with any of 64 trillion (8 million x 8 million) diploid combinations.

Crossing over also adds to the variability established by meiosis. The exchange of segments of nonsister chromatids results in new genetic combinations of maternal and paternal genes on the same chromosome. This *genetic recombination* results in gene combinations different from those inherited from the previous generation.

Independent assortment of chromosomes, random fertilization, and genetic recombination all contribute to genetic variability in a sexually reproducing population. *Mutations*, changes in the DNA of genes, are the ultimate source of this genetic diversity, which is passed on and continually reordered by sexual reproduction.

Variety of Sexual Life Cycles

In most animals, meiosis occurs in the formation of gametes, which are the only haploid cells in the life cycle. In most fungi and some protoctists, the only diploid stage is the zygote. Meiosis occurs right after the gametes fuse, and mitosis creates a multicellular haploid organism. Gametes will be produced by mitosis in these organisms.

Plants and some species of algae have a third type of life cycle, called an *alternation of generations*, that includes both diploid and haploid multicellular stages. The multicellular diploid *sporophyte* stage produces haploid spores by meiosis. These spores develop into a multicellular haploid plant, the *gametophyte*, which produces gametes by mitosis. Gametes fuse to form a diploid zygote that develops into the next sporophyte generation.

STRUCTURE YOUR KNOWLEDGE

1. Create a diagram of an animal sexual life cycle, including the major events and processes.
2. Fill in the following table, listing the key events that occur in the stages of meiosis.

STAGE	KEY EVENTS OF MEIOSIS
INTERPHASE I	
PROPHASE I	
METAPHASE I	
ANAPHASE I	
TELOPHASE I	
INTERKINESIS	
PROPHASE II	
METAPHASE II	
ANAPHASE II	
TELOPHASE II	

TEST YOUR KNOWLEDGE

MULTIPLE CHOICE: *Choose the one best answer.*

1. The most important difference between sexual and asexual reproduction is that
   a. two parents are needed for sexual reproduction.
   b. the offspring of sexual reproduction have a unique combination of genes inherited from two parents.
   c. sexual reproduction requires parents that are diploid.
   d. asexual reproduction does not involve meiosis.

2. A karyotype is a
   a. genotype of an individual.
   b. unique combination of chromosomes found in a gamete.
   c. blood type determination of an individual.
   d. picture of homologous chromosomes of an individual.

3. Autosomes are
   a. sex chromosomes.
   b. chromosomes that occur singly.
   c. chromosomes that move independently during meiotic cell divisions.
   d. none of the above.

4. Syngamy
   a. occurs during prophase I of meiosis.
   b. is the process of fertilization or fusion of gametes.
   c. results in crossing over and genetic variation in offspring.
   d. is the production of similar gametes.

5. During the first meiotic division (meiosis I)
   a. homologous chromosomes separate.
   b. the chromosome number becomes haploid.
   c. crossing over between nonsister chromatids occurs.
   d. all of the above occur.

6. A cell with a diploid number of 6 could produce gametes with how many different combinations of chromosomes?
   a. 6
   b. 8
   c. 12
   d. 64

7. Genetic recombination involves
   a. crossing over between nonsister chromatids.
   b. the random fertilization of gametes.
   c. independent assortment of chromosomes at metaphase I.
   d. all of the above.

8. In fungi and some protoctists,
   a. the zygote is the only haploid stage.
   b. gametes are formed by meiosis followed by mitosis.
   c. the adult organism is haploid.
   d. the gametophyte generation produces gametes by mitosis.

9. In the alternation of generations found in plants,
   a. the sporophyte generation produces spores by meiosis.
   b. the gametophyte generation produces gametes by mitosis.

   c. the zygote will develop into a sporophyte generation by mitosis.
   d. all of the above are correct.

10. Which of the following is not a source of genetic variation in sexually reproducing organisms?
    a. chiasmata
    b. synapsis
    c. independent assortment of chromosomes
    d. crossing over

11. Meiosis II is similar to mitosis because
    a. sister chromatids separate.
    b. homologous chromosomes separate.
    c. DNA replication precedes the division.
    d. they both take the same amount of time.

12. Homologous chromosomes
    a. have identical genes.
    b. have genes for the same traits at the same loci.
    c. pair up in prophase II.
    d. are found in haploid cells.

CHAPTER 13

MENDEL AND THE GENE IDEA

FRAMEWORK

Through his work with garden peas, Mendel developed the fundamental principles of inheritance and the laws of segregation and independent assortment. This chapter describes the basic monohybrid and dihybrid crosses that Mendel performed to establish that inheritance involves particulate genetic factors (genes) that segregate independently in the formation of gametes and recombine to form offspring. The laws of probability can be applied to predict the outcomes of genetic crosses.

Various complications in the expression of genotypes as phenotypes include intermediate inheritance, multiple alleles, pleiotropy, penetrance and expressivity, epistasis, and polygenic traits. Genetic screening and counseling for recessively inherited genetic disorders use Mendelian principles in the analysis of human pedigrees.

CHAPTER SUMMARY

From ancient Greece through the nineteenth century, *pangenesis* was the prevailing theory of inheritance. According to this theory, particles, called pangenes, from all parts of the body come together to form eggs or semen, and changes to various parts of the body during an organism's life could be passed on to the next generation. Two additional theories, developed in the seventeenth century with the early microscopic observations of sperm and ovarian follicles, were that the sperm contained a miniature human being, the homunculus, with the mother serving only as an incubator, or that the egg contained the miniature being with the semen's purpose being to stimulate growth.

By the early nineteenth century, biologists believed that both parents contribute to the characteristics of off-spring with the blending or irreversible mixing of hereditary materials. In the 1860s, Gregor Mendel discovered the fundamental principles of inheritance that refuted this "blending" theory.

Mendel's Model of Inheritance

Mendel's work demonstrated that parents pass on to their offspring discrete, particulate heritable factors (genes) that retain their individuality and ability to be separated in the formation of gametes. Mendel worked with easily distinguishable characteristics of garden peas. Peas were a good choice of study organism because it was easy to grow them, to control their fertilization, and to quantify their offspring.

The seven characteristics Mendel chose occurred in two alternative forms that could be developed into *true breeding* varieties—self-fertilizing parents always produced offspring with the parental form of the characteristic. To follow the transmission of these well-defined traits, Mendel cross-pollinated plants that were true breeding for alternate forms of the same characteristic, and then self-pollinated the next generation. The true-breeding parental plants are known as the *P generation* (for parental); the results of the first *cross* are called the F_1 *generation* (for first filial); and the next generation, from the self-cross of the F_1, is known as the F_2 *generation*.

Mendel found that the F_1 offspring did not show a blending of the parental characteristics. Instead, only one of the parental forms of the trait was found in the hybrid offspring. In the F_2 generation, however, the missing parental form reappeared in the ratio of 3:1, three offspring with the parental trait shown by the F_1 to one offspring with the reappearing trait.

Mendel's explanation for this phenomenon contains four parts: (1) there are alternate forms, now called *al-*

leles, for the heritable factors called genes, (2) an organism has two genes (alleles) for each inherited trait, one inherited from each parent, (3) gene pairs separate (segregate) during the formation of gametes, so an egg or sperm only carries one allele or gene for each inherited trait, and (4) when two different alleles occur together, one allele (called the *dominant allele*) may be completely expressed while the other *recessive allele* is masked.

This hypothesis explains the 3:1 ratio observed in the F_2 plants. During the separation of genes in the formation of gametes in the F_1 generation, one-half of the gametes would receive one allele while the other half would receive the alternate allele. Random fertilization of gametes would result in one-fourth of the plants having two dominant alleles, one-half having one dominant and one recessive allele, and one-fourth receiving two recessive alleles. The dominant allele would be the one expressed in the half of the offspring that are hybrid; so the ratio of plants showing the dominant or recessive trait would be 3:1.

The mechanism of inheritance that accounts for these results is stated by Mendel's *law of segregation*: gene pairs segregate during gamete formation, and the paired condition is reestablished by the random fusion of gametes at fertilization.

Dominant alleles are often symbolized by a capital letter; recessive alleles by a small letter. An organism that has a pair of identical alleles for a trait is said to be *homozygous*. If the organism has two different alleles, it is said to be *heterozygous* for that trait. Homozygotes are true breeding; heterozygotes are not, since they produce gametes with one or the other allele that can recombine to produce homozygous-dominant, heterozygous, and homozygous-recessive offspring, in the *genotypic ratio* of 1:2:1. The *phenotype* is an organism's expressed traits; its *genotype* is its genetic makeup.

The *phenotypic ratio* is often different from the genotypic ratio, since homozygous-dominant and heterozygotes may have the same phenotype. A *testcross*, breeding a recessive homozygote with an organism of unknown genotype, can be done to determine the genotype of organisms expressing the dominant phenotype. The recessive homozygote can contribute only a recessive allele, so that if all the offspring express the dominant allele, the unknown parent must have been homozygous dominant. If there is a 1:1 phenotypic ratio of offspring, the parent must have been a heterozygote.

General rules of probability apply to the segregation of genes and the reconstitution of pairs at fertilization. The probability of an event occurring is the ratio of the number of possible events that fulfill that criterion over all the possible events; for example, there is a 50% probability of getting a head in a coin toss.

The outcome of *independent events* is not affected by previous events. For the probability of independent events occurring simultaneously, however, the *rule of multiplication* states that the probability of a compound event is equal to the product of the separate probabilities of the independent events. The probability of a particular genotype being formed by fertilization is equal to the product of the probability of the formation of each type of gamete needed to produce that genotype. The probability of getting a homozygous-recessive offspring from a cross of heterozygotes is equal to the probability of getting one gamete with the recessive allele (1/2) times the probability of getting the other gamete with the recessive allele or $1/2 \times 1/2 = 1/4$.

If the genotype can be formed in more than one way, then the *rule of addition* states that the probability of an event is equal to the sum of the separate probabilities of the different ways in which the event can occur. Thus, a heterozygote offspring can occur if the egg contains the dominant allele and the sperm the recessive (1/4 probability) or if the egg contains the recessive and the sperm the dominant (1/4). A heterozygote offspring should be formed from a hybrid cross 1/4 + 1/4, or 1/2 (50%) of the time.

These rules of probability account for the 1:2:1 ratio of genotypes seen in the offspring of a hybrid cross. Large samples will usually show these expected distributions more precisely because the influence of individual events is averaged out.

Mendel deduced the law of segregation from his work with *monohybrid crosses*, crosses that involved parents that differed in only a single trait. He used *dihybrid crosses*, involving parents that differed in two traits, to determine whether the two traits were transmitted together (in the same combination in which they were inherited from the parental plants) or independently of each other.

If the two pairs of genes segregate independently, then gametes from an F_1 hybrid generation (AaBb) should contain four combinations of genes in equal quantities (AB, Ab, aB, ab). The random fertilization of these four classes of gametes should result in 16 (4 × 4) gamete combinations that produce four phenotypic categories in a ratio of 9:3:3:1 (nine offspring showing both dominant traits, three showing dominant of one and recessive of one, three showing recessive and dominant traits reversed, and one showing both recessive traits). Mendel obtained these ratios when he scored the F_2 progeny of a dihybrid cross, and they provided evidence for his law of independent assortment. This principle states that genes assort independently from each other when they segregate in the formation of gametes.

Fairly complex genetics problems can be solved by

applying the rules of probability to segregation and *independent assortment*. The probability of a particular genotype arising from a cross can be determined by considering each gene involved as a separate monohybrid cross and then multiplying the probabilities of all the independent events involved in the final genotype.

From Genotype to Phenotype: Some Complications

Mendel's laws were established to explain inheritance in terms of discrete factors, now called genes, that are transmitted from generation to generation. Not all genes follow the case of simple dominance that Mendel discovered; nor do genotypes always predict phenotypes in a rigid manner. However, more complicated inheritance patterns do not alter the particulate theory of inheritance, with genes being transmitted according to rules of chance, or Mendel's two laws.

Some characteristics show intermediate inheritance, in that the F_1 hybrids have a phenotype intermediate between that of the parents. In *incomplete dominance*, neither allele masks the expression of the other, and the phenotype of heterozygotes is distinguishable from the two homozygous conditions. In the case of a monohybrid cross with incomplete dominance, the phenotypic ratio is equal to the genotypic ratio.

Multiple alleles may exist for some genes, in which more than two allelic forms are found. Some of the alleles may be *codominant*, meaning that both alleles are expressed. Human blood groups include three alleles in which the alleles I^A and I^B are dominant over i, but codominant with each other. Individuals with the ii genotype have type O blood. Blood type is critical in transfusions because the antigens found on the red blood cells of unmatched donated blood may interact with the recipient's antibodies and cause agglutination, or clumping, of blood cells within the blood vessels.

Pleiotropy is the characteristic of a single gene having multiple phenotypic effects in an individual. Many hereditary diseases, with complex sets of symptoms, are caused by a single gene. Siamese cats are often cross-eyed because the gene that affects fur pigmentation also influences the connections between the eyes and brain of the cat.

The phenotype of an individual is the result of complex interactions between its genotype and the environment. *Penetrance* is the proportion of individuals that express the phenotype for a particular genotype. A gene with incomplete penetrance is not expressed in the phenotype of all individuals that carry the gene. The *expressivity*, or degree to which a gene is expressed, can also vary. Occasionally, a strong environmental in-

fluence will produce a phenotype, called a *phenocopy*, which simulates the effect of a gene that is not present in the genotype.

In *epistasis*, one gene may mask the expression of another, unrelated gene. An F_2 phenotypic ratio of 9:7 can be explained as the result of one gene being epistatic to another. Unless the first gene is present in its dominant form, the other gene cannot be expressed.

Quantitative traits, such as height or skin color, may vary in a population in a continuous way. In *polygenic inheritance*, two or more genes have an additive effect on a single phenotypic trait. Each dominant allele contributes one "unit" to the expressed trait. A polygenic trait, combined with the influence of environmental factors, may result in a *normal distribution*, forming a bell-shaped curve, of the trait within a population.

Mendelian Inheritance in Humans

A *family pedigree* is a family tree with the history of a particular trait shown across the generations. By convention, circles represent females, squares are used for males, and solid symbols indicate individuals that express the phenotype in question. Parents are joined by a horizontal line and offspring are listed below parents from left to right in order of birth. The genotypes of individuals in the pedigree can often be deduced by following the patterns of inheritance.

For genetic disorders inherited as simple Mendelian traits, geneticists, physicians, and genetic counselors use pedigree analysis to make statistical predictions concerning the probability that a trait may be inherited.

For the 1000 or so genetic disorders that are inherited as simple recessive traits, only homozygous-recessive individuals express the phenotype. *Carriers* of the disease are heterozygotes who are phenotypically normal but may transmit the recessive allele to their offspring. A mating between two heterozygotes has a 1/4 chance of producing an offspring with the homozygous recessive disorder. A normal child produced by this mating has a 2/3 chance of being a carrier.

Genetic disorders are rarely evenly distributed among all racial or cultural groups due to the different genetic histories of such groups. *Cystic fibrosis* is the most common genetic disease in the United States; it is found in 1 in every 2500 Caucasians, but more rarely in other races. This recessive allele causes excessive secretions of mucus in various organs, leading to blockage of the digestive tract, cirrhosis of the liver, pneumonia, or other infections. Most afflicted children die before adolescence.

Tay-Sachs disease is a lethal disorder in which the brain cells are unable to metabolize a type of lipid that then accumulates in and damages the brain, resulting in an early death. There is a disproportionately high incidence of this disease among Jewish people whose ancestors lived in Central Europe.

Sickle-cell anemia is the most common inherited disease among blacks; in the United States it affects 1 in 500 black children. Due to a single amino acid substitution in the hemoglobin protein, the molecules tend to crystallize, causing the red blood cells to deform into a sickle shape. The sickled cells block blood vessels and cause severe pain. Worldwide, about 100,000 people die from this disease annually. Heterozygous individuals are said to have *sickle-cell trait*, but are usually healthy. The resistance to malaria that may accompany the sickle-cell trait may explain why this lethal recessive allele remains in relatively high frequency in areas where malaria is common.

The likelihood of two mating individuals carrying the same rare deleterious allele increases when the individuals are close relatives. People with common ancestors are more likely to carry the same recessive genes, and *consanguineous matings* are more likely to produce offspring homozygous for a lethal trait. Matings between "blood" relatives are indicated on pedigrees with double lines.

There is some debate over the extent to which human consanguinity increases the risk of inherited diseases. Many societies have laws or taboos forbidding marriages between close relatives, perhaps as a consequence of the observed increase in stillbirths and birth defects resulting from such matings. In some populations, however, marriage between close relatives is common and has no observable ill effect. Inbreeding among zoo animals and domesticated animals, and in some small natural populations of animals, has resulted in a higher incidence of harmful recessive traits, but has enabled some endangered species to avoid extinction.

A few human disorders are due to dominant genes. For example, in achondroplasia, dwarfism is due to a single copy of a mutant allele. Homozygosity for this dominant allele is lethal. This rare allele illustrates that the dominance or recessiveness of an allele is not determined by its frequency in the population.

Dominant lethal alleles are more rare than are recessive lethals because the harmful allele cannot be masked in the heterozygote. Most dominant lethal alleles are the result of mutations and kill the developing organism before it can reproduce and pass on the new form of the gene. A late-acting lethal dominant allele, however, can be passed on if the symptoms do not develop until after an organism is old enough to reproduce. *Huntington's chorea* is a degenerative disease of the nervous system that does not develop until the individual is 35 to 45 years old. Children of a person who develops Huntington's have a 50% chance of having inherited the dominant lethal allele and thus of developing the disease later in life. Medical researchers are working to develop a method to detect the lethal gene without waiting for the symptoms to begin.

Scientists are working to remedy inherited diseases by adding, removing, or altering genes in somatic cells. The severity of some inherited diseases can be lessened by treating the symptoms. Prevention is sometimes possible by assessing the risk of a genetic disorder through genetic counseling before a child is conceived or in the early stages of pregnancy.

The probablilty of a child having a genetic defect can be determined by considering the family history of the disease. If two prospective parents both have siblings who had the disorder, both sets of prospective grandparents must have been carriers. The parents each have a 2/3 chance that they are carriers. The probability that *both* parents are carriers is 2/3 x 2/3; the chance that two heterozygotes will have a recessive homozygous child is 1/4. The overall chance that a child will inherit the disease is 2/3 x 2/3 x 1/4, or 1/9. Should this couple have a baby that has the disease, this establishes that they are both carriers, and the chance that a subsequent child will have the disease is 1/4.

Determining whether parents are carriers is important for determining the risk of a genetic disorder. Biochemical tests that permit carrier recognition are being developed. The use of new methods for analyzing DNA should help in the detection of harmful genes.

Amniocentesis, a procedure done between the fourteenth and sixteenth week of pregnancy, allows detection of several genetic disorders. Amniotic fluid removed from the sac surrounding the baby in the uterus is analyzed biochemically, and fetal cells present in the fluid are cultured for karyotyping to check for certain chromosomal defects.

Chorionic villi sampling is a new technique in which a small amount of the fetal tissue making up the placenta is suctioned off. These rapidly growing cells can be karyotyped immediately, giving much faster results than amniocentesis. The procedure can also be performed at only 8 to 10 weeks of pregnancy.

Ultrasound is a simple noninvasive procedure that can reveal major abnormalities. It uses sound waves to produce an image of the fetus. Fetoscopy, the insertion of a needle-thin viewing scope and light into the uterus, also allows the fetus to be checked for anatomical problems. Fetoscopy and amniocentesis can result in complications or fetal death, and should be reserved for situations in which the risk of genetic disorders or other defects outweighs the risk of the procedure.

Some genetic disorders can be detected at birth. The recessively inherited disorder *phenylketonuria* (PKU) is caused by the lack of an enzyme to break down properly the amino acid phenylalanine. An accumulation of phenylalanine and its by-product causes severe mental retardation. If the disease is detected in the newborn by a routine screening test, the diet of the child can be adjusted and normal development is possible.

STRUCTURE YOUR KNOWLEDGE

1. Relate Mendel's two laws of inheritance to the behavior of chromosomes in meiosis that you studied in Chapter 12.

2. Tall red-flowered plants are crossed with short white-flowered plants. The resulting F_1 generation consists of all tall pink-flowered plants. Assuming that height is a simple case of dominance and flower color involves incomplete dominance, determine the results of an F_1 cross of TtRr plants. Fill in the gametes and genotypes in the following Punnett square and the phenotypes and numbers of the F_2 generation.

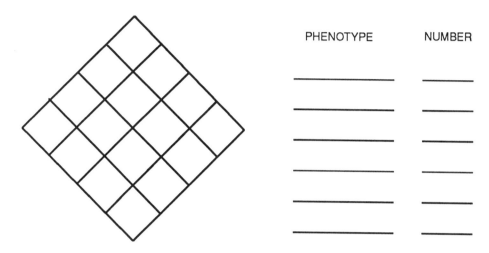

PHENOTYPE	NUMBER
_____	____
_____	____
_____	____
_____	____
_____	____
_____	____

3. Albinism (lack of skin pigmentation) is caused by a recessive allele. Consider the following human pedigree for this trait (solid symbols represent individuals who are albinos). From your knowledge of Mendelian inheritance, determine the probable genotypes of the parents in generation I, the mates in generation II, and son 4 in generation III. Can you determine the genotype of son 3 in generation II? Why or why not? (Let AA and Aa represent normal pigmentation and aa be the albino genotype.

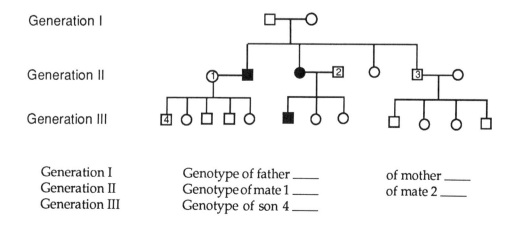

Generation I
Generation II
Generation III

Genotype of father ____   of mother ____
Genotype of mate 1 ____   of mate 2 ____
Genotype of son 4 ____

TEST YOUR KNOWLEDGE

MATCHING: *Match the definition with the correct term.*

1. ___ codominance

2. ___ homozygous

3. ___ heterozygous

4. ___ expressivity

5. ___ polygenic trait

6. ___ pleiotropy

7. ___ incomplete dominance

8. ___ testcross

9. ___ dihybrid cross

10. ___ penetrance

11. ___ epistasis

12. ___ phenocopy

A. cross with homozygous recessive to determine genotype of unknown

B. cross between two hybrids

C. cross that involves two different gene pairs

D. the degree to which a gene is expressed

E. the expressed characteristics of an individual

F. containing two different alleles for same locus

G. containing multiple alleles

H. heterozygote is intermediate between two parental homozygous phenotypes

I. both alleles are expressed in heterozygote

J. single gene with multiple phenotypic effects

K. proportion of individuals that express genotype

L. phenotype induced by environment, not coded for in genotype

M. one gene masks the expression of another gene

N. two or more genes with additive effect on phenotype

O. true breeding variety

MULTIPLE CHOICE: *Choose the one best answer.*

1. The genotypic ratio of a cross
   a. will be equal to the phenotypic ratio.
   b. will change if the gene involved exhibits incomplete dominance.
   c. will change if the gene involved has multiple alleles.
   d. will reflect all possible combinations of gametes.

2. The phenotypic ratio of a cross
   a. will be equal to the genotypic ratio.
   b. will change if the gene involved is pleiotropic.
   c. will change if the gene involved has multiple alleles.
   d. will include all the different phenotypic combinations of gametes.

3. Dominant alleles
   a. are expressed even when present only in one allele of a gene pair.
   b. will mask the expression of another unrelated gene.
   c. are the most frequently found alleles in a population.
   d. have pleiotropic effects when present in the heterozygote.

4. According to the theory of pangenesis,
   a. the characteristics of the offspring are a blending of parental traits.
   b. the sperm contains a miniature human being called the homunculus.
   c. changes to an organism's body can be passed to the next generation.
   d. particulate heritable factors separate in the formation of gametes.

5. The F_2 generation
   a. has a phenotypic ratio of 3:1.
   b. is the result of the self-fertilization or crossing of F_1 individuals.

c. can be used to determine the genotype of individuals with the dominant phenotype.

d. will show up in a dihybrid cross.

6. According to Mendel's law of segregation,
   a. there is a 50% probability that a gamete will get a dominant allele.
   b. gene pairs segregate independently of other genes in gamete formation.
   c. allele pairs separate in gamete formation.
   d. the laws of probability determine gamete formation.

7. A 1:1 phenotypic ratio in a testcross
   a. indicates that the alleles are codominant.
   b. indicates that one parent must have been homozygous recessive.
   c. indicates that the dominant phenotype parent was a heterozygote.
   d. indicates that the alleles segregated independently.

8. According to Mendel's law of independent assortment,
   a. an individual heterozygous for three genes should produce eight different gametic combinations of alleles.
   b. gene pairs segregate independently of other genes in gamete formation.
   c. the F_2 phenotypic ratio of a dihybrid cross of two true breeding varieties should be 9:3:3:1.
   d. all of the above are correct.

9. Quantitative traits
   a. are found in large numbers of offspring.
   b. may exist in a normal distribution within a population.
   c. may be the result of pleiotropic genes.
   d. may be the result of varying penetrance of a gene.

10. The probability that a particular genotype may result from a cross
    a. can be determined as the product of the probabilities of the formation of the gametes needed to produce the genotype.
    b. can be determined from the genotypic ratio for the cross.
    c. will depend on the genotypes of the parents.
    d. is related to all of the above.

11. Carriers of a disease
    a. are indicated by solid symbols on a family pedigree.
    b. are heterozygotes for the gene that can cause the disease.

c. will produce children with the disease.

d. usually are involved in consanguineous matings.

12. A lethal recessive allele is more likely to be maintained in a population
    a. if it is somehow beneficial in the heterozygous condition.
    b. if it is not expressed until late in life.
    c. if it can be identified by genetic screening.
    d. both a and b are correct.

13. Dominant lethal alleles
    a. are expressed late in life.
    b. are lethal in both the homozygous and heterozygous condition.
    c. are more likely to occur in consanguineous matings.
    d. are responsible for cystic fibrosis, Tay-Sachs disease, and sickle-cell anemia.

14. With chorionic villi sampling,
    a. a couple is tested to see if they are carriers of a genetic disorder.
    b. a fetus is tested for anatomical problems.
    c. fetal cells of the placenta can be karyotyped.
    d. amniotic fluid is biochemically tested.

15. If both parents are carriers of a lethal recessive gene, the probability that their child will inherit and express the disorder is
    a. $2/3 \times 2/3 \times 1/4$, or $1/9$.
    b. $1/2$.
    c. $1/4$.
    d. $1/2 \times 1/2 \times 1/4$, or $1/16$.

GENETIC PROBLEMS

1. Blood typing is often used as evidence in paternity cases, when the blood type of the mother and child may indicate that a man alleged to be the father could not possibly have fathered the child. For the following mother and child combinations, indicate which blood groups of potential fathers would be exonerated.

Blood group of mother	Blood group of child	Man is exonerated if he belongs to blood group(s)
AB	A	_____
O	B	_____
A	AB	_____
O	O	_____
B	A	_____

2. Polydactyly (extra fingers and toes) is due to a dominant gene. A father is polydactyl, the mother has the normal phenotype, and they have had one normal child. What is the genotype of the father? Of the mother? What is the probability that a second child will have the normal number of digits?

3. For the following genotypes, indicate what proportion of the gametes will have the indicated genes.

Parental genotype	Type of gamete	Proportion expected
AABb	AB	_____
AaBb	ab	_____
AABbcc	ABc	_____
AaBbCc	ABc	_____

4. For the following crosses, indicate the probability of obtaining the indicated genotype in an offspring. Remember it is easiest to treat each gene separately as a monohybrid cross and then combine the probabilities.

Cross	Offspring	Probability
AAbb x AaBb	AAbb	_____
AaBB x AaBb	aaBB	_____
AABbcc x aabbCC	AaBbCc	_____
AaBbCc x AaBbcc	aabbcc	_____

5. In dogs, black (B) is dominant to chestnut (b), and solid color (S) is dominant to spotted (s). What are the genotypes of the parents that would produce a cross with 3/8 black solid, 3/8 black spotted, 1/8 chestnut solid, and 1/8 chestnut spotted puppies? (Hint: first determine what genotypes the offspring must have before you deal with the fractions.)

6. The height of spike weed is a result of polygenic inheritance involving 3 genes, each of which can contribute 5 cm to the plant. The base height of the weed is 10 cm, and the tallest plant can reach 40 cm. If a tall plant (AABBCC) is crossed with a base-height plant (aabbcc), what is the height of the F_1 plants? How many phenotypic classes will there be in the F_2?

7. In guinea pigs, the gene for production of melanin is epistatic to the gene for the deposition of melanin. The dominant allele M causes melanin to be produced; mm individuals cannot produce the pigment. The dominant allele B causes the deposition of a lot of pigment and produces a black guinea pig, whereas only a small amount of pigment is laid down in bb animals, producing a light-brown color. Without an M allele, no pigment is produced so the allele B has no effect and the guinea pig is white. A homozygous black guinea pig is crossed with a homozygous recessive white: MMBB x mmbb. Give the phenotypes of the F_1 and F_2 generations.

8. When hairless hamsters are mated with normal-haired hamsters, about one-half the offspring are hairless and one-half are normal. When hairless hamsters are crossed with each other, the ratio of normal-haired to hairless is 1:2. How can you account for the results of the first cross? Explain the unusual ratio obtained in the second cross.

9. If two medium-tailed pigs were mated and the litter produced included three stub-tailed, six medium-tailed, and four long-tailed piglets, what would be the simplest explanation of these results?

10. The ability to taste phenylthiocarbamide (PTC) is controlled in humans by a single dominant allele (T). A woman nontaster married a man taster and they had three children, two boy tasters and a girl nontaster. All the grandparents were tasters. Create a pedigree for this family for this trait. (Solid symbols should signify nontasters.) Where possible, indicate whether tasters are TT or Tt.

CHAPTER 14

THE CHROMOSOMAL BASIS OF INHERITANCE

FRAMEWORK

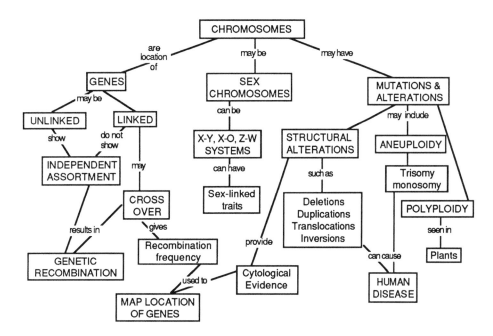

CHAPTER SUMMARY

Chromosome Theory of Inheritance

By 1900, three botanists had independently arrived at the same genetic principles that Mendel had discovered 35 years previously. These principles, combined with the cytological evidence of the processes of mitosis and meiosis developed in the late 1800s, led to the development of the *chromosome theory of inheritance*. According to this theory, Mendelian genes are located on chromosomes that undergo segregation and independent assortment in the process of gamete formation.

T. H. Morgan, working with the fruit fly, *Drosophila melangaster*, first associated a specific gene with a specific chromosome. Fruit flies are excellent organisms for genetic studies because they are prolific breeders and have only four pairs of chromosomes, which are easily distinguishable with a microscope. The sex chromosomes occur as XX in female and XY in male flies.

Traits that are the normal phenotypes found most commonly in nature are called *wild type*, whereas alternative traits, assumed to have arisen as mutations in a wild-type gene, are called *mutant phenotypes*. The genetic notation commonly used by *Drosophila* geneticists employs small letters to signify the mutant allele and a superscript plus sign for the wild allele.

Morgan discovered a mutant white-eyed male fly that he mated with a wild-type red-eyed female. The F_1 were all red-eyed and the F_2 showed the 3:1 phenotypic ratio typical of a simple dominant trait. In the F_2, however, all female flies were red-eyed and one-half of the males were red-eyed and one-half were white-eyed.

Morgan deduced that the gene for eye color was *sex-linked*, occurring on one sex chromosome but not the other. In this case, the gene for eye color is located on the X chromosome; males only have one X and their phenotype is determined by the eye-color gene they inherit from their mother. The association of a specific gene with a chromosome provided evidence for the chromosome theory of inheritance.

Linked genes are genes that are located on the same chromosome. Since chromosomes are inherited as a unit, the law of independent assortment usually does not apply to linked genes.

In 1908, Bateson and Punnett observed an unusual pattern of inheritance of two traits in the sweet pea. When they crossed two heterozygotes, formed from parents homozygous for either the dominant or recessive traits, they did not get the expected 9:3:3:1 ratio. Each trait individually appeared in the expected 3:1 ratio, but a large number of offspring showed either both dominant or both recessive traits. The explanation for this phenomenon emerged later from the work of Morgan: the genes for the two traits were linked on the same chromosome and inherited together, unless a crossover event occurred in the formation of eggs or pollen.

Chromosomal Basis of Recombination

The production of offspring that combine traits from the two parents is called *genetic recombination*. Mendel observed that, in a dihybrid cross between a heterozygote and a homozygous recessive individual, one-half of the offspring, called *parental types*, will have phenotypes like one or the other parent, and one-half the offspring, called *recombinants*, will have combinations of the two traits that are unlike the parents. This 50% frequency of recombination is observed when two genes are located on different chromosomes. It is the result of the random alignment of homologous chromosomes at metaphase I and the resulting independent assortment of alleles.

Linked genes do not assort independently, and one would not expect to see recombination of parental alleles in the offspring. In a testcross (cross using homozygous recessive individuals) with flies heterozygous for two genes, Morgan found that most of the offspring resembled the parental phenotypes, indicating that the genes were linked. Seventeen percent of the offspring, however, showed a recombination of parental traits. An exchange of segments between homologous chromosomes, called *crossing over*, accounts for the recombination of linked genes. The reciprocal trade between nonsister chromatids during prophase I transfers the same genetic loci and results in new combinations of alleles.

Morgan's group was the first to work out methods to map genes in sequence on particular chromosomes. Sturtevant suggested that recombination frequencies reflect the relative distances between genes. If genes are located farther apart on the chromosome, there is a greater probability that a crossover event will occur between them. Sturtevant used recombination data to locate genes on a map and defined one *map unit* as equal to a 1% recombination frequency.

The sequence of genes on a chromosome can be determined by finding the recombination frequency between different pairs of genes. Thus, if a and b are 12 map units apart, b and c are 3 map units apart, and c and a are 9 units apart, the sequence of genes must be a-c-b.

In a dihybrid testcross with a heterozygote, one would predict a 1:1 ratio if the genes were linked, and a 1:1:1:1 phenotypic ratio if the genes assorted independently. A 50% recombination of linked genes would also result in a 1:1:1:1 ratio. Thus, it is impossible to determine linkage if genes are 50 or more map units apart. Distant genes on the same chromosome may be mapped by determining and then adding the recombination frequencies determined with intermediate genes.

Sturtevant and his co-workers clustered the genes for various mutations of *Drosophila* into four groups of linked genes. Combined with the fact that there are four sets of chromosomes in *Drosophila*, these results provided additional evidence for the location of genes on chromosomes.

Crossover data provide only relative distances for the location of genes on chromosomes. *Cytological mapping* is a technique that associates a mutant phenotype with a visible chromosomal defect and can thus pinpoint the location of a particular gene.

X-rays can cause *point mutations*, in which a single spot of a gene is altered, or *chromosomal mutations*, in which breaks in the chromosome may be visible in the microscope. Muller discovered in the 1920s that the use of X-rays greatly increased the number of mutants of *Drosophila*, providing more raw material for genetic studies, but also indicating the dangers of X-rays.

The *polytene chromosomes*, found in the salivary glands of *Drosophila* larvae, consist of hundreds of aligned chromatids on which differences in staining density produce characteristic patterns of dark and light bands. Chromosomal mutations actually can be seen in these chromosomes. Homologous chromosomes closely pair in these cells, a rare event in non-meiotic cells. When one homologue has a structural alteration, the pairing is incomplete and the normal homologue loops out at the unpaired segment.

The chromosomal location of a gene can be determined by matching a specific mutant phenotype with a localized deformity in polytene chromosome pairing. By inducing mutations with X-rays and other mutagens and associating mutant phenotypes with chromosomal defects, geneticists have developed extensive cytological maps of the genes of *Drosophila* and other organisms. These maps differ from those determined from crossover data in the spacing between genes. Recombinational frequency does not translate into a fixed length on a chromosome because the frequency of crossing over varies for different regions of a chromosome.

Chromosomal Basis of Sex

Sex is a phenotypic character usually determined by the presence or absence of special chromosomes. Humans, other mammals, and fruit flies all share an *X-Y system* of sex determination. Males produce two kinds of gametes, sperm with either an X or a Y chromosome, and are called the *heterogametic sex*. Females are the *homogametic* sex; all eggs contain an X chromosome.

An *X-O system* is found in grasshoppers, crickets, roaches, and some other insects. The female has two X chromosomes. There is no Y chromosome—males are XO, having only one sex chromosome. In the *Z-W system* seen in birds, some fishes, and butterflies and moths, the female is the heterogametic sex, and males are designated as ZZ.

Bees, wasps, and ants have no sex chromosomes. Males develop from unfertilized eggs (called *parthenogenesis*) and are haploid. Females develop from fertilized eggs and are diploid.

The chromosomal basis of sex determination of plants that produce male and female flowers on two separate plants (*dioecious*) is usually an X-Y system. Most plant species and some animals (such as earthworms and snails) are *monoecious*; a single individual produces both sperm and eggs. In these cases, all individuals of the species have the same complement of chromosomes.

Sex chromosomes may carry genes for traits that are not related to maleness or femaleness. In humans, the X chromosome is much larger than the Y one, and most X-linked genes (contained on the X chromosome) do not have corresponding loci on the Y chromosome. Males inherit their sex-linked traits from their mothers; fathers can pass sex-linked traits to only their daughters.

Recessive sex-linked traits are seen more often in males. Females need to inherit recessive alleles on both their X chromosomes to express a recessive trait. *Hemophilia* is a sex-linked recessive trait that is characterized by excessive bleeding due to the lack or malfunction of a blood-clotting factor. Color-blindness and two types of muscular dystrophy are also sex-linked traits.

Most genes on the Y chromosome code for traits found only in males and have no counterparts on the X. The H-Y antigen, found on sperm and other specialized cells of the male and thought to play a role in sexual development, is a Y-linked trait.

Normal diploid cells have two copies of autosomal genes and thus a double dose of each gene's product. The *Lyon hypothesis*, formulated by geneticist Mary F. Lyon, states that one of the X chromosomes is present as a *Barr body* in each female somatic cell. With only one functional X chromosome, the dosages of sex-linked genes are equal in both females and males. Which of the two X chromosomes is inactivated is a random event occurring in embryonic cells. A female who is heterozygous for a sex-linked trait will express one allele in approximately one-half her cells and the alternate in the other cells. This genetic mosaicism is seen in the coloration of a tortoiseshell cat.

Sex-limited traits, such as beard growth in humans, are expressed exclusively in one sex, although the genes for them may be carried on autosomes, and thus be present in and transmitted by both sexes.

With *sex-influenced traits*, the penetrance or expressivity of autosomal genes is sex-dependent. Baldness may be expressed in a man having only one "baldness" allele, whereas a woman will express baldness only if she is homozygous for the allele. The different hormonal conditions of males and females determine the penetrance or expressivity of sex-influenced traits.

Chromosomal Alterations

Chromosomal mutations, caused by errors in meiosis or mutagens, can alter either the number of chromosomes in the cell or the structure of individual chromosomes. In *nondisjunction*, a pair of homologous chromosomes does not separate properly in meiosis I, or sister chromatids do not separate in meiosis II, and a gamete may receive either two or none of one type of chromosome. A zygote formed with one of these aberrant gametes has a chromosomal alteration known as *aneuploidy*, a nontypical number of chromosomes. If one chromosome is present in triplicate, the organism is said to be *trisomic* for that chromosome, and the chromosome number is $2N + 1$. A $2N - 1$ chromosome number indicates that an individual is *monosomic* for a chromosome. Aneuploid organisms usually have a set of symptoms caused by the abnormal dosage of genes on the extra or missing chromosome.

Polyploidy is a chromosomal alteration in which an organism has more than two complete chromosome sets, as in *triploid* ($3N$) or *tetraploid* ($4N$) organisms. Polyploidy is common in the plant kingdom and has played an important role in the evolution of plants. In animals, complete polyploids are less common than mosaic polyploids, in which patches of tetraploid tissue grow from a cell in which the chromosomes did not separate in mitosis.

Breakage of chromosomes can lead to *deletion*, in which a fragment of the chromosome is lost; *duplication*, in which a broken fragment may join to the homologous chromosome; *translocation*, in which the fragment may join a nonhomologous chromosome; or *inversion*, in which the fragment may rejoin the original chromosome in the reverse orientation. Errors in crossing over can result in deletions and duplications caused by nonequal exchange of chromatids.

A homozygous deletion is usually lethal; most genes are necessary for an organism's existence. Duplications and translocations are typically harmful. With inversions, even though all the genes are present in proper quantities, the phenotype may be altered due to the position effects of neighboring genes on the expression of the relocated genes.

The frequency of aneuploid zygotes may be fairly high in humans, but development is usually so altered that the embryos spontaneously abort long before birth. Some genetic diseases, expressed as "syndromes" of characteristic traits, are the result of aneuploidy.

Down's syndrome, affecting 1 out of every 600 children born, is the most common serious birth defect in the United States. Trisomy of chromosome 21 results in characteristic facial features, short stature, heart defects, and mental retardation. Meiotic nondisjunction seems to become more common in older women, and the incidence of trisomy 21 and other major chromosomal defects increases for older mothers.

Patau's syndrome, caused by trisomy of chromosome 13, is characterized by harelip, cleft palate, and eye, brain and circulatory defects. Edward's syndrome, caused by a trisomy of chromosome 18, severely affects every organ system of the body. Victims of both these syndromes rarely live more than 1 year.

Most sex-chromosome aneuploidies upset the genetic balance less than do autosomal aneuploidies, perhaps because so few genes are located on the Y chromosome and extra X chromosomes are inactivated as Barr bodies. One in every 400 live births results in an XXY male who will exhibit *Klinefelter's syndrome*, a condition in which the individual has abnormally small testes, is sterile, may have feminine body contours, and is usually of normal intelligence. Additional X or Y chromosomes (XXXY, XXYY, and so on) usually result in individuals with Klinefelter's syndrome who are mentally retarded.

Males with a single extra Y chromosome may be somewhat taller than average males, but they do not exhibit any well-defined syndrome. Trisomy X results in *metafemales* with limited fertility and intelligence. Monosomy X individuals exhibit *Turner's syndrome*,

phenotypically female, sterile individuals with short stature, no secondary sex characteristics, and usually normal intelligence.

In humans, the presence of a Y chromosome determines "maleness," and the absence of a Y chromosome determines "femaleness," regardless of the number of X chromosomes. In *Drosophila*, the ratio of X chromosomes to autosomes determines sex. An XXY fruit fly is female; an XO fly is male.

Structural alterations of chromosomes, such as deletions, are associated with specific human disorders and often severe physical and mental defects. The *cri du chat* syndrome of mental retardation, small head, and unusual cry is caused by a deletion in chromosome 5.

Mapping Human Chromosomes

Sex-linked traits and syndromes associated with observable chromosomal alterations have resulted in the mapping of some genes to particular chromosomes. Methods of cell culture and the formation of hybrid cells between two different cell lines (even different species) have allowed human genes to be assigned to particular chromosomes. Hybridized cells kept in cell culture for many generations lose most of the chromosomes of one of the parental lines. The presence or absence of a particular gene product can then be related to the presence or absence of a particular chromosome. It is now possible to determine the exact location of a gene using recombinant DNA technology.

Extranuclear Inheritance

Exceptions to Mendelian inheritance are found in the case of extranuclear genes located in cytoplasmic organelles that are usually transmitted to offspring in the cytoplasm of the egg cell. Cytoplasmic genes were first observed by Correns in 1909 in the maternal inheritance of the plastids determining leaf coloration. Maternal inheritance of mitochondrial genes in mammals is due to the large cytoplasmic contribution of the egg cell.

STRUCTURE YOUR KNOWLEDGE

1. Two of the genes Mendel used to support his law of independent assortment were actually located on the same chromosome. Explain why genes located more than 50 map units apart behave as though they are not linked. How can one determine whether these genes are linked and what the relative distance between them is?

2. Fill in the following table of the various chromosomal systems of sex determination.

SYSTEMS OF SEX DETERMINATION					
	X-Y	X-O	Z-W	HAPLO-DIPLOID	NO SEX CHROMOSOMES
FEMALE					
MALE					
EXAMPLES					

3. You have found a new mutant phenotype in fruit flies that you suspect is recessive and sex-linked. What is the best cross you could make to confirm your predictions?

4. Fill in the following table concerning human diseases or syndromes related to chromosomal abnormalities. What explanation can you give for the adverse phenotypic effects associated with these chromosomal alterations?

NAME OF DISEASE OR SYNDROME	CHROMOSOMAL ALTERATION INVOLVED	SYMPTOMS OR ASSOCIATED TRAITS
	Trisomy 21	
	Trisomy 13	Harelip, cleft palate, eye, brain, circulatory defects
Edwards Syndrome		
		Sterility, small testes, enlarged breasts or feminine body contours, normal intelligence if XXY, mental retardation if additional X or Y
Turner Syndrome		
Metafemale	Trisomy X	
		Mental retardation, small head, unusual facial features, unusual cry

TEST YOUR KNOWLEDGE

MULTIPLE CHOICE: *Choose the one best answer.*

1. The chromosomal theory of inheritance states that
    a. genes are located on chromosomes.
    b. chromosomes and their associated genes undergo segregation during meiosis.
    c. chromosomes and their associated genes undergo independent assortment in gamete formation.
    d. all of the above are correct.

2. A wild type is
    a. the phenotype found most commonly in nature.
    b. the dominant allele.
    c. designated by a small letter.
    d. your basic party animal.

3. Sex-linked traits
    a. are found less frequently in female individuals.
    b. are coded for by genes located on a sex chromosome.
    c. are found in only one or the other sex, depending on the sex-determination system of the species.
    d. are always inherited from the mother in mammals and fruit flies.

4. Which of the following is not true of cytological maps?
    a. They are more accurate than are maps based on recombination data.
    b. They are most easily made with polytene chromosomes.
    c. They rely on chromosome aberrations and mutant phenotypes for their construction.
    d. They rely on hybrid cells and cell culture to determine gene loci.

5. Genetic recombination
    a. results in recombinant offspring.
    b. occurs in the fertilization process.
    c. occurs by independent assortment and crossing over in the formation of gametes.
    d. involves all of the above.

6. A 1:1:1:1 ratio of offspring from a dihybrid testcross indicates that
    a. the genes are linked.
    b. the genes are not linked.
    c. crossing over has occurred.
    d. the genes are 25 map units apart.

7. In grasshoppers, the heterogametic sex is
    a. females that produce gametes with either a Z or a W chromosome.
    b. males that produce gametes with either an X or no sex chromosome.
    c. males that produce gametes with either an X or a Y chromosome.
    d. neither sex, since sex depends on haploidy or diploidy.

8. Dioecious plants
    a. are usually polyploid.
    b. produce male and female flowers on separate plants.
    c. do not have a chromosomal basis for sex determination.
    d. all of the above are true.

9. A son inherits color-blindness
    a. from his mother.
    b. from his father.
    c. from his mother only if she is color-blind.
    d. from his father only if he is color-blind.

10. Baldness is a sex-influenced trait in that
    a. its expressivity varies between men and women.
    b. it is expressed exclusively in men even though its gene is carried on an autosome.
    c. it is inherited only from fathers.
    d. it is carried on the X chromosome and thus shows up more often in men.

11. Nondisjunction
    a. occurs when homologous chromosomes do not separate properly in meiosis I.
    b. occurs when sister chromatids do not separate in meiosis II.
    c. can result in both trisomic and monosomic offspring.
    d. may involve all of the above.

12. In translocation,
    a. a fragment of a chromosome joins to its homologue.
    b. a fragment of a chromosome joins to a nonhomologous chromosome.
    c. the breakage and refusion of a chromosome does not affect phenotype.
    d. the offspring shows aneuploidy.

13. A triploid individual
    a. is the result of a duplication.
    b. has a $2N + 1$ chromosome number.
    c. has a $3N$ chromosome number.
    d. results from cells in which chromosomes did not separate during mitosis.

14. Extra dosages of genes
    a. usually have deleterious effects.
    b. can be caused by duplications.
    c. are found in Down's, Patau's, and Edwards's syndromes.
    d. all of the above.

15. Sex-chromosome aneuploidies upset genetic balance less than do autosomal aneuploidies because
    a. extra X chromosomes are inactivated as Barr bodies.
    b. the Y chromosome only determines sex; it carries no genes.
    c. many more of these individuals lead normal lives.
    d. all of the above.

16. Which of the following is not true of maternal inheritance?
    a. It is the result of the larger contribution of egg cytoplasm.
    b. It gave evidence for the existence of cytoplasmic genes.
    c. Mitochondrial, ribosomal, and plant plastid genes come from the mother.
    d. It does not follow Mendelian principles of inheritance.

17. Human chromosomes can be mapped using
    a. loops in polytene chromosomes that correspond to mutant syndromes.
    b. new techniques of cell culture and hybridization of cells.
    c. amniocentesis.
    d. recombination frequency data.

18. According to the Lyon hypothesis,
    a. females are a genetic mosaic due to random nonseparation of chromatids during mitosis.
    b. males inherent their sex-linked traits from their mothers.
    c. equal dosages of x-linked genes for males and females occur because of the formation of Barr bodies.
    d. maternal inheritance is a result of the large amount of egg cytoplasm.

19. A sex-linked lethal recessive gene occurs in pigeons. What would be the sex ratio in the offspring of a cross between a male heterozygous for the lethal gene and a normal female?
    a. 2:1 male to female
    b. 1:2 male to female
    c. 1:1 male to female
    d. 4:3 male to female

20. The sex-determination system of *Drosophila* resembles that of humans in that
    a. an XXX is a viable female in both species.
    b. an XO is a viable female in both species.
    c. sex is determined by the number of X chromosomes present.
    d. male sex is determined by the presence of the Y chromosome.

GENETICS PROBLEMS

1. Two normal color-sighted individuals produce the following children and grandchildren. Fill in the probable genotype of the indicated individuals in this pedigree. Solid symbols represent color-blindness. Choose an appropriate notation for the genotypes.

GENOTYPES

1 _____   5 _____

2 _____   6 _____

3 _____   7 _____

4 _____

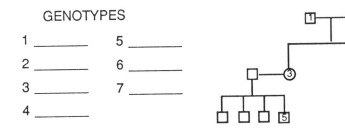

2. The following recombination frequencies were found. Determine the order of these genes on the chromosome.

a – c	10%	b – c	4%	d – c	20%
a – d	30%	b – d	16%		
a – e	6%	b – e	20%		

3. In guinea pigs, black (B) is dominant to brown (b), and solid color (S) is dominant to spotted (s). A heterozygous black, solid-colored pig is mated with a brown-spotted pig. The total offspring for several litters are

black solid	16
black spotted	5
brown solid	5
brown spotted	14

Are these genes linked or nonlinked? If they are linked, how many map units are they apart?

CHAPTER 15

THE MOLECULAR BASIS OF INHERITANCE

FRAMEWORK

This chapter outlines the key evidence that was gathered to establish DNA as the molecular basis of inheritance. Watson and Crick's double helix, with its rungs of specifically paired nitrogenous bases and twisting side ropes of phosphate and sugar groups, provided the three-dimensional model of DNA that explained DNA's ability to encode a great variety of information and produce exact copies of itself through semiconservative replication. The replication of DNA is an extremely fast and accurate process involving many enzymes and proteins, including: proteins that bind to origins of replication; helicases, topoisomerases, and single-strand binding proteins that unwind and unkink DNA and keep it apart; primases that form the RNA primer needed to begin replication; DNA polymerases and DNA ligase that direct the addition of nucleotides and link fragments of DNA together; and various enzymes that proofread and repair the DNA molecule.

CHAPTER SUMMARY

Deoxyribonucleic acid, DNA, is the genetic material, the substance of genes, the basis of heredity. Nucleic acids' unique ability to direct their own reproduction allows for the precise replication and transmission of DNA to all the cells in the body and from one generation to the next. DNA encodes the blueprints for the proteins and enzymes that direct and control the developmental, biochemical, anatomical, physiological, and behavioral traits of organisms.

The Search for the Genetic Material

The role of DNA in heredity was first established through work with microorganisms—bacteria and viruses. By the 1940s, chromosomes were known to carry hereditary information and to consist of proteins and DNA. Most scientists believed that the proteins carried the genetic program because of the known specificity and heterogeneity of proteins. The simple, repetitious nature of nucleic acids did not seem capable of coding the wealth of information needed for heredity.

The work of Griffith in 1928 provided the first evidence that the genetic material was some sort of heat-stable chemical. Griffith worked with two strains of *Diplococcus pneumoniae*—a smooth strain (S) that synthesized a mucous coat or capsule and a rough strain (R) that did not form a capsule. Only live S strain cells injected in mice caused pneumonia; mice injected with heat-killed S cells or R cells did not develop pneumonia and die. However, when Griffith mixed heat-killed S cells and live R cells and then injected the mixture, the mice died. Griffith was able to isolate live S cells from the blood of these mice, even though only dead S cells had been used. R cells had somehow acquired the ability to make polysaccharide coats from dead S cells. This transfer of genetic material is called *transformation*. Griffith's use of heat to destroy the S cells indicated that the genetic material might not be proteins since heat denatures most proteins.

Avery worked for a decade to identify the transforming agent by purifying chemicals from heat-killed S cells. In 1944, he and his colleagues McCarty and MacLeod announced that DNA was the molecule that transferred the genetic information.

Viruses consist of little more than DNA, or sometimes RNA, contained in a protein coat. They can reproduce only by infecting another cell and commandeering that cell's metabolic machinery. *Bacteriophages*, or *phages* for short, are viruses that infect bacteria. In 1952, Hershey and Chase showed that DNA was the genetic material of a phage known as T2 that infects the bacterium *Escherichia coli* (E. coli).

It was known that the T2 virus consisted of DNA and protein and was able to somehow inject its genetic material into the bacterium so that the *E. coli* cell turned into a T2-manufacturing cell. Hershey and Chase devised an experiment using radioactive isotopes to tag the phage's DNA and proteins to determine which was transferred to the bacteria. One batch of T2 was grown with radioactive sulfur to become incorporated into and tag protein; another was grown with radioactive phosphorus that labeled the DNA.

The labeled T2 cells were allowed to infect separate samples of *E. coli*. The cultures were blended to shake off parts of the phages that remained outside the bacteria cells and then centrifuged to separate the heavier bacterial cells from the lighter viral particles. Radioactivity was measured and compared for the pellet and supernatant.

In the samples with the labeled proteins, the radioactivity was found in the supernatant, indicating that the phage protein did not enter the bacterial cells. In the samples with the labeled DNA in the T2 phage, most of the radioactivity was contained in the bottom of the tube in the bacterial cell fraction. When these *E. coli* cells were returned to culture, they lysed and released phages.

Circumstantial evidence that DNA was the genetic material came from the observation that, prior to mitosis, a eukaryotic cell doubles its DNA content, and diploid cells have twice as much DNA as do haploid cells.

Chargaff, in 1947, reported that DNA composition is species-specific. Using paper chromatography to separate bases, he found that each species he studied had a different ratio of nitrogenous bases. Chargaff also determined that the number of adenines and thymines were equal and the number of guanines and cytosines were equal in the DNA from all the organisms he studied. The A=T and G=C property of DNA, known as *Chargaff's Rules*, was not explained until the double helix was discovered.

Discovery of the Double Helix

By the early 1950s, the arrangement of covalent bonds in a nucleic acid polymer was established and the race was on to determine the three-dimensional structure of DNA. Linus Pauling in California and Maurice Wilkins and Rosalind Franklin in London were working on the problem; but Watson and Crick were the first to solve the DNA puzzle.

Crick was studying protein structure using a technique called *X-ray crystallography*. An X-ray beam passed through a crystal can expose photographic film to produce a pattern of spots that a crystallographer can interpret into information about the three-dimensional atomic structure of the crystal. While visiting Wilkins, Watson saw an X-ray photo produced by Franklin that clearly showed the basic shape of DNA to be a helix.

He and Crick deduced that the helix had a width of 2-nanometers with its purine and pyrimidine bases stacked 0.34 nanometers apart. This width suggested that the helix consisted of two strands, thus the term *double helix*.

Watson and Crick developed models of wire to build a double helix that would conform to the X-ray measurements and the known chemistry of DNA. They finally arrived at a model that paired the nitrogenous bases on the inside of the helix with the sugar-phosphate chains on the outside. Franklin's X-ray data indicated that the helix makes one full turn every 3.4 nanometers; thus ten layers of nucleotide pairs, stacked 0.34 nanometers apart, are present in each turn of the helix. The sugar-phosphate side chains are oriented in opposite directions, said to be *antiparallel*.

The idea that there was specific base pairing meant that one strand of the double helix would complement the information contained on the other strand. At first Watson assumed that bases paired with themselves, but the diameter of the double helix would not permit such an arrangement. To produce the molecule's 2-nanometer width, a purine must pair with a pyrimidine. The molecular arrangement of the side groups of the bases permitted two hydrogen bonds to form between adenine and thymine, and three hydrogen bonds between guanine and cytosine. This complementary pairing explained Chargaff's rules (A=T and G=C). The cumulative effect of a multitude of these weak interactions, coupled with the van der Waals forces between the stacked bases, contributes to the stability of the double helix.

The *sequence* of nucleotides along the length of a DNA was not controlled by any pairing rules. The infinite variety of sequences possible would support the necessity of encoding unique and detailed genetic information.

In April 1953, Watson and Crick published a paper in *Nature* reporting the double helix as the molecular model for DNA. This model set the stage for the recent exciting advances in molecular biology.

DNA Replication

A template theory had already been advanced for the explanation of how genetic material could be copied. When a gene replicates, a "negative image" could be created that would then serve as a template for synthesis of copies of the original "positive image." The specific base pairing of the Watson and Crick model immediately suggested a template mechanism of DNA replication.

In a second paper, Watson and Crick suggested that DNA replication was accomplished by the separation of the two DNA strands and the creation of two new complementary strands as nucleotides paired up along the two exposed strands. This *semiconservative* model predicted that the two daughter DNA molecules would each have one old strand from the parent DNA and one newly formed strand. A conservative model would mean that the parent strand would remain intact and the duplicate molecule would be totally new.

Meselson and Stahl provided the evidence for the semiconservative model. They grew *E. coli* in a medium with ^{15}N, a heavy isotope that the bacteria incorporated into their nitrogenous bases. When centrifuged, this DNA could be separated by density from nonlabeled DNA. Cells with labeled DNA were transferred to a medium with the lighter isotope, ^{14}N. After one generation of bacterial growth, the DNA extracted from the culture was all of intermediate density; it was a hybrid of the parental heavier DNA strand and the newly formed lighter DNA strand. Had DNA replication been conservative, one-half of the DNA would have been parental and heavy, and one-half would have been newly made from ^{14}N and light.

DNA replication is extremely rapid (50 bases added per second in mammals, 500 per second in bacteria) and accurate (only 1 base mispaired per 1 billion nucleotides). More than a dozen enzymes and other proteins are involved in this intricate process.

Replication seems to begin at special sites, called *origins of replication*, where specific proteins that initiate replication bind. Replication spreads in both directions from these origins. Bacterial and viral DNA molcules have only one replication origin, but the large eukaryotic DNA molecules may have hundreds to thousands of origin sites. The points along the DNA molecule where replication is occurring are called *replication forks* because of their Y shape.

Enzymes called *helicases* unwind the helix and *single-strand binding proteins* help keep the separated strands apart. The kinks formed from separating the strands of the double helix are untangled by enzymes called *topoisomerases* that temporarily break or nick the molecule so that it can rotate freely to "unkink."

A primer of RNA is needed to initiate DNA replication. An enzyme called *primase* pairs about five RNA nucleotides to a short portion of the DNA strand that serves as the primer template. Enzymes called *DNA polymerases* then catalyze the synthesis of a new DNA strand, pairing bases and connecting the nucleotides of the new strand. The energy for creating the covalent bonds between nucleotides is provided by the release of a pyrophosphate group from each incoming nucleotide.

DNA is replicated in a 5'—>3' direction. The 5' end of the strand is the one with either a phosphate group or a free (nonphosphorylated) hydroxyl group attached to the number five carbon of the terminal sugar. At the 3' end, a free -OH is attached to the number three carbon of the end sugar.

Because the DNA strands run in an antiparallel direction, the simultaneous synthesis of both strands presents a problem. The *leading strand* is the new 5'—>3' strand being polymerized as a single polymer. The *lagging strand*, which runs in the 3'—>5' direction, is created as a series of short segments, each formed in the 5'—>3' direction as the DNA molecule unzips. These segments are known as *Okazaki fragments* after the scientist who discovered them. Each fragment requires an RNA primer. A continuous strand of DNA is produced after a DNA polymerase removes the RNA primer and replaces it with DNA, and a linking enzyme called *DNA ligase* joins the 3' end of each new fragment to the 5' end of the growing chain.

Pairing errors in nucleotide placement may occur as often as 1 per 10,000 bases. The amazing accuracy of DNA replication is achieved by the proofreading function of DNA polymerase. In bacteria, DNA polymerase checks each nucleotide against its template after insertion and backs up and replaces incorrect nucleotides. It is not known whether, in eukaryotic cells, DNA polymerase or other enzymes perform this proofreading function.

DNA Repair

DNA molecules may be accidentally altered by the action of reactive chemicals, radioactive emissions, X-rays, and ultraviolet light (UV). These changes, or mutations, may be corrected through the action of many types of DNA repair enzymes. In *excision repair*, the damaged strand is cut out by a repair enzyme and the gap is correctly filled through the action of a DNA polymerase and DNA ligase. This type of repair is particularly important in skin cells to repair the DNA damage done by ultraviolet rays of sunlight.

Alternative Forms of DNA

The elegantly simple architecture of the double helix reaffirmed the fundamental structure-fits-function theme of biology. Recent evidence indicates that not all DNA corresponds to the double helix model. DNA molecules may be linear or circular, supercoiled, or occasionally single-stranded. Even linear DNA may show variations in its helical structure, with different three-dimensional forms known as *A-DNA*, *C-DNA*, and *Z-DNA*. Z-DNA twists in the left-hand direction, opposite that of the most common *B-DNA* represented by the Watson–Crick double helix. The biological functions of these nonB forms of DNA are being investigated.

STRUCTURE YOUR KNOWLEDGE

1. Fill in the following chart about these key investigators and the evidence they provided as to the structure and function of DNA.

INVESTIGATOR	ORGANISMS USED	TECHNIQUES, EXPERIMENTS	CONCLUSIONS
Griffith			
Avery			
Hershey & Chase			
Chargaff			
Franklin			
Watson & Crick			
Meselson & Stahl			

2. Create a concept map to illustrate the key participants and events involved in DNA replication.

TEST YOUR KNOWLEDGE

MULTIPLE CHOICE: *Choose the one best answer.*

1. One of the reasons most scientists believed proteins were the carriers of genetic information was that
   a. proteins were more heat stable than nucleic acids.
   b. DNA was not always double stranded.
   c. proteins were much more complex molecules than were nucleic acids.
   d. early experimental evidence pointed to proteins as the hereditary material.

2. Transformation involves
   a. the transfer of genetic material, often from one bacterial strain to another.
   b. the creation of a strand of RNA from a DNA molecule.
   c. the infection of bacterial cells by bacteriophages.
   d. the type of replication shown by DNA.

3. According to Chargaff's rules,
   a. heavier labeled fractions will be found in the pellet of a centrifuge tube.
   b. DNA replication must be semiconservative.
   c. A=T and G=C.
   d. each species has a different ratio of nitrogenous bases.

4. For his work with DNA, Chargaff used
   a. X-ray crystallography.
   b. paper chromatography.

c. heavy isotopes of sulfur and phosphorus.

d. ultracentrifugation and ^{15}N.

5. In his work with pneumonia-causing bacteria and mice, Griffith found
   a. that DNA was the transforming agent.
   b. that the R and S strains mated.
   c. that heat-killed S cells could cause pneumonia when mixed with heat-killed R cells.
   d. that some heat-stable chemical was transferred to R cells to transform them into S cells.

6. When T2 phage are grown with radioactive sulfur,
   a. their DNA is tagged.
   b. their proteins are tagged.
   c. their DNA is found to be of medium density in a centrifuge tube.
   d. they transfer their radioactivity to *E. coli* when they infect them.

7. Meselson and Stahl
   a. provided evidence for the semiconservative replication of DNA.
   b. were able to separate phage protein coats from *E. coli* by using a blender.
   c. found that DNA labeled with ^{15}N was of intermediate density.
   d. grew *E. coli* on labeled phosphorus and sulfur.

8. Watson and Crick concluded that each base could not pair with itself because
   a. there would not be room for the helix to make a full turn every 3.4 nanometers.
   b. the width of 2 nm would not permit two purines to pair together.
   c. the bases could not be stacked 0.34 nm apart.
   d. identical bases could not hydrogen bond together.

Use these centrifuge tubes showing density bands of DNA to answer questions 9 and 10.

a.　　b.　　c.　　d.

9. *E. coli* grown on ^{15}N medium are transferred to ^{14}N medium for one generation of growth. DNA extracted from these cells is mixed with cesium chloride and placed in an ultracentrifuge. What density distribution of DNA would you predict for this experiment from the possible DNA density bands illustrated above?
   a. ___
   b. ___
   c. ___
   d. ___

10. In an experiment similar to that in question 9, the cells are allowed to grow on the ^{14}N medium for another generation. What density distribution would you predict for this experiment from those illustrated above?
    a. ___
    b. ___
    c. ___
    d. ___

11. The energy for the polymerization of DNA comes from
    a. DNA polymerase.
    b. the hydrolysis of ATP.
    c. the loss of a pyrophosphate from nucleoside triphosphate monomers.
    d. the mitochondria.

12. The continuous elongation of DNA along one strand of DNA
    a. occurs on the leading strand.
    b. occurs because DNA polymerase can elongate only in the 5'—>3' direction.
    c. does not produce Okazaki fragments.
    d. all of the above.

13. Topoisomerases
    a. unwind DNA strands prior to replication.
    b. unkink single-strand DNA by nicking and resealing the polymer.
    c. provide excision repair of damaged DNA.
    d. create the RNA primer to initiate replication.

14. DNA ligase is needed
    a. to join Okazaki fragments on the leading strand.
    b. to replace the RNA primer.
    c. to join the 3' end of DNA repaired segments to the 5' end of the gap.
    d. all of the above.

15. Which of the following statements about DNA polymerase is incorrect?
  a. It is found only in eukaryotes.
  b. It is able to proofread and correct for errors in its base pairing.
  c. It is unable to join free nucleotides unless an RNA primer is present.
  d. It only works in the 5'—>3' direction.

16. Thymine dimers, covalent links between adjacent thymine bases in DNA, may be induced by UV light. When they occur, they are repaired by
  a. excision enzymes that cut out the damaged region.
  b. DNA polymerase.
  c. ligase.
  d. all of the above.

17. Okazaki fragments are produced by
  a. RNA primase.
  b. DNA polymerase working in 5' —> 3' direction.
  c. excision enzymes.
  d. Both a and b are correct.

18. Not all DNA conforms to the double helix model because
  a. some DNA may be single-stranded.
  b. some DNA may have a different three-dimensional structure.
  c. Z-DNA twists in the left-hand direction.
  d. all of the above.

Use this diagram of replicating DNA to answer questions 19 through 25. Each letter is used only once.

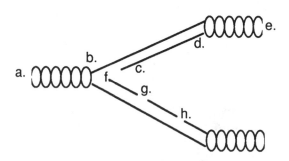

19. A helicase would be found at letter ____ .
20. An RNA primase would be found at letter ____ .
21. An Okazaki fragment is indicated by letter ____ .
22. A ligase would be found at letter ____ .
23. On the leading strand, DNA polymerase would be moving from letter ____ to ____ .
24. The parental strand of DNA is indicated by letter ____ .
25. One of the daughter strands is indicated by letter ____ .

Use the following diagram to answer questions 26 through 28.

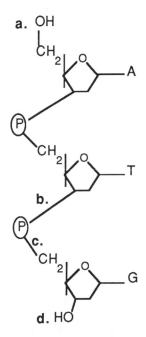

26. Which letter (a, b, c, or d) indicates the 5' end of this single DNA strand?
27. Which letter (a, b, c, or d) indicates a phosphodiester bond formed by DNA polymerase?
28. The base sequence of the DNA strand made from this template would be (from top to bottom)
  a. ATC
  b. CGA
  c. TAC
  d. UAC

CHAPTER 16

FROM GENE TO PROTEIN

FRAMEWORK

This chapter deals with the central dogma of molecular biology, the sequence of DNA to RNA to proteins. The steps involved in the synthesis of mRNA and proteins are detailed, and the effect of DNA mutations on the functional products of genes is described. The following outline lists the major concepts involved in the pathway from gene to protein.

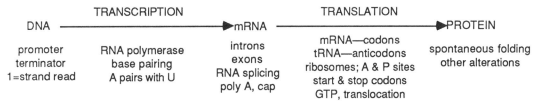

CHAPTER SUMMARY

The *central dogma* of molecular biology describes the sequence of DNA—>RNA—>protein. The DNA inherited by an organism directs the activities of each cell by controlling the synthesis of enzymes and other proteins through messenger RNA.

Genes Control Metabolism

In 1909, British physician Garrod first suggested that genes determine phenotype through the action of enzymes that control chemical processes in the cell. Using the example of alkaptonuria, an inherited condition in which the urine is dark red, Garrod reasoned that inherited diseases were attributable to an inability to make certain enzymes.

In the 1930s, Beadle and Ephrussi speculated that the mutations causing the various eye colors in *Drosophila* were a result of a nonfunctioning enzyme at some point in the metabolic pathway leading to pigment formation. A short time later, Beadle and Tatum did work with mutants of the orange bread mold *Neurospora crassa* that demonstrated the relationship between genes and enzymes.

Neurospora is a useful organism for genetic studies for several reasons. Its phenotype is a direct expression of its genotype—there are no wild-type alleles present in this haploid organism to mask the effect of a gene that is present in a mutant form. The diploid zygote formed from sexual mating undergoes meiosis to produce four haploid spores, which then each divide mitotically. The resulting eight *ascospores* are contained within a thin-walled sac called an *ascus* from which they can be removed and individually grown on artificial growth medium.

Beadle and Tatum studied several nutritional mutants, called *auxotrophs*, that could not grow on the minimal medium of inorganic salts, sucrose, and biotin used by wild-type *Neurospora*. The mutants were unable to synthesize the compounds they needed due to a nonfunctional enzyme in one of their metabolic pathways. By transferring bits of fungi growing from spores of irradiated fungi to various combinations of minimal medium and added nutrients, Beadle and Tatum were able to isolate auxotrophs with different metabolic defects along various metabolic pathways. Assuming that each mutant was defective in a single gene and that its particular metabolic pathway was blocked by lack of a specific enzyme, they formulated the *one gene–one enzyme* hypothesis: the function of a gene is to control the production of a specific enzyme.

Molecular biologists revised this to the one gene–one protein hypothesis, since all enzymes are proteins but not all proteins are enzymes. Many proteins consist of more than one polypeptide chain, so the axiom has been further revised to the one gene–one polypeptide hypothesis.

The Languages of Macromolecules

Nucleic acids and proteins are similar in that they are constructed of many monomers linked together in specific linear sequences. DNA and RNA are long sequences of only four different monomers: four nucleotides differing in their nitrogenous bases. The base sequence of DNA in a gene encodes the sequence of the 20 different amino acid monomers that may be linked together to form a polypeptide chain.

Transcription is the transfer of information from DNA to RNA—the same "language" of nucleic acids is used. *Translation* is the term for the transfer of information from RNA to a polypeptide—the language changes from nucleotides to amino acids.

The translation of nucleotides into amino acids involves a sequence of three nucleotides, called a *codon*, to specify each amino acid. Since there are only four possible bases in a nucleotide sequence, a two-base code provides only 16 (4^2) unique arrangements. A *triplet code* provides 64 (4^3) possible codons, more than enough to specify the 20 amino acids.

Transcription: The Synthesis of mRNA

The genetic information in DNA is transcribed by the synthesis of *messenger RNA (mRNA)* along a DNA template. As in DNA replication, the double helix unwinds and separates, and the RNA molecule grows along one of the DNA strands in the 5'—>3' direction. The ribonucleotides take their places along the DNA template by forming hydrogen bonds with the nucleotides there, following the same base-pairing rules except that U, rather than T, pairs with A. *RNA polymerase* links the ribonucleotides together.

A promoter sequence signals the RNA polymerase to start transcribing, and a terminator sequence signals the polymerase to stop synthesis and release the RNA molecule. As the RNA is made, it peels away from the DNA template and the hydrogen bonds between DNA strands reform. For any gene, only one strand of DNA is signaled to be read by a promoter region. Different genes may be read from different strands.

RNA molecules average 6000 nucleotides in length; although some may be as long as 20,000 nucleotides. Several molecules may be transcribed simultaneously from a single gene through the action of multiple molecules of RNA polymerase. The mRNA may or may not be modified before it crosses the nuclear membrane to serve for protein synthesis in the cytoplasm.

Translation: The Synthesis of Protein

Transfer RNA (tRNA) pairs the appropriate amino acid with its codon on mRNA. These single-stranded, short, RNA molecules are arranged into a cloverleaf shape by four regions of hydrogen bonding between complementary base sequences and then folded into a three-dimensional, roughly L-shaped structure. The loop protruding from one end of the L holds the specialized base triplet called the *anticodon*, which recognizes, by base pairing, a particular codon on mRNA. The 3' end on the tRNA molecule on the other end of the L is the site of attachment of its amino acid.

Amino acid activating enzymes, or *aminoacyl-tRNA synthetases*, have specific active sites that bind one type of amino acid with its appropriate tRNA molecule. Each enzyme first binds its amino acid to its active site along with an ATP molecule that loses a pyrophosphate and joins to the amino acid as AMP. The appropriate tRNA then displaces the AMP and bonds to the amino acid. This activated amino acid–tRNA complex can furnish its amino acid to the growing polypeptide chain according to the sequence of codons on mRNA.

Ribosomes, consisting of two subunits composed of proteins and a specialized form of RNA, called *ribosomal RNA (rRNA)*, have a binding site for mRNA, a *P site* that holds the tRNA carrying the growing polypeptide chain, and an *A site* that binds to the tRNA carrying the next amino acid. The transfer of this amino acid from its tRNA to the carboxyl end of the growing polypeptide chain is catalyzed by the ribosome.

The three stages of protein synthesis—chain initiation, chain elongation, and chain termination—all require enzymes. The first two stages also need phosphate-bond energy provided by guanosine triphosphate (GTP), an energy compound closely related to ATP.

Initiation requires several proteins, called *initiation factors*, GTP, the mRNA to be read, the two subunits of a ribosome, and the first amino acid attached to its tRNA. Translation must begin with the correct nucleotide so that the grouping of bases into codons, called the *reading frame*, produces the proper sequence of amino acids.

The mRNA and an initiator tRNA, usually carrying methionine, are bound to the small subunit of the ribosome. The tRNA is bound to the start codon, usually AUG, on the mRNA. The nucleotides to the 5' side of the start codon constitute a recognition signal and base-pairing region for rRNA in the ribosome. Next, the large subunit of the ribosome attaches to the small one. The initiator tRNA fits into the P site of the now functional ribosome.

The addition of amino acids, involving the use of several proteins called *elongation factors*, occurs in a three-step cycle: (1) with the hydrolysis of a high-energy phosphate bond of GTP, the mRNA codon in the A site forms hydrogen bonds with the appropriate anticodon of a tRNA carrying its amino acid; (2) an enzyme called *peptidyl transferase* catalyzes the formation of a peptide bond between the polypeptide held in the P site and the amino acid in the A site—the polypeptide is released from the tRNA that was holding it and is now held by the tRNA of the amino acid in the A site; and (3) the tRNA, carrying the growing polypeptide, is translocated to the P site. This process requires the hydrolysis of a GTP molecule and makes the next mRNA codon available in the A site. Translation and translocation occur in the 5'——>3' direction.

Termination occurs when a termination codon—UAA, UAG, or UGA— reaches the A site of the ribosome. A protein called *release factor* binds to the termination codon and causes peptidyl transferase to attach a water molecule to the polypeptide chain, freeing the completed polypeptide from the tRNA in the P site. The ribosome then separates into its small and large subunits.

During and following translation, a polypeptide undergoes several changes in preparation for its function in the cell. One or more amino acids at the beginning, amino end of the chain may be enzymatically removed; whole segments of the polypeptide may be excised; or the chain may be cleaved into several pieces. Disulfide bridges between cysteine molecules may form as the chain spontaneously folds and coils as dictated by the primary sequence of amino acids. The attachment of sugars or phosphate groups may modify certain amino acid residues; or several polypeptides may associate into a quaternary structure.

Proteins destined to become part of membranes or to be exported from the cell are produced by ribosomes bound to the ER. Proteins that are to function in the cytoplasm are usually produced by free ribosomes. A ribosome may produce an average-sized protein in less than 1 minute. Usually several ribosomes are reading a mRNA at one time, creating a cluster known as a *polyribosome*.

The Genetic Code

In the early 1960s, the genetic code was "cracked" by a series of experiments that determined the amino acid translations of each of the codons of nucleic acids. Nirenberg synthesized artificial mRNA by linking together uracil ribonucleotides creating multiple UUU-UUU codons. Adding this "poly-U" to an *in vitro* system of a test tube containing all the biochemical ingredients necessary for protein synthesis, Nirenberg obtained a polypeptide containing the single amino acid phenylalanine. The codons AAA, GGG, and CCC were deciphered the same way.

Using more elaborate techniques to decode the triplets with mixed bases, scientists had deciphered all 64 triplets by the mid-1960s. Sixty-one of the triplets code for amino acids. The three remaining codons function as stop signals. AUG both codes for methionine and functions as the start signal for translation.

The code is often redundant; more than one codon may specify a single amino acid. Often this redundancy, called "degeneracy", occurs as differences in the third base of the triplet. The code is never ambiguous; no codon specifies two different amino acids.

In general, a particular nucleotide sequence on DNA is read in only one reading frame—starting at a start triplet and reading each triplet sequentially. An exception is found in a small virus in which two genes share some of the same nucleotides that are grouped in different sets of triplets by the two genes.

Sixty-four codons can be read from mRNA, but there are only about 40 different tRNA molecules. In a phenomenon known as *wobble*, the third nucleotide (5' end) of some tRNA molecules can form hydrogen bonds with more than one kind of base in the codon. In several tRNA, the unusual base inosine (I) is found in the third position and can pair with U, C, or A. Thus one tRNA can recognize three mRNA codons, all of which code for the same amino acid carried by that tRNA.

The genetic code of codons and their corresponding amino acids is universal for almost all organisms. A bacterial cell can translate the genetic messages of human cells and vice versa. This universality lends compelling evidence to the evolutionary connection of all living organisms.

Recent exceptions to the constancy of the genetic code have been found in several single-celled ciliates in which the RNA codons UAA and UAG code for glutamine instead of stop signals. The genetic code found in mitochondrial DNA and protein synthesis machinery varies with different organisms.

Split Genes and RNA Processing in Eukaryotes

A surprising and recent development in molecular biology involves *split genes*, in which much of the DNA does not code for amino acid sequences of proteins. Long segments of noncoding sequences of bases, known as *introns* or intervening sequences, have been found within the boundaries of eukaryotic genes. The remaining coding regions are called *exons*, since they are the regions that are expressed. An entire mRNA transcript is made of the gene, and then the introns are removed and the exons joined to produce a mRNA with a continuous coding sequence. This process, known as *RNA splicing* or *RNA processing*, is also present in the production of tRNA and mRNA.

Signals for RNA splicing are sets of a few nucleotides at either end of each intron. Several nuclear enzymes may be involved, although some splicing occurs without enzymes—the intron RNA catalyzes the process itself.

Before leaving the nucleus, eukaryotic mRNA is modified by the addition of a *cap* of modified guanosine triphosphate at the 5' end and a string of 150 to 200 adenine nucleotides, called *poly A*, at the 3' end. The cap appears to enhance translation, and both additions apparently protect the ends of the mRNA.

There are several hypotheses concerning the functions of introns. One is that they are somehow involved in regulation of gene activity or in the flow of mRNA into the cytoplasm. Another hyothesis is that they facilitate recombination between exons to create a diversity of proteins. It is believed that exons may represent *domains* that code for functional segments of a protein, such as a binding site. The crossing over between domains within a gene, coupled with mutational changes, can give rise to new genes.

More complex organisms have more and longer introns. Simpler organisms, such as yeast, have few genes containing introns. The evolutionary significance of these differences remains an interesting question.

Mutations and Their Effects on Proteins

Mutations, changes in the nucleotide sequence of DNA in a gene, may be divided into two general types: base-pair substitutions and base-pair insertions or deletions. *Base-pair substitutions* involve the replacement of one nucleotide and its complementary partner with another pair of nucleotides. When a single nucleotide pair is involved, the mutation is called a *point mutation*.

Due to the redundancy of the genetic code, some base-pair substitutions have no effect on the translation of the gene. An exchange in the third nucleotide of a codon may still result in the insertion of the same amino acid.

A substitution may result in the insertion of a different amino acid without altering the character of the protein, if the new amino acid is similar in properties or is located in an area of the protein not crucial to that protein's function.

A base-pair substitution that results in the insertion of a different amino acid in a critical portion of a protein, such as the active site of an enzyme, may significantly affect protein function. Occasionally such a mutation results in an improved protein, but much more frequently the mutation is harmful to the organism. An incorrectly coded amino acid is called a "missense" mutation. When the point mutation changes a codon for an amino acid into a stop codon, the translation of the polypeptide chain is prematurely halted. Such "nonsense" mutations almost always lead to nonfunctional proteins.

Base-pair insertions or *deletions* usually have a more substantial effect than do base-pair substitutions because they may alter the reading frame of the nucleotide triplets. Whenever insertions or deletions are not in a multiple of three nucleotides, all nucleotides downstream from the mutation will be improperly grouped into codons, creating extensive missense and ending usually in nonsense—premature termination. These *frameshift mutations* almost always produce nonfunctional proteins.

A *conditional mutation* is harmful or fatal under some conditions but not others. Temperature-senstitive mutations involve an amino acid substitution that is not harmful at a *permissive* temperature, but changes the stability of the protein at a different, usually higher, temperature and inactivates the protein. Geneticists make use of temperature-sensitive mutations to grow mutants at permissive temperatures and then study the effects of their altered enzymes at nonpermissive temperatures.

Mutagenesis, the generation of mutations, may occur in a number of ways. *Spontaneous mutations* include base-pair substitutions, insertions, and deletions that may occur during DNA replication or repair. Physical and chemical agents called mutagens can cause mutations in DNA. X-rays can cause double-strand breaks in DNA resulting in chromosomal rearrangements and deletions. UV radiation creates pyrimidine dimers. Chemical mutagens include *base analogues* that substitute for normal bases in DNA synthesis and result in mispairing and base-pair substitutions. Reactive chemicals alter bases and also lead to mispairing. Regions of DNA that tend to move from one DNA section to another are called *transposons* and may act as mutagens by disrupting DNA sequences.

Redefining the Gene

The definition of a gene has evolved from Mendel's inheritable factors, to Morgan's loci along chromosomes, to Beadle and Tatum's one gene–one polypeptide theory. Research continually refines our understanding of the structural and functional aspects of genes, to include introns and exons, mRNA splicing, and some overlapping genes. The best working definition of a gene may be James Watson's suggestion that a gene is a sequence of nucleotides with a specific functional product such as a polypeptide or an RNA molecule.

STRUCTURE YOUR KNOWLEDGE

1. Describe transcription and translation, including the molecules involved, enzymes needed, and products made. Make a chart of these processes if you wish. How are these two processes similar? How are they different?

2. What is the genetic code? Explain redundancy and the wobble hypothesis. What is meant by saying that the genetic code is almost universal?

3. Prepare a concept map dealing with the topic of mutation.

TEST YOUR KNOWLEDGE

MULTIPLE CHOICE: *Choose the one best answer.*

1. Beatle and Tatum's study of *Neurospora* showed that
   a. auxotrophs could not grow on minimal medium.
   b. mutants had defective enzymes in their metabolic pathways.
   c. the occurrence of different mutants with defective metabolic pathways supported a one gene–one enzyme hypothesis.
   d. all of the above.

2. Transcription involves
   a. the transfer of information from DNA to mRNA.
   b. the transfer of information from DNA to tRNA.
   c. the transfer of information from mRNA to an amino acid sequence.
   d. the transfer of information from DNA to an amino acid sequence.

3. Which of the following is not true of a codon?
   a. It consists of three nucleotides.
   b. It codes for a specific amino acid or a stop signal.
   c. It extends from one end of a tRNA molecule.
   d. It may code for the same amino acid as another codon does.

4. RNA polymerase
   a. works on both DNA strands, but always in a 5′ —> 3′ direction.
   b. creates hydrogen bonds between nucleotides on the DNA strand and their complementary ribonucleotides.
   c. starts transcribing at an AUG triplet on one DNA strand.
   d. transcribes both introns and exons.

5. Transfer RNA
   a. is the nucleic acid that forms the small subunit of the ribosome.
   b. binds to its specific amino acid by replacing AMP in the active site of an aminoacyl-tRNA synthetase.
   c. uses GTP as the energy source to bind its amino acid.
   d. forms hydrogen bonds with the anticodon in the A site of a ribosome.

6. Translocation involves
   a. the hydrolysis of a GTP molecule.
   b. the movement of the tRNA in the A site to the P site.
   c. the movement of mRNA over one triplet in the A site.
   d. all of the above.

7. Changes in a polypeptide following translation may involve
   a. the formation of disulfide bridges between cysteine molecules.
   b. the action of enzymes to add amino acids at the beginning of the chain.
   c. the removal of poly A from the end of the chain.
   d. all of the above.

8. A single mRNA may produce several proteins at a time by
   a. the action of several ribosomes in a cluster called a polyribosome.
   b. several RNA polymerase molecules working sequentially.
   c. association with rough ER.
   d. containing several promoter regions.

9. The phenomenon known as wobble refers to
   a. the movement of a tRNA from the A to the P site.
   b. the redundancy of the genetic code.
   c. the degeneracy of the genetic code.
   d. the ability of a tRNA to pair with several codons.

10. Which of the following is not true of RNA splicing?
    a. Exons are excised before the mRNA is translated.
    b. Nuclear enzymes may be involved in the splicing.
    c. The existence of exons and introns may facilitate crossing over between domains within a gene.
    d. Simpler organisms exhibit less RNA splicing.

11. Base-pair substitutions may have little effect on the resulting protein for all of the following reasons except which one?
    a. The redundancy of the code may result in no change in translation.
    b. As long as the substitution is three nucleotides the reading frame is not altered.
    c. The missense mutation may not occur in a critical part of the protein.
    d. The new amino acid may have properties similar to those of the replaced one.

12. A conditional mutation
    a. may occur only under high temperatures.
    b. may not harm the organism at nonpermissive temperatures.
    c. may revert back under permissive conditions.
    d. may involve temperature-sensitive mutations.

13. Base analogues
    a. may be created by UV radiation.
    b. are regions of DNA that can move from one section of DNA to another.
    c. may substitute for normal bases and result in base-pair substitutions.
    d. are chemical mutagens that spontaneously occur.

14. A base deletion early in the coding sequence of a gene most likely will result in a protein that is
    a. functionally disrupted by the mutation.
    b. prematurely terminated.
    c. largely missense coding due to a frameshift mutation.
    d. all of the above.

15. The bonds between the anticodon of a tRNA molecule and the complementary codon of mRNA are
    a. formed by the input of energy from GTP.
    b. catalyzed by peptidyl transferase.
    c. hydrogen bonds.
    d. formed by the input of energy from ATP.

16. A prokaryotic gene 600 nucleotides long can code for a polypeptide chain of about how many amino acids?
    a. 200
    b. 300
    c. 600
    d. 1800

17. Which of the following mutations would be expected to have the most harmful effect on the resulting protein?
    a. a base deletion near the start of the coding sequence
    b. a base deletion near the end of the coding sequence
    c. a three-base addition near the start of the coding sequence
    d. a base substitution

18. Peptidyl transferase
    a. translocates the tRNA holding the polypeptide chain from the A to the P site.
    b. hydrolyzes a GTP to pair the codon and anticodon in the A site.
    c. catalyzes the formation of a peptide bond between a polypeptide and the amino acid in the A site.
    d. binds the initiator tRNA to the start codon in the P site.

19. How many amino acids could be coded for if there were six instead of four different nucleotide bases and codons consisted of doublets (two nucleotides) instead of triplets?
    a. 6
    b. 12
    c. 216
    d. 432

20. Frameshift mutations can be caused by
    a. base-pair substitution.
    b. transposons.
    c. base analogues.
    d. a and b.

FILL IN THE BLANKS: *Complete the following table for the DNA and mRNA sequences, the anticodons and their amino acid. Use the following portion of the genetic code to help you answer these questions.*

Codon	Amino Acid	Codon	Amino Acid
AUG	Methionine	GGG	Glysine
AAG	Lysine	GCA	Alanine
CCA	Proline	UGU	Cysteine
GUC	Valine	UAG	Stop

DNA	1	4		T T C	10
mRNA	2	5	7		U A G
Anticodon	3	C A G	8	11	
Amino acid	methionine	6	9	12	

CHAPTER 17

THE GENETICS OF VIRUSES AND BACTERIA

FRAMEWORK

This chapter introduces the genetic characteristics of viruses and bacteria. Viruses consist of a *genome* (single- or double-stranded RNA or DNA) enclosed in a protein *capsid*. They replicate by using the metabolic machinery of their specific host cell. *Bacteriophages* may be virulent and follow the *lytic* cycle or temperate with a *lysogenic* replication cycle. Tumor viruses may introduce or turn on *oncogenes* and thus transform host cells into cancerous cells.

The bacterial genome consists of a *circular DNA molecule* and various *plasmids*. Transfer of genetic material between bacteria may occur by transformation, transduction, or conjugation. The *F factor* plasmid that controls conjugation may be integrated into the bacterial chromosome to produce *Hfr* cells. *R plasmids* confer antibiotic resistance to bacterial cells. Structural genes with related functions may occur in *operons* controlled by a single *promoter* region. The expression of *repressible* and *inducible* operons is controlled by *repressor proteins*. *Activating proteins* exert positive control over gene expression.

CHAPTER SUMMARY

The study of viruses and bacteria has provided much information about the molecular principles of the genetics of all organisms, an appreciation for the special genetic features of microbes (with implications for understanding how viruses and bacteria cause disease), and new powerful techniques of manipulating genes that have had an effect on basic research and biotechnology.

Discovery of Viruses

The search for the cause of tobacco mosaic disease led to the discovery of viruses. Mayer, in 1883, found that he could transmit the disease by spraying sap from an infected plant onto a healthy plant. Since he could not resolve the infectious agent from the sap, he concluded that this contagious disease was caused by an unusually small bacteria. Ivanowsky used a filter designed to remove bacteria to filter sap from infected plants and found the sap still transmitted the disease.

The possibility of a filterable bacterial toxin causing tobacco mosaic disease was eliminated when, in 1897, Beijerinck discovered that the infectious agent in the filtered sap could reproduce. A plant that was infected with filtered sap could pass on the disease to other plants through its own filtered sap. The infectious agent could not be cultivated on nutrient media, and it was not killed by alcohol. In 1935, Stanley crystallized the infectious particle, now known as *tobacco mosaic virus* (TMV). Since that time, many viruses have actually been seen with the electron microscope.

Viral Structure

A *virion*, or viral particle, may consist simply of nucleic acid enclosed in a protein shell. Viral genomes may be single- or double-stranded DNA, or single- or double-stranded RNA. As few as four or as many as several hundred genes may be contained on a linear or circular single molecule of nucleic acid.

The *capsid*, or protein shell built from a large number of protein subunits, may be rod-shaped (helical), polyhedral, or complex in structure. Membranous *envelopes*, derived from membranes of the host cell and including viral proteins and glycoproteins, may cloak the capsid of viruses found in animals.

The most complex capsids are found among viruses that infect bacteria. The seven phages that infect the bacterium *Escherichia coli* were the first *bacteriophages* to be discovered. The three "T-even" phages—T2, T4, and T6—have a similar capsid structure consisting of an icosahedral (20-sided) head enclosing the genetic material and a protein tailpiece with tail fibers for attaching to a bacterium.

Replication of Viruses

Viruses are *obligate intracellular parasites* that lack the metabolic equipment needed to express their genes and reproduce. They produce hundreds or thousands of progeny in each generation by directing the host cell's enzymes, ribosomes, and other resources to make copies of the viral genome and capsid proteins.

The replication of the viral genome depends on the form of the viral nucleic acid. The replication of double-stranded DNA is similar to that of cellular genes. Most RNA viruses have a gene for *RNA replicase*, an enzyme that uses viral RNA as a template for making complementary RNA strands. Some RNA viruses have a gene for an enzyme called *reverse transcriptase*, which uses viral RNA as a template for DNA synthesis, which is later transcribed into viral RNA. The three patterns of viral genome replication are DNA—>DNA, RNA—>RNA, and RNA—>DNA—>RNA.

Each virus type has a *host range*, a limited group of host cells that it can infect. Proteins on the outside of the virion recognize *receptor sites* on the surface of the host's cell.

Bacterial Viruses

Bacteriophages are the best understood of viruses. The study of the T phages helped demonstrate that DNA is the genetic material. The study of lambda, another phage of *E. coli*, found in 1951 by Lederberg, led to the discovery of the lytic and lysogenic cycles of double-stranded DNA viruses.

Virulent bacteriophages cause their host cells to *lyse* during their replication cycle known as the *lytic cycle*. The T4 phage uses its tail fibers to stick to a receptor site on the surface of an *E. coli* cell. The sheath of the tail contracts through the expenditure of ATP stored in the tailpiece, and thrusts the viral DNA into the cell. The empty capsid is left attached to the outside of the cell. The *E. coli* cell begins to transcribe and translate the 100 or so genes of the phage. One of the first genes translated codes for an enzyme that chops up the host cell's DNA, sparing the phage DNA because it contains a modified form of cytosine. Nucleotides from the host cell's degraded DNA are used to create viral DNA. Three sets of capsid proteins are made and assembled into phage tails, tail fibers, and polyhedral heads. The viral components spontaneously assemble into 100 to 200 phage particles that are released after a lysozyme is produced that digests the bacterial cell wall.

The lytic cycle takes only 20 to 30 minutes at 37° C, during which time the T4 population increases more than a 100-fold. The concentration of phage particles in a sample can be determined by spreading a dilute sample onto a cloudy "lawn" of bacteria growing in a Petri dish. Each *plaque*, or clear hole in the lawn, indicates the lysis of bacterial cells by successive generations of a phage particle.

Bacteria defend against viral infection when mutations change the receptor sites used by a phage or when viral DNA is broken down after entering the cell by bacterial *restriction enzymes*. Bacterial hosts and their viral parasites are continually *co-evolving* in response to the new defenses and offenses of each other.

A *temperate virus* can reproduce without killing its host when it follows a *lysogenic cycle*. When the phage lambda binds to the surface of a bacterium, it can follow either a lytic cycle or a lysogenic cycle, in which its injected DNA inserts into the bacterial chromosome.

Most of the genes of the inserted phage genome, known as a *prophage*, are inactive. One prophage gene coding for a *repressor protein* remains active and keeps the other genes switched off. Reproduction of the host cell replicates the prophage along with the cellular DNA. The prophage may be excised from the bacterial chromosome spontaneously or through the action of environmental stimuli, and start the phage's lytic cycle.

A host cell carrying a prophage is called a *lysogenic cell*. Expression of a few of the prophage's genes may cause a change in the phenotype of the bacterium, called *lysogenic conversion*. Several disease-causing bacteria would be harmless except for the expression of prophage genes coding for toxins.

Plant Viruses and Viroids

Most plant viruses are RNA viruses, often having rod-shaped capsids with spirally arranged capsomeres. Plant viral diseases may spread through *horizontal transmission*, in which the virus invades from an external source. Injuries to the plant increase susceptibility to viral infections. Insects can acts as carriers or *vectors* of viruses, passing disease from plant to plant. The inheritance of a viral infection from a parent plant is called *vertical transmission*.

Viral particles spread easily throughout a plant through the plasmodesmata. Reducing the spread of disease and breeding resistant varieties are the best preventions for plant viral infections.

Viroids are very small molecules of naked RNA that can disrupt the metabolism of a plant cell and severely stunt plant growth. It is not known where viroids come from or how they function in disrupting the regulation of genes in the host cell.

Animal Viruses

A membranous envelope surrounding the capsid is present in several groups of animal viruses but not in plant viruses or phages. Paramyxoviruses, including the viruses that cause measles and mumps, have single-stranded RNA genomes. Glycoproteins extending from the membranous envelope of a paramyxovirus attach to receptor sites on the host cell's plasma membrane, and the envelope fuses with the plasma membrane, transporting the capsid into the cell. Viral enzymes replicate the RNA genome and make mRNA, which the host cell's ribosomes use for protein synthesis. New capsids form around viral genomes and leave the cell by budding off within an envelope from the host's plasma membrane. This process is called a *productive cycle* because the viruses exit without destroying the host cell.

The envelopes of herpesviruses come from the host's nuclear membrane. Clinical evidence indicates that the herpesvirus' double-stranded DNA can integrate into the cell's genome as a *provirus*, similar to a bacterial prophage.

The symptoms of a viral infection may be caused by toxins produced by infected cells, toxic components of the virions themselves, cells killed or damaged by the virus, or the body's defense mechanisms fighting the infection. The ability of the infected tissue to regenerate or repair itself may determine the severity of a viral infection.

Vaccines are harmless derivatives of pathogens that induce the immune system to produce antibodies against the actual disease agent. Vaccinations have greatly reduced the incidence of many viral diseases.

Since viruses use the host's cellular machinery to replicate, few drugs have been found to cure viral infections. Adenine arabinoside is one of the drugs that is an analogue of purine nucleosides and thus interferes with viral nucleic acid synthesis. It is effective against a number of human viruses.

Tumor viruses, which can cause cancer in animals, have been found. When tumor viruses infect cells growing in tissue culture, the cells are transformed, assume rounded shapes, and lose the contact inhibition regulation of growth. There is strong circumstantial evidence that links viruses to human cancer.

Tumor viruses transform cells through integration of viral nucleic acid into the host cell genome. Retroviruses must use reverse transcription to transcribe DNA from their RNA genetic material. DNA viruses can be directly inserted into a chromosome.

Oncogenes are genes responsible for triggering cancerous transformation in cells. Surprisingly, these genes have been found not only in tumor viruses but also within the genomes of normal cells of many species. These oncogenes code for cellular growth factors or proteins involved in growth-factor action. Some tumor viruses may lack oncogenes but transform cells simply by turning on the cells' oncogenes. Carcinogens may also act by turning on cellular oncogenes.

Origin of Viruses

It is unlikely that viruses are degenerate cells or relics of a precellular stage of evolution. Different families of viruses are genetically more similar to their host cells than to each other. It is most likely that viruses evolved from fragments of cellular nucleic acid that acquired special packaging. Viruses inhabit a gray area between life and nonlife—they are inert molecules containing a genetic program they can express only within other living cells.

The Bacterium and Its Genome

The single, circular *bacterial chromosome* is simpler in structure than and has one-thousandth as much DNA as the chromosomes of a eukaryotic cell. Transcription and translation can occur simultaneously because the chromosome is not separated from the rest of the cell. Smaller rings of DNA, called *plasmids*, carry accessory genes and are found in many bacteria.

Replication proceeds bidirectionally from a single origin during cellular reproduction. Binary fission can occur as rapidly as once every 20 minutes. Rapid growth, easily observable phenotypes, and mechanisms for the transfer of genes make bacteria ideal subjects for genetic studies.

Some bacteria can take up segments of naked DNA in a process called *transformation*. If the foreign DNA is integrated into the bacterial chromosome by recombination, a new combination of genes is produced.

Bacteriophages can transfer genes from one bacterium to another by *general transduction*, when a random piece of host DNA is accidentally packaged within a phage capsid, or by *restricted transduction*, when a prophage includes some bacterial genes as it excises from the bacterial chromosome. In restricted transduction, most of the phage genes are also included in the capsid, and the bacterial genes transduced are those adjacent to the prophage site on the chromosome.

Lederberg and Tatum discovered the third type of genetic transfer between bacteria. In *conjugation*, two cells temporarily join by appendages called *sex pili* and DNA transfers from one cell to the other. A plasmid called the fertility factor, or *F factor*, carries the genes for pili production and the processes involved with the transfer of DNA from donor to recipient. Cells containing the F factor are called F^+. The F^- factor plasmid replicates along with the bacterial chromosome, and F^+ cells pass on the trait to daughter cells. The plasmid also replicates before conjugation and is transferred to the recipient cell, changing it from an F^- to an F^+ cell.

Cells in which the F factor inserts into the bacterial chromosome are called *Hfr* for High-Frequency Recombination. When these cells undergo conjugation, the F factor is transferred along with the attached bacterial chromosome. Movement may disrupt the mating, resulting in a partial transfer of genes. Recombination between the new DNA and the recipient cell's chromosome produces new combinations of genes.

Hfr bacteria of a given strain will always transfer chromosomal genes in the same order. The site of the F factor on the chromosome determines the sequence of gene transfer. The loci of genes can be determined by experiments using interrupted mating. Hfr and F⁻ strains with different alleles are mixed together, and samples are removed and agitated to disrupt conjugation at different time intervals. Using various nutritional selective media and an antibiotic-resistance marker, the number and order of transferred genes can be determined.

Plasmids are small, circular DNA molecules that replicate separately from the bacterial chromosome. Plasmids that can integrate into the bacterial chromosome, such as F factor, are called *episomes*.

R plasmids carry genes that code for antibiotic-destroying enzymes. As many as seven genes for resistance to different antibiotics can be carried on one R plasmid. R plasmids can be transferred to nonresistant cells, creating the medical problem of antibiotic-resistant pathogens.

Transposons

Transposons, mobile segments of DNA, include two types of nucleotide sequences: a sequence for transposase, the enzyme responsible for the cutting and ligating of DNA, and a pair of "inverted repeat" sequences at the ends of the transposon that serve as recognition sites for transposase. *Insertion sequences* (IS) are the simplest transposons, consisting of only a transposase gene and inverted repeats. Their insertion into a gene as they move about the chromosome usually serves to inactivate the gene.

Complex transposons contain additional DNA and may serve to move genes from one chromosome to another, and occasionally from one species to another. Transposon insertion is not dependent on DNA sequence homology and can occur almost anywhere.

Control of Gene Expression in Prokaryotes

A cell is able to adjust its metabolism in response to environmental conditions by controlling either enzyme activity, through feedback inhibition, or enzyme synthesis, through regulation of gene expression.

Constitutive genes are unregulated and are continually involved in RNA synthesis. The rate at which they are transcribed, however, is determined by the efficiency of their promoter region for binding RNA polymerase, which is related to the relative need for their encoded proteins. *Regulated genes* can be switched on and off as metabolic needs change. The mechanism

for gene regulation was first described by Jacob and Monod in 1961 for the lactose genes in *E. coli*.

Operons are clusters of genes on a chromosome or phage genome that have related functions. The *structural genes* within an operon may code for the different enzymes of a single metabolic pathway. The tryptophan *(trp)* operon of *E. coli* contains five structural genes for the enzymes of the tryptophan pathway. Because a single promoter region serves all the genes of an operon, an RNA polymerase will tend to transcribe all the genes at one time, producing a *polycistronic* mRNA molecule with the transcript of all the enzymes for the pathway.

An *operator* is a segment of DNA that overlaps the promoter region and controls the movement of RNA polymerase onto the promoter and along the structural genes of the operon. A *repressor* is a protein with an "active site" that binds to a specific operator and blocks attachment of RNA polymerase. An *activator* is a protein that binds near or within the promoter and promotes transcription.

Regulator genes code for repressor (and activator) proteins. Most regulator genes are constitutive, constantly producing repressor proteins at a slow rate. The activity of the repressor protein is determined by the presence or absence of a metabolite (a precursor, intermediate, or end product of a metabolic pathway). In the *trp* operon, tryptophan is the metabolite that binds to an allosteric site of the repressor protein, changing its conformation into its active state that has a high affinity for the operator and switches off the *trp* operon.

In this regulatory system, tryptophan is a *co-repressor*. Both the repressor protein and tryptophan are needed to turn off the operon. Should the tryptophan concentration in the cell fall too low, the *trp* operon is no longer repressed, and mRNA for the enzymes for tryptophan synthesis is produced. Enzymes are called *repressible enzymes* when a metabolite inhibits or represses their synthesis.

The synthesis of *inducible enzymes* is stimulated by a metabolite. The *lac* operon, controlling lactose metabolism in *E. coli*, is an inducible operon that contains three structural genes. Unlike the *trp* repressor, which is inactive until bound with its co-repressor, the *lac* repressor is innately active, binding to the *lac* operator and switching off the operon. The presence of lactose inactivates the repressor protein.

Repressible enzymes are usually found in anabolic pathways, where the pathway's end product serves to switch off enzyme synthesis and prevent overproduction of metabolic products. Inducible enzymes are found in catabolic pathways, where the presence of nutrient molecules stimulates the production of enzymes necessary for their breakdown. Both types of regulation are examples of *negative control*: they involve a repressor protein that, when active, binds with the operator and blocks transcription.

A regulatory system that uses *positive control* is one in which an activator molecule turns on transcription. Even when lactose inactivates the repressor of the *lac* operon, little mRNA is produced because the *lac* promoter has a low affinity for RNA polymerase. The binding of a *catabolite-activating protein (CAP)* to the promoter region stimulates gene expression by increasing the promoter's ability to associate with RNA polymerase. *Cyclic AMP (cAMP)*, derived from ATP, accumulates in the cell when glucose is missing and binds with the activating protein. The cAMP–CAP complex attaches to the *lac* promoter region and stimulates transcription. A low level of cAMP in the cell signifies that sufficient glucose is present for an energy source; the cAMP–CAP complex disassociates; the activating protein releases from the promoter; and transcription of the *lac* operon slows down.

The regulation of the *lac* operon includes negative control by the repressor protein that is inactivated by the presence of lactose, and positive control by the catabolite-activating protein that functions when complexed with cAMP. *E. coli* is able to control its gene expression depending on the secondary energy sources (such as lactose) available to the cell and the presence of glucose, its primary energy and carbon source. Prokaryotic cells have complex genetic and metabolic regulatory mechanisms.

STRUCTURE YOUR KNOWLEDGE

1. Create a concept map or diagram that describes the lytic and lysogenic cycles of bacteriophage.
2. Develop a concept map to illustrate your understanding of the fascinating mechanisms by which bactera are able to regulate their gene expression in response to varying needs.

TEST YOUR KNOWLEDGE

MATCHING: *Match the term with its description.*

1. ___ viral enzyme that uses RNA as a template for DNA synthesis

2. ___ viral enzyme that uses RNA as a template for RNA synthesis

3. ___ bacterial enzyme that chops up foreign DNA

4. ___ change in phenotype of bacterium as result of expression of prophage genes

5. ___ transfer of genes through bridge joining two bacteria

6. ___ enzyme that cuts and ligates DNA at inverted repeats

7. ___ enzyme that dissolves bacterial cell wall

8. ___ involves budding of viruses from host cell, surrounded by membraneous envelope

9. ___ injection of host cell DNA incorporated into phage capsid into a new cell

10. ___ a mRNA molecule that is a transcript of several genes

A. conjugation

B. lysogenic conversion

C. reverse transcriptase

D. lysozyme

E. lysogenic cycle

F. plaque

G. polycistronic

H. restriction enzyme

I. RNA replicase

J. productive cycle

K. transduction

L. transformation

M. transposase

MULTIPLE CHOICE: *Choose the one best answer.*

1. The study of the genetics of viruses and bacteria has done all the following except
   a. provide information on the molecular biology of all organisms.
   b. illuminate the sexual reproductive cycles of viruses.
   c. develop new techniques for manipulating genes.
   d. develop an understanding of the causes of cancer.

2. Beijerinck concluded that the cause of tobacco mosaic disease was not a filterable toxin because
   a. the infectious agent could not be cultivated on nutrient media.
   b. a plant sprayed with filtered sap would develop the disease.

c. the infectious agent could be crystallized.
d. the infectious agent reproduced in and could be transmitted in filtered sap from a plant infected with filtered sap.

3. Viral genomes may be any of the following except
    a. several molecules of single-stranded DNA.
    b. double-stranded RNA.
    c. a circular DNA molecule.
    d. a linear single-stranded RNA molecule.

4. Retroviruses have a gene for reverse transcriptase that
    a. uses viral RNA as a template for making complementary RNA strands.
    b. uses viral RNA as a template for DNA synthesis.
    c. destroys the host cell's RNA.
    d. translates RNA into proteins.

5. Virus particles assemble from capsid proteins and nucleic acid molecules
    a. spontaneously by the formation of weak bonds.
    b. at the direction of viral enzymes.
    c. by using host cell enzymes.
    d. through the addition of ATP stored in the tailpiece.

6. The study of lambda, a phage of *E. coli,*
    a. led to the discovery of the *lac* operon.
    b. showed that DNA was the genetic material.
    c. led to the discovery of the lytic and lysogenic replication cycles.
    d. revealed the structure of an icosahedral head and tail fibers attached to the tailpiece.

7. Horizontal transmission of a plant viral disease may involve
    a. the movement of viral particles through the plasmodesmata.
    b. the inheritance of an infection from a parent plant.
    c. the spread of an infection by vegetative propagation.
    d. insects as vectors carrying viral particles between plants.

8. A provirus is similar to a prophage because both
    a. can cause lysogenic conversion.
    b. involve a viral genome integrated into a host chromosome.
    c. produce toxins that can cause disease.
    d. are early evolutionary stages of viruses.

9. Drugs that are effective in treating viral infections
    a. induce the body to produce antibodies.
    b. inhibit the action of viral ribosomes.

c. interfere with viral nucleic-acid synthesis.
d. change the cell-recognition sites on the host cell.

10. Tumor viruses
    a. integrate viral nucleic acid into the host cell genome.
    b. transform cells growing in tissue culture into rounded cells that lose their contact inhibition.
    c. may not always contain oncogenes but may turn on the host cell's oncogenes.
    d. may do all of the above.

11. Transcription and translation can occur simultaneously in bacteria because
    a. the mRNA does not have to leave the nucleus to reach the ribosomes.
    b. the same enzyme serves both functions.
    c. replication proceeds bidirectionally around the circular chromosome.
    d. one promoter region serves all the structural genes in an operon.

12. Restricted transduction occurs when
    a. restriction enzymes cut up viral DNA.
    b. naked pieces of DNA are picked up by bacterial cells.
    c. a prophage includes some bacterial genes when it excises from the bacterial chromosome.
    d. a Hfr cell mates with an F$^-$ cell and transfers part of the bacterial chromosome.

13. The F factor of bacteria
    a. carries genes for sex pili production.
    b. may be transferred and convert a cell from F$^-$ to F$^+$.
    c. is an episome.
    d. all of the above

14. R plasmids
    a. carry genes for antibiotic-destroying enzymes.
    b. may be transferred and convert a cell from F$^-$ to F$^+$.
    c. are episomes.
    d. all of the above

15. Tiny molecules of naked RNA are
    a. retroviruses.
    b. transposons.
    c. viroids.
    d. episomes.

16. Regulator genes are genes that
    a. code for repressor and activator proteins.
    b. are usually constitutive.

c. are not contained in the operon they control.

d. all of the above

17. Inducible enzymes
    a. are usually involved in anabolic pathways.
    b. are produced when a metabolite inactivates the repressor protein.
    c. are produced when an activator molecule enhances the attachment of RNA polymerase with the promoter.
    d. all of the above

18. In *E. coli*, tryptophan switches off the *trp* operon by
    a. binding to the promoter.
    b. binding to the operator.
    c. binding to the repressor and increasing the latter's affinity for the operator.
    d. inactivating the repressor protein.

19. A mutation that renders the regulator gene of an inducible operon inactive would result in
    a. continuous transcription of the structural genes.
    b. inhibition of transcription of the structural genes.
    c. irreversible binding of the repressor to the promoter.
    d. no difference in transcription rate when an activator protein was present.

20. A repressor protein
    a. may be activated or inactivated by a metabolite.
    b. binds to the operator region and blocks attachment of RNA polymerase to the promoter.
    c. provides a means of negative control over gene transcription.
    d. all of the above

MATCHING: *Match these components of the* lac *operon with their functions:*

1. ___ beta-galac-tosidase

2. ___ cAMP–CAP complex

3. ___ lactose

4. ___ operator

5. ___ promoter

6. ___ regulator gene

7. ___ repressor

8. ___ structural gene

A. is inactivated when attached to lactose

B. codes for synthesis of repressor

C. hydrolyzes lactose

D. stimulates gene expression

E. repressor attaches here

F. RNA polymerase attaches here

G. acts as inducer that inactivates repressor

H. codes for an enzyme

CHAPTER 18

CONTROL OF GENE EXPRESSION AND DEVELOPMENT IN EUKARYOTES

FRAMEWORK

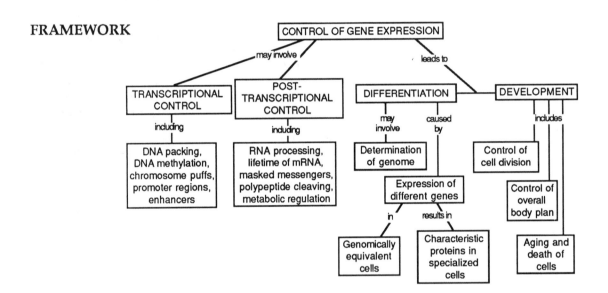

CHAPTER SUMMARY

Eukaryotic cells in multicellular organisms must control their gene expression, not only in response to their external and internal environments, but also to direct the cellular differentiation necessary to create specialized cells. A cell expresses only about 1% of its genome at one time. Humans have 50,000 to 100,000 genes active in different cells at different times.

The mechanisms for the control of gene expression began to be known in the mid-1970s, aided by the techniques of gene cloning and nucleic acid sequencing. As James Watson pointed out in the interview, the understanding of the molecular basis of development is just beginning.

In eukaryotes, the greater complexity of chromosome structure, gene organization, and cellular structure provides more avenues for control of gene activity than are present in prokaryotes.

Packing of DNA in Eukaryotic Chromosomes

Each chromosome consists of a single, extremely long DNA molecule (about 2 meters of DNA in the 46 chromosomes of human cells). Multilevel folding compacts the DNA in the chromosomes and helps to determine the activity of its genes.

Histones are small, positively charged proteins, bound tightly to DNA, that contribute to the folding. Partially unfolded chromatin appears as a string of *nucleosome* "beads," consisting of DNA wound around a protein core of histone. Nucleosomes may limit the access of transcription proteins to DNA.

The *30-nanometer chromatin fiber* (about 30 nm in diameter) is the next order of DNA packing. It apparently consists of a repeating array of six nucleosomes organized around a histone molecule, called H1.

The *looped domain* is a fold in the 30-nm chromatin

fiber, containing 20,000 to 100,000 base pairs, that may contain genes that are expressed in a coordinated fashion. Looped domains may also coil and fold, further compacting the chromatin. Chromatin visible with the light microscope during interphase is called *heterochromatin*. This highly compacted DNA is not actively transcribed; however when it is in the more open form of chromatin, called *euchromatin*, it can be transcribed.

Control of Gene Expression

One of the two X chromosomes in female mammalian somatic cells is present as a highly compacted Barr body, an example of heterochromatin that is not transcribed. Its inactivity may be related to the *DNA methylation* of cytosine residues. Genes are more heavily methylated in cells in which they are not expressed; drugs that inhibit methylation can induce gene reactivation. DNA methylation may be a long-term control mechanism for gene expression.

DNA methylation may also cause the double helix to assume the left-hand Z form, which some molecular biologists speculate functions in gene regulation.

As in prokaryotes, some control of gene expression occurs at transcription. The polytene chromosomes seen in the salivary glands of *Drosophila* larvae consist of hundreds of parallel chromatids. Chromosome puffs appear along the polytene chromosomes at specific stages in development. These DNA loops may make the DNA region more accessible to RNA polymerase. Autoradiography has shown that the puffs correspond to regions of intense RNA synthesis. The shifting locations of puffs, especially during critical times in development, indicate that genes are turned on and off. Ecdysone, the insect hormone that initiates molting, can induce these changes in puff patterns, showing that gene regulation is responsive to chemical signals.

Because so little DNA is expressed in most eukaryotic cells, scientists speculate that genes are usually inactive and that most gene-regulatory proteins interact with the nucleotide sequences called regulatory sites to activate transcription of genes.

Some promoter regions that are specifically bound by RNA polymerase have been located. The three different RNA polymerases, which transcribe genes for rRNA, tRNA, and mRNA, each have their own type and location of promoter to which they bind.

Enhancers are DNA sequences that increase the activity of nearby genes several hundred-fold. Enhancer sequences, which can move about within regions of DNA, are thought to be recognition sites for proteins that increase the DNA's accessibility to RNA polymerase.

Posttranscriptional Control

Gene expression, measured by the amount of proteins (or tRNA or rRNA) that is made, can be blocked or stimulated at any posttranscriptional step. Translation is separated from transcription by the nuclear envelope.

RNA processing involves the removal of the coded intron DNA segments and the splicing together of exons. Different splicing patterns of the same RNA transcript can create different proteins, providing extra genetic flexibility. The addition of nucleotides to both ends of mRNA during processing influences its exit from the nucleus and its lifetime and translation in the cytoplasm.

Prokaryotic organisms can vary their protein synthesis rapidly in response to environmental changes because their mRNA molecules are degraded after only a few minutes of activity. Eukaryotic mRNA, on the other hand, can last hours or even weeks; the quantity of protein synthesis they direct is related to the length of their lifetime. Hemoglobin mRNA is particularly stable; mature red blood cells (lacking nuclei) continue to make hemoglobin using mRNA that had accumulated in the cytoplasm while the nucleated cell was differentiating.

The translation of certain mRNA can be delayed until a control signal initiates it. A great deal of mRNA is synthesized and stored in egg cells as "masked messenger," waiting to be translated during the active protein synthesis period following fertilization. Initiation factors may be necessary for translation within eukaryotic cells and may serve as control factors for gene expression.

Following translation, polypeptides are often extensively cleaved to produce an active protein. The hormone insulin does not become active until a large center portion is removed from the polypeptide chain, leaving two shorter chains, connected by disulfide bridges. The selective degradation of proteins and metabolic regulation of enzymes also serve as control mechanisms in the cell.

Gene Organization in Eukaryotes and Its Evolution

Unlike prokaryotic genes, which are often organized into operons controlled by the same regulatory sites and transcribed together into a single polycistronic mRNA, eukaryotic genes coding for enzymes in the same metabolic pathway are often scattered throughout the genome and separately transcribed. A proposed mechanism for the integrated control of these scattered genes involves specific nucleotide sequences associated with each gene that may serve as recognition signals for regulatory molecules, which then transcribe or repress all the genes in synchrony.

Multigene families are collections (often clustered but occasionally scattered in the genome) of similar or identical genes, presumably of common origin, that usually code for RNA products. The identical genes coding for the major rRNA molecules are arranged one after another, hundreds to thousands of times, forming huge *tandem arrays* that enable cells that are actively synthesizing proteins to produce the millions of ribosomes they need.

An example of a multigene family of nonidentical genes is the two-part family of genes that code for the α and ß polypeptide chains of hemoglobin. Different versions of each subunit are clustered together in the order in which they are expressed during development.

Families of identical genes most likely arose by repeated gene duplication, a phenomenon that seems to occur due to mistakes in DNA replication and recombination. Nonidentical gene families probably arose from mutations in duplicated genes. *Pseudogenes*, sequences of DNA similar to real genes but lacking signals for gene expression, provide evidence for gene duplication and mutation.

Highly repetitive short sequences may make up 10% to 25% of the total DNA in complex eukaryotes. Called *satellite DNA* because their base compositions may be sufficiently different from the rest of the cell's DNA to isolate them by centrifugation, this DNA is primarily located at the centromeres and may serve a structural rather than a genetic role in the cell.

The Program of Development

A major challenge of biology is to understand the developmental processes that produce a complex organism from a single fertilized egg. The long-term regulation of gene expression leads to the *differentiation* or specialization of cells that express a characteristic subset of the organism's genes to produce their particular set of proteins.

The genetic program for development is contained in the DNA. The zygote is said to be *totipotent* because it can give rise to a complete adult organism. For many species, embryonic cells remain totipotent through the first few cell divisions. Separation of two-celled embryos may give rise to identical (monozygotic) twins. As the embryo develops, the cells lose the ability to develop into all the tissues of the adult. *Determination*, the restriction of developmental potential, occurs progressively as the cells develop the particular molecular and structural characteristics that determine their functions.

Differentiation of cells is signaled by changes in cellular structure, the appearance of characteristic proteins, or, initially, by the accumulation of specific mRNA molecules. Environmental cues or chemicals turn on the genes that allow cells to become specialists at making the particular proteins associated with their functions.

Even though they do not express most of their genes, all cells contain the full complement of genes; they exhibit *genomic equivalence*. Experiments involving the transplantation of nuclei from differentiated cells into zygotes support genomic equivalence and the reversibility of differentiation. In the 1950s, Briggs and King transplanted nuclei from embryonic and tadpole cells into enucleated eggs. The ability of the transplanted nucleus to direct normal development was inversely related to the age of the cell from which it was taken. Determination affects the ability of a nucleus to express all its genome, but this change may be reversible, indicating that all the genes are still present in the genome.

Although nuclei from differentiated cells can produce whole organisms when placed into the cytoplasmic environment of an egg cell, no one has been able to induce a differentiated cell from a vertebrate to form an embryo. This feat is done routinely, however, with differentiated plant cells. Carrots were the first plants to be produced from mature somatic cells. This technique is used to clone plants, and is an important technique in genetic engineering work.

Regeneration is the replacement of lost parts in an organism. Many invertebrates have extensive abilities to regenerate complete organisms from fragments of their bodies. Salamanders are able to regenerate lost limbs through the formation of a mass of undifferentiated cells, called a blastema, that can develop into the tissues of a new limb. The blastema cells, derived from muscle and epithelial cells, undifferentiate and then become redetermined along new developmental pathways.

As supported by evidence from nuclear transplantation, somatic cell development of plants, and regeneration, the specialized cells of an organism contain all the genes present in the zygote. The mechanism for differentiation of cells must then rely on the selective and differential expression of identical genomes.

Genes That Control Development

When normal cells differentiate, control mechanisms, such as contact inhibition, limit their growth and division. A cell transformed to a cancer cell escapes from these control mechanisms and may form a mass called a *tumor* within the body. In culture, tumor cells continue to divide indefinitely. The HeLa cell line has been reproducing in culture for 30 years.

Benign tumors remain at their original site in the body and can be completely removed by surgery. *Malignant tumors* are formed by cells that not only multiply excessively, but also have abnormal cell surfaces that cause them to lose the recognition and sticking qualities of normal cells. These cells may spread into tissues around the tumor or travel by way of blood and lymph vessels to form tumors in new locations in the body.

This spread of cancer cells is called *metastasis*.

Cancer may be caused by physical or chemical agents, called *carcinogens*, or induced by certain viruses. The cancer mechanism involves the activation of oncogenes, either native to the cell or introduced by viruses. The cellular genes corresponding to oncogenes (called *proto-oncogenes* when the cell is noncancerous) are thought to be key genes that control cell growth and differentiation. The study of cancer provides insight into the control of normal development and vice versa.

A gene that has an essential function in a normal cell may become an oncogene if it is present in more copies than normal, undergoes translocation that separates it from its normal control regions, or has a mutated nucleotide sequence that creates a more active or resilient protein. These changes may remove the gene from its normal control mechanisms.

Homeotic genes, first identified in fruit flies, appear to control the developmental fate of groups of cells. A sequence of about 200 nucleotides, called a *homeo-box*, was found repeated at least seven times in each of the fruit fly homeotic genes. Similar sequences have been identified in many segmented animals, but not in animals lacking a segmented body pattern.

The homeo-box, located within the coding sequence of genes and presumably translated into protein, contains sequences that resemble genes for DNA-binding proteins. The protein produced by the homeotic gene may regulate genes involved in complex morphogenetic developments.

Genome Modification

Multiple copies of rRNA genes are present in the genomes of vertebrates. In the oocyte of the frog, 1 million or more extra copies of the rRNA genes are synthesized and contained as extrachromosomal circles of DNA, enabling the developing egg to make the huge numbers of ribosomes needed for protein synthesis after fertilization. The selective synthesis of extra DNA is called *gene amplification*.

Chromosome diminution is the elimination of whole or parts of chromosomes from some cells early in development. At the 16-cell stage of a developing gall midge, 14 of the cells lose 32 of their initial 40 chromosomes. The remaining two cells are the germ cells that give rise to gametes.

Rearrangements of the DNA on chromosomes can activate or inactivate genes. The two mating types found in yeasts are determined by the presence of the *alpha* (α) or *a* gene at the *mating-type locus* on a chromosome. Yeast cells can change their mating type by switching the gene with a "silent" copy of the other gene. Unexpressed copies of both genes are kept within the genome so that mating type can be changed by this *cassette mechanism*.

Immunoglobulin genes develop permanent rearrangement of DNA segments in their differentiation. B-lymphocytes are white blood cells that produce proteins, known as *antibodies* or *immunoglobulins*, that recognize and bind to viruses, bacteria, and other invading molecules. Each differentiated B-lymphocyte and its descendants produce one specific antibody. During differentiation, the antibody gene is pieced together from a number of different, widely separated, DNA sequences for the constant, junction, and variable regions of an immunoglobulin polypeptide chain. Much of the variation in antibodies arises from different combinations of variable and constant regions, and different combinations of the four polypeptides that form the complete antibody molecule. The rearrangment and deletion of DNA in the formation of antibody genes is a unique example of genomic nonequivalence.

Aging as a Stage in Development

Following embryonic development, organisms may continue to change by such processes as growth and maturation, metamorphosis, wound healing, and regeneration of body parts. Aging and death may be considered forms of developmental change.

Genes program the aging and death of selected parts of organisms throughout development. The death of cells in the webbing between the fingers of an embryo, the production of mature xylem vessels by the digestion of the living cell, and the loss of tadpole tails are examples of the programmed death of cells.

Evidence that cells may be subject to programmed obsolescence was provided by the work of Hayflick and others at Stanford University. They found that human embryonic fibroblasts in cell culture always stopped reproducing, degenerated, and died after a specific number of mitotic divisions. Other cell types also stopped dividing after a characteristic number of divisions.

One explanation of the predictability of aging is that the cumulative effects of mutations and other insults cause the functional decline of cells within an average number of cell divisions. Another hypothesis suggests that aging and death are programmed either by the expression of specific genes or by scheduled changes, such as a decline in the ability of DNA to replicate or repair itself, that affect all genes. All somatic cells die. Gametes, however, may be thought of as immortal in that their DNA may continue to replicate through the generations.

Epigenesis Vs. Preformation

The idea of *preformation* included the notion that the egg or sperm contains a preformed, miniature individual, needing only to grow during its development, which contains future preformed, miniature individuals. *Epigenesis* is the view that form emerges gradually through progressive development. With the aid of the microscope, epigenesis replaced preformation as the theory of embryology.

The embryo's form does indeed emerge gradually as it develops from the egg. The concept of preformation, however, is valid in that the total plan for development is present in the combination of the genome of the zygote and the cytoplasm of the egg. Nuclear–cytoplasmic interactions are fundamental to differentiation. The selected control of gene expression involves the chemical dialogue between nucleus and cytoplasm—chemical messages influence expression of genes and the products of genes influence the chemical environment. Somehow these complex and intricate control mechanisms direct the development of multicellular eukaryotes.

STRUCTURE YOUR KNOWLEDGE

1. Develop a concept map concerning the major mechanisms that have been identified that control the expression of eukaryotic genes.

2. You have learned that the DNA chromosome is far from a simple linear sequence of genes. This chapter contains a great deal of information about the physical structure and the organizational and regulatory nucleotide sequences of the eukaryotic genome. Create a concept map or table that helps you to organize your knowledge of the structure and functions of the complex eukaryotic DNA.

3. Discuss development, determination, and differentiation, and relate them to genomic equivalency and the control of gene expression.

TEST YOUR KNOWLEDGE

MULTIPLE CHOICE: *Choose the one best answer.*

1. The control of gene expression is a more complicated process in eukaryotic cells because
   a. genes are separated from ribosomes by the nuclear envelope.
   b. gene expression is involved in the differentiation of specialized cells as well as the response to the environment.
   c. the chromosomes are linear and more numerous.
   d. operons are controlled by more than one promoter region.

2. Nucleosomes are
   a. small, positively charged proteins that bind tightly to DNA.
   b. small bodies in the nucleus involved in rRNA synthesis.
   c. basic units of DNA packing consisting of DNA wound around a core of histone.
   d. repeating arrays of six histone units organized around an H1 molecule.

3. Heterochromatin is
   a. chromatin visible with the light microscope.
   b. the most highly compacted form of DNA, consisting of coiled and folded looped domains.
   c. a nontranscribed form of DNA.
   d. all of the above.

4. DNA methylation of cytosine residues
   a. can be induced by drugs that reactivate genes.
   b. may contribute to the formation of the inactive Barr bodies found in female somatic cells.
   c. helps to produce the right-hand coiling found in the double helix.
   d. all of the above.

5. Chromosome puffs along the polytene chromosomes of *Drosophila*
   a. are loops of the hundreds of parallel chromatids that may make the DNA more accessible to RNA polymerase.
   b. are looped domains, or folds in the chromatin fiber containing genes that are expressed in a coordinated fashion.
   c. are regions of DNA methylation causing highly compacted and inactive areas on the chromosome.
   d. contain collections of multigene families.

6. Which of the following is not true of enhancers?
   a. They can move about within regions of DNA.
   b. They greatly increase the activity of nearby genes.
   c. They are promoter regions that specifically bind one of the three different RNA polymerases.
   d. They are thought to be recognition sites for proteins that make DNA more accessible to RNA polymerase.

7. Masked messenger refers to
   a. mRNA that is not translated until a control signal initiates it.
   b. the addition of nucleotides to both ends of mRNA.
   c. RNA processing with different splicing patterns that can result in mRNA for different proteins.
   d. the blocking of mRNA from translation by regulatory proteins.

8. Multigene families are
   a. equivalent to the operons of prokaryotes.
   b. collections of similar or identical genes that usually code for RNA.
   c. controlled by the same promoter region.
   d. collections, often scattered in the genome, of the genes that code for enzymes within the same metabolic pathway.

9. Pseudogenes are
   a. tandem arrays of rRNA genes that enable actively synthesizing cells to create enough ribosomes.
   b. highly repetitive short sequences found around the centromeres.
   c. genes that can become oncogenes when induced by carcinogens.
   d. sequences of DNA that are similar to real genes but lack signals for gene expression.

10. Satellite DNA
    a. is a structural component of chromosomes involved in replication and separation of chromatids.
    b. contains multiple copies of rRNA genes in extrachromosomal circles.
    c. comprises repeated sequences of nucleotides, called homeo-boxes, found in homeotic genes.
    d. helps to make the promoter more accessible to RNA polymerase.

11. Differentiation is brought about by
    a. the genomic equivalency of developing cells.
    b. the restriction of genetic potential during development.
    c. the selective expression of genes.
    d. homeotic genes in segmented animals.

12. A tumor that metastasizes
    a. is not malignant.
    b. contains proto-oncogenes.
    c. spreads cancer cells to new locations in the body.
    d. can be completely removed by surgery.

13. The formation of blastema cells
    a. is part of the development of a complete plant from a somatic cell.
    b. provides evidence for the irreversible determination of cells.
    c. is the result of nuclei transplant experiments.
    d. requires the dedifferentiation and redetermination of cells, providing evidence for genomic equivalency.

14. Which of the following is not true?
    Proto-oncogenes
    a. are thought to be genes that control cell growth and differentiation.
    b. may be activated by carcinogens.
    c. are introduced into the cell by viruses.
    d. are noncancerous.

15. An example of genomic nonequivalence is
    a. chromosome diminution.
    b. cassette mechanism of gene replacement.
    c. rearrangements of immunoglobulin genes.
    d. both a and c are correct.

16. A gene can develop into an oncogene by
    a. being present in more copies than normal.
    b. undergoing a translocation that removes it from its control region.
    c. developing a mutation that creates a more active or resistant protein.
    d. all of the above.

17. Gene amplification involves
    a. the production of loud genes.
    b. the enlargement of chromosomal areas undergoing transcription.
    c. the synthesis of extra DNA to meet particular metabolic needs.
    d. the transformation of genes into oncogenes.

18. The specific number of mitotic divisions observed in fibroblasts in cell culture
    a. provides evidence of the programmed obsolescence of cells.
    b. indicates that the cells have been transformed by cancer viruses.
    c. provides evidence for the determination of the genetic program.
    d. is equal to the number of mitotic divisions found in most cells in culture.

19. Epigenesis
    a. is the theory that egg or sperm contains a miniature individual.
    b. is the view that form emerges gradually in development.
    c. is the result of programmed obsolescence.
    d. is the expression of extrachromosomal DNA.

20. Gene expression
    a. controls the chemical environment of the cytoplasm.
    b. is controlled by the chemical environment of the cell and cytoplasm.
    c. is controlled by nuclear–cytoplasmic interactions.
    d. is a result of preformation.

CHAPTER 19

RECOMBINANT DNA TECHNOLOGY

FRAMEWORK

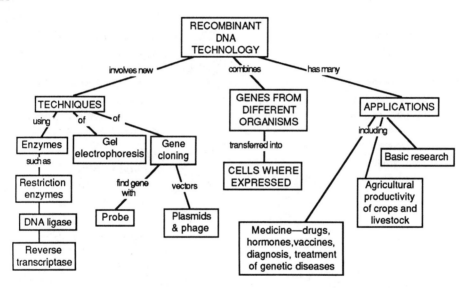

CHAPTER SUMMARY

Recombinant DNA technology involves a set of newly developed techniques for recombining genes from different organisms and then introducing this recombinant DNA into living cells in which the genes are expressed. Using techniques developed in basic research on the biochemistry and molecular biology of bacteria, scientists can now isolate and produce large quantities of particular genes and their products. This methodology has begun an industrial revolution in biotechnology—the use of living systems to manufacture practical products.

Basic Strategies of Gene Manipulation

Before 1975, the technology for altering genes in organisms involved finding natural mutants or creating new mutations with mutagenic radiation or chemicals. The *selection* of the desired mutants involved methods that would permit only the desired organism to survive and grow. Bacteria and their phages could transfer genes from one bacterial strain to another by the processes of transformation, conjugation, and transduction by phage. Geneticists were able to use these processes for the detailed molecular study of prokaryotic and phage genes. The detailed study of eukaryotic genes has been made possible with recombinant DNA techniques.

Restriction enzymes protect bacteria against foreign DNA by cutting up nonbacterial DNA in a process called *restriction*. Most restriction enzymes recognize short nucleotide sequences and cut at specific points within them. The cell protects its own DNA from restriction by the methylation of nucleotide bases within recognition sequences, called *modification*. Recognition sequences are symmetrical sequences of four to eight nucleotides running in opposite directions on the two strands, such as GAATTC on one strand and CTTAAG on the other. The restriction enzyme usually cuts phosphodiester bonds in a staggered way, such as between the G and A on both strands. "Sticky ends" of short, single-stranded sequences occur on both sides of the cut.

DNA from different sources can be recombined in the laboratory when they are cut by the same restriction enzyme and the complementary bases on the sticky ends of the restriction fragments pair by hydrogen bonding and are then joined together by *DNA ligase*.

Restriction fragments can be separated by *gel electrophoresis*, a technique that separates macromolecules on the basis of their size and electrical charge. Negatively charged nucleic acids move through the electric field produced in a slab of gel at a rate inversely proportional to their size. The band patterns of restriction fragments produced in the gel can be used to identify specific viral or plasmid DNA's. Individual fragments can be removed from the gel and retain their biological activity, permitting the isolation and purification of restriction fragments.

Small DNA fragments can be inserted into plasmid DNA that can then serve as *vectors* for moving recombinant DNA from test tubes back into cells. The plasmid method of gene cloning involves isolating plasmids that carry identification genes, such as antibiotic-resistance genes, treating the plasmids with a restriction enzyme that will cut the DNA ring at a single site, and mixing them with "foreign" DNA that has been treated with the same restriction enzyme and has complementary sticky ends. Recombinant molecules are formed by DNA ligase after sticky ends hydrogen bond. The plasmids are introduced into bacterial cells by transformation and reproduced as the bacteria develop clones of cells. Clones carrying the recombinant DNA are identified by the use of selective media (usually for antibiotic resistance).

Bacteriophages can also serve as vectors for introducing recombinant DNA into cells. Recombinant phage DNA is created using restriction enzymes and restriction fragments of foreign DNA. Once inside the cell, the phage DNA replicates itself to form new phage particles, simultaneously cloning the inserted genes.

The genes used for cloning can be isolated directly from a particular organism by cutting the host DNA into thousands of pieces with restriction enzymes and inserting them into plasmids or viral DNA. The collection of the huge number of DNA segments from a genome, derived in this *shotgun approach*, is called a *gene library*.

Copy DNA eliminates the problem of the large introns that may make a eukaryotic gene too large to clone easily. Using the enzyme reverse transcriptase, obtained from retroviruses, mRNA is used as a template to produce cDNA. Since cDNA has the introns already removed, it will produce mRNA that can be more easily translated by bacterial cells, which lack RNA-processing machinery.

Finding the bacterial clone that contains the gene of interest from the gene library formed by the shotgun or cDNA technique may be difficult. If the gene is translated into protein, the presence of the protein can be determined by its activity or its structure (using antibodies that combine with it). Techniques for detecting the gene itself involve the use of a nucleic acid sequence called a *probe*, which has complementary sequences to the gene. The radioactively labeled probe will hybridize with the gene of interest by base-pairing and thus will locate bacterial clones with the gene.

Genes for cloning may be chemically synthesized when the exact nucleotide sequence is known. The sequencing of genes is made possible by the use of restriction enzymes, which cut long DNA sequences into fragments, and gel electrophoresis, which can separate the fragments according to their lengths.

In the Sanger method of DNA sequencing, single-stranded restriction fragments are incubated with radioactively labeled nucleotides, included in which are modified nucleotides (dideoxy forms) that block further synthesis when they are incorporated into a growing DNA strand. The result is a set of radioactive strands of varying lengths that can be separated and identified on the basis of the bands they produce in gel electrophoresis.

Thousands of DNA sequences are being collected in computer data banks where they can be analyzed for genetic control elements and similarities among genes of different organisms, and automatically translated into amino acid sequences.

Differences in both the control of and mechanisms for transcription and translation between prokaryotes and eukaryotes present problems in getting bacteria to express eukaryotic genes. Large-scale commercial production of a gene product in bacteria has been accomplished, however, with the aid of genetic tricks such as using plasmid vectors that produce many gene copies per cell, changing the gene's promoter to a highly active one, and attaching a bacterial gene that is produced in large quantities to the start of the eukaryotic gene. When bacterial cells are engineered to secrete the protein product, the task of purification is simplified.

Yeast present an excellent vehicle for recombinant DNA technology because of their ease of culture, their ability to take up DNA by transformation, and their capacity to express eukaryotic genes since they are eukaryotic cells. Cultured animal and plant cells must be used in some genetic research and in commercial applications that require synthesis of complex products (such as antibodies).

Applications of Recombinant DNA Technology

Recombinant DNA technology has opened up the study of the molecular details of eukaryotic gene structure and function. Genes can be studied directly, without having to infer genotype from phenotype.

Genes can be produced in large amounts and used as labeled probes to locate similar DNA segments in the same or other genomes, providing information on evolutionary relationships. Copy DNA probes allow location and study of the natural form of the gene, including its regulatory and noncoding sequences.

DNA probes also can be used to map genes on eukaryotic chromosomes with an *in situ hybridization* technique. Labeled DNA probes base-pair (hybridize) with intact chromosome on a microscope slide, and autoradiography and chromosome staining show the location of the gene.

Many molecules controlling cell metabolism and development are produced in such small quantities that they cannot be purified and characterized by normal biochemical techniques. The production of cDNA from the mRNA molecules in the cells of an organism and their cloning in bacteria have provided large enough quantities of these proteins to investigate their structures and functions.

Recombinant DNA techniques are providing means to improve the productivity of agricultural plants and animals. The production of vaccines, growth factors, and hormones may raise livestock productivity. Plant crops with engineered genes may be made resistant to diseases, harsh growth conditions, or herbicides.

The regeneration of plants from single cells growing in tissue culture has made plant cells easier to manipulate than are the cells of mammals. The search for genetic vectors for plants cells, however, has been more difficult than that for animal cells. A plasmid carried by the bacterium *Agrobacterium tumefaciens* is the best-developed vector.

This bacterium produces crown gall tumors in the plants it infects when the *Ti plasmid* (tumor inducing) integrates a segment of its DNA into the plant cells' chromosomes. Foreign genes are inserted into the Ti plasmid, and *Agrobacterium* with recombinant plasmids are used to infect plant cells growing in culture. When these cells regenerate whole plants, the foreign gene is included in the plant genome.

Early results of genetic engineering in plants have been positive. Plant strains that carry a bacterial gene for resistance to a herbicide have been developed. Crop yields may be improved by the production of bacterial species, found surrounding plant roots, that are genetically engineered to produce a toxin harmful to crop pests. The production of bacteria with increased nitrogen-fixing potential and the engineering of plants that can fix nitrogen themselves are examples of a valuable use of recombinant DNA technology in improving plant productivity.

Gene splicing has been used to produce hormones and other mammalian proteins in bacteria. Insulin and human growth hormone were the first two polypeptide hormones made by recombinant DNA techniques to be approved for use in the United States. Before 1982, insulin for treating diabetics was obtained from pig and cattle pancreatic tissues. Now there are several ways of producing insulin that is chemically identical to human insulin in genetically engineered bacteria.

Because growth hormones are species-specific, human growth hormone, used for treating children with hypopituitarism that results in dwarfism, had to be obtained in the small quantities available from human cadavers. A genetically engineered version of human growth hormone was developed in 1979 and approved for use in 1985.

Traditional vaccines against viral diseases are either particles of a virulent virus that have been *inactivated* or active virus particles of an *attenuated* (nonpathogenic) viral strain. These vaccines induce an immune response in the organism in which antibodies specific for the invading pathogen are developed. Recombinant DNA techniques have been used to make large amounts of protein molecules from disease-causing viruses and bacteria. If the protein, called a *subunit*, triggers an immune response against the pathogen, it can be used as a vaccine. Genetic engineering methods can also be used to modify directly the genome of the pathogen so as to attenuate it.

The live vaccinia virus, which is the basis of the smallpox vaccine, has been modified to carry genes that induce immunity to other diseases. The inclusion of genes for immunity to several diseases may allow a single live vaccinia innoculation to protect people from multiple diseases.

Diagnosis of genetic diseases is becoming possible using gene cloning and the normal gene as a probe for locating the corresponding gene in the cells being tested. Comparison of the probe and gene make use of restriction enzymes and gel electrophoresis to compare the patterns formed by the two sets of restriction fragments on the gel.

Genetic engineering may provide the means for correcting genetic disorders in individuals by replacing or supplementing defective genes. New genes would presumably be introduced into a few somatic cells of types that actively reproduce within the body so that the gene would be replicated when the cells are reinserted into the individual. The protein product of the introduced normal gene would need to be able to alleviate the mutant phenotype despite the absence of the gene and protein in most other tissues of the body.

Some of the technical problems involved with human gene therapy include how to get the proper control mechanisms to operate on the transferred gene, and when in development and into which tissues the gene should be introduced. An ethical and social question is whether the germ cells should be treated in order to correct defects in future generations. Opponents fear that tampering with human genes will eventually lead to the practice of eugenics, the deliberate effort to control the distribution of human genes.

Safety and Policy Matters

Scientists have worried that dangerous consequences might be involved in recombinant DNA manipulations, in particular that hazardous new pathogens might be produced and released into the environment. Early scientists working in this field developed a set of guidelines to deal with policy and safety issues. The National Institutes of Health (NIH) adopted these recommendations as part of their formal program for guidelines for biomedical research.

The safety measures were of two types: the first to protect scientists from infection by engineered microbes and to prevent the microbes from accidentally escaping from the laboratory, and the second to develop mutant strains of microbes that could not possibly survive outside the laboratory.

A new genetic engineering industry began to develop, with many of its scientists recruited from academic posts. Concerns surfaced about university-based biologists who were now part of both the academic and business worlds. Critics feared that safety and ethical questions would not be dealt with in the rush to build profitable business enterprises.

As organisms were developed to be introduced into the environment, it became necessary to decide which federal agencies were best suited to evaluate proposals for the organisms' release. The NIH has a standard procedure for considering gene-splicing research proposals, but it is not a regulatory agency. In several instances, the federal courts have become involved to answer questions of which federal regulations apply. It is important that considered and practical regulations be developed to guide the use of this powerful new technology.

STRUCTURE YOUR KNOWLEDGE

1. Fill in the following table on the basic strategies of gene manipulation used in recombinant DNA technology.

Technique or Tool	Brief Description of Technique or Tool	Use in Recombinant DNA Technology
Restriction enzymes		
DNA ligase		
Gel electrophoresis		
Reverse transcriptase		
Labeled probe		
Gene cloning		

2. There are many applications for recombinant DNA technique in basic research, in agriculture, and in medicine. Describe four or five examples in each of these categories.

TEST YOUR KNOWLEDGE

MULTIPLE CHOICE: *Choose the one best answer.*

1. The role of restriction enzymes in recombinant DNA technology is to
   a. provide a vector for the transfer of the recombinant DNA.
   b. produce cDNA from mRNA.
   c. produce a staggered cut at specific recognition sequences on DNA.
   d. reseal "sticky ends" after base pairing of complementary bases.

2. Molecules are separated by gel electrophoresis on the basis of
   a. their charge.
   b. their charge and size.
   c. their polarity.
   d. their shape.

3. Plasmid DNA and bacteriophages
   a. are involved in cloning genes.
   b. are both vectors for transferring recombinant DNA into cells.
   c. can be treated so that they carry foreign genes in their DNA.
   d. all of the above.

4. Which of the following is not true of cDNA?
   a. It will produce mRNA more easily transcribed by bacterial cells.
   b. It is produced from mRNA using reverse transcriptase.
   c. It can be used as a probe to locate a gene.
   d. It can eliminate the problem of eukaryotic genes that are too large, due to the presence of introns, to clone easily in bacterial cells.

5. Genes for cloning may be chemically synthesized
   a. when the exact sequence of nucleotides is known.
   b. through the use of restriction enzymes and gel electrophoresis to separate restriction fragments.
   c. by the Sanger method.
   d. using cDNA.

6. Yeast is a good organism for recombinant DNA technology because
   a. it can synthesize complex products such as antibodies.
   b. it can express eukaryotic genes even though it is a prokaryote.
   c. it can take up recombinant DNA by transformation.
   d. all of the above.

7. *Agrobacterium*
   a. is the best-known vector for transferring recombinant DNA into plant cells.
   b. carries the Ti plasmid that can integrate into a host cell's genome.
   c. causes tumors in the plants it affects.
   d. all of the above.

8. An attenuated virus
   a. is a live virus that is nonpathogenic.
   b. consists of viral particles that have been inactivated.
   c. can transfer recombinant DNA to its host cell.
   d. will not produce an immune response.

9. Difficulties in getting prokaryotic cells to express eukaryotic genes include the fact that
   a. the signals that control gene expression are different.
   b. the genetic code differs between the two.
   c. prokaryotic cells cannot transcribe introns.
   d. all of the above.

10. Copy DNA cannot create as complete a gene library as the shotgun approach can because
    a. it eliminates introns from the genes.
    b. a cell produces mRNA for only a small portion of its genes.
    c. the shotgun approach produces more restriction fragments.
    d. cDNA is not as easily integrated into plasmids or phage genomes.

11. Which of the following is not true of recognition sequences?
    a. Modification by methylation of bases within them prevents restriction of bacterial DNA.
    b. They are usually symmetrical sequences of four to eight nucleotides.
    c. They signal the attachment of RNA polymerase.
    d. There is a specific restriction enzyme for each recognition sequence.

12. Recombinant DNA technology contributes to basic research in
    a. the evolutionary relationships of organisms.
    b. the control of gene expression in eukaryotes.
    c. the biochemical analysis of growth factors and molecules produced in small quantities.
    d. all of the above.

ANSWER SECTION

CHAPTER 12: MEIOSIS AND SEXUAL LIFE CYCLES

Suggested Answers to Structure Your Knowledge

1.

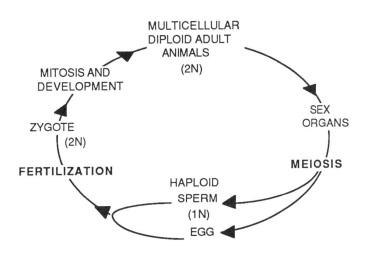

2.

STAGE	KEY EVENTS OF MEIOSIS
INTERPHASE I	Chromosome replication, sister chromatids attached at centromere
PROPHASE I	Synapsis of homologous pairs; crossing over, spindle forms
METAPHASE I	Tetrads align at metaphase plate; independent assortment
ANAPHASE I	Homologous chromosomes separate, move toward opposite poles
TELOPHASE I	Haploid sets of double-stranded chromosomes reach poles, cytokinesis
INTERKINESIS	Nuclear membranes may form, no replication of chromosomes
PROPHASE II	Spindle forms
METAPHASE II	Chromosomes align at metaphase plate
ANAPHASE II	Centromeres of sister chromatids separate, chromosomes move apart
TELOPHASE II	Cytokinesis, 4 haploid cells with single-stranded chromosomes

Answers to Test Your Knowledge

Multiple Choice:

1. b	5. d	9. d
2. d	6. b	10. b
3. d	7. a	11. a
4. b	8. c	12. b

CHAPTER 13: MENDEL AND THE GENE IDEA

Suggested Answers to Structure Your Knowledge

1. The segregation of gene pairs (alleles) takes place at anaphase I when homologous chromosomes move to opposite poles of the cell. The two cells formed from this division have one-half the number of chromosomes and one copy of each gene. Independent assortment relates to the lining up of synapsed chromosomes at the equatorial plate in a random fashion during metaphase I. Chromosomes initially received from the mother and father may orient in either direction. The number of different chromosome arrangements in gametes is equal to 2^N where N is the haploid number of chromosomes. The number of gametic arrangements of genes in crosses involving more than one gene is equal to 2^n where n is the number of genes (as long as they are on different chromosomes).

2.

TtRr x TtRr

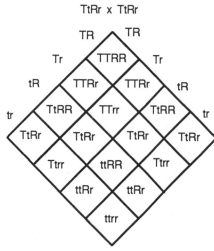

PHENOTYPE	NUMBER
tall, red	3
tall, pink	6
tall, white	3
short, red	1
short, pink	2
short, white	1

3.
Generation I
Genotype of father = Aa   of mother = Aa
Generation II
Genotype of mate 1   of mate 2 = Aa
          = AA (probably)
Generation III
Genotype of son 4 = Aa

   The genotype of son 3 could be AA or aa. If his wife is AA, then he could be Aa (both his parents are carriers) and the recessive allele never would be expressed in his offspring. Even if he and his wife were both carriers, there would be an 81/256 or 31.6% chance that all four children would be normally pigmented.

Answers to Test Your Knowledge

Matching:

1. I	5. N	9. C
2. O	6. J	10. K
3. F	7. H	11. M
4. D	8. A	12. L

Multiple Choice:

1. d	6. c	11. b
2. d	7. c	12. d
3. a	8. d	13. b
4. c	9. b	14. c
5. b	10. d	15. c

Genetics Problems:

1.

Mother	Child	Man exonerated if
AB	A	no groups exonerated
O	B	A and O
A	AB	A and O
O	O	AB
B	A	B and O

2. Father's genotype must be Pp because polydactyly is dominant and he has had one normal child. Mother's genotype is pp. The chance of the next child having normal digits is 1/2 or 50% because the mother can only donate a p allele and there is a 50% chance that the father will donate a p allele.

3. AB gamete from AABb parent = 1/2
ab gamete from AaBb parent = 1/4
ABc gamete from AABbcc parent = 1/2
ABc gamete from AaBbCc parent = 1/8

4. Probability of AAbb offspring = 1/2
(to get AA) x 1/2 (bb) = 1/4
Probability of aaBB offspring = 1/4
(aa) x 1/2 (BB) = 1/8
Probability of AaBbCc = 1 (Aa) x 1/2
(Bb) x 1 (Cc) = 1/2
Probability of aabbcc = 1/4 (aa) x 1/4
(bb) x 1/2 (cc) = 1/32

5. The genotypes of the puppies were 3/8 B-S- (since B and S are dominant alleles, the "-" indicates that the second gene can be either B or b, S or s, and still produce a dominant phenotype), 3/8 B-ss, 1/8 bbS-, and 1/8 bbss. Because recessive traits show up in the offspring, both parents had to be at least heterozygous for both genes. Black occurs in a 6:2 or 3:1 ratio, indicating a heterozygous cross. Solid occurs in a 4:4 or 1:1 ratio,

indicating a cross between a heterozygote and homozygous recessive. Parental genotypes were BbSs x Bbss.

6. The F_1 cross produced AaBbCc plants with 3 units of 5 cm added to the base height of 10 cm, to produce 25-cm tall F_1 plants. Of the 64 possible combinations of gametes in the F_2, there will be 7 different phenotypic classes, varying from all dominant alleles, 5 dominant and 1 recessive, 4 dominant and 2 recessive, and so on, through all 6 recessive alleles.

7. It is good to decide first what phenotypes are produced by each genotype. M-B- individuals will be black, because both genes are dominant. M-bb individuals are brown. mm- individuals will all be white, because no melanin is produced to be laid down (the B genotype does not matter). All F_1 guinea pigs will be MmBb and black. The normal 9 M-B-, 3 M-bb, 3 mmB-, 1 mmbb phenotypic ratio will be 9 black, 3 brown, and 4 white due to the epistatic effect of mm on the B gene.

8. The 1:1 ratio of the first cross indicates a cross between a heterozygote and a homozygote, Hh x hh, or Hh x HH. The second cross indicates that the hairless hamsters cannot be hh, or only hairless hamsters should be produced. If the hairless hamsters are Hh, then one would expect a 3:1 phenotypic ratio and a 1:2:1 genotypic ratio. The 1:2 ratio indicates that the homozygous-recessive genotype may be lethal, so that embryos that are hh never develop.

9. The 1:2:1 ratio indicates a case of incomplete dominance. Medium-tailed pigs are probably Tt, and the heterozygous cross produced three tt stub-tailed pigs, six Tt medium-tailed pigs, and four TT long-tailed pigs.

10.

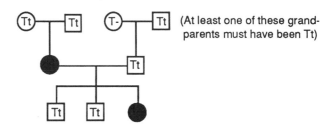

(At least one of these grandparents must have been Tt)

CHAPTER 14: CHROMOSOMAL BASIS OF INHERITANCE

Suggested Answers to Structure Your Knowledge

1. Genes that are not linked assort independently, and the ratio of offspring from a testcross with a dihybrid heterozygote should be 1:1:1:1 (AaBb x aabb gives AaBb, Aabb, aaBb, aabb offspring). Genes that are linked and do not cross over should produce a 1:1 ratio (AaBb and aabb). If crossovers occur 50% of the time, then the heterozygote will produce equal quantities of

AB, Ab, aB, and ab gametes and the genotypic ratio of offspring will be 1:1:1:1, the same as for unlinked genes. Crosses with intermediate genes on the chromosome could establish both that the genes A and B are on the same chromosome and that they are a certain distance apart.

2.

SYSTEMS OF SEX DETERMINATION					
	X - Y	X - O	Z - W	HAPLO-DIPLOID	NO SEX CHROMOSOMES
FEMALE	XX	XX	ZW	Diploid	Both gametes produced by same organism
MALE	XY	XO	ZZ	Haploid	
EXAMPLES	Humans, mammals, fruit flies, dioecious plants	Crickets, roaches, grasshoppers	Birds, some fish, moths, butterflies	Bees, ants, wasps	Worms, snails, monoecious plants

3. If the gene for the mutant is sex-linked, you could assume it was on the X chromosome. A cross between mutant female flies and normal males should produce

all normal female and mutant male flies.

X^mX^m x X^+Y produces X^+X^m and X^mY offspring.

4.

NAME OF DISEASE OR SYNDROME	CHROMOSOMAL ALTERATION INVOLVED	SYMPTOMS OR ASSOCIATED TRAITS
Down's syndrome	Trisomy 21	Characteristic facial features, short stature, heart defects, respiratory infections, mental retardation
Patau syndrome	Trisomy 13	Harelip, cleft palate, eye, brain, circulatory defects
Edwards syndrome	Trisomy 18	Serious effects on almost every organ system
Klinefelter's syndrome	XXY (or XXYY, XXXY, XXXXY, XXXXXY)	Sterility, small testes, enlarged breasts or feminine body contours, normal intelligence if XXY, mental retardation if additional X or Y
Turner's syndrome	Monosomy X	Short stature, sex organs do not mature, no secondary sex characteristics, normal intelligence
Metafemale	Trisomy X	Limited fertility, often mentally retarded
Cri du chat syndrome	Deletion in chromosome 5	Mental retardation, small head, unusual facial features, unusual cry

Answers to Test Your Knowledge

Multiple Choice:

1. d	6. b	11. d	16. c
2. a	7. b	12. b	17. b
3. b	8. b	13. c	18. c
4. d	9. a	14. d	19. a
5. d	10. a	15. a	20. a

Genetics Problems:

1. The gene is sex-linked, so a good notation is X^c, X^c, and Y so that you will remember that the Y does not carry the gene. Genotypes are:

1. $X^C Y$	4. $X^C X^c$	6. $X^c Y$
2. $X^C X^c$	5. $X^C Y$	7. $X^C X^c$
3. $X^C X^C$ or $X^C X^c$		

2. e, a, c, b, d

3. The genes appear to be linked, because the parental types appear most frequently in the offspring. Recombinant offspring represent 10 out of 40 total offspring for a recombination frequency of 25%, indicating that the genes are 25 map units apart.

CHAPTER 15: THE MOLECULAR BASIS OF INHERITANCE

Suggested Answers to Structure Your Knowledge

1.

INVESTIGATOR	ORGANISMS USED	TECHNIQUES, EXPERIMENTS	CONCLUSIONS
Griffith	S and R strains of bacteria, S caused pneumonia in mice	Inject mice with heat-killed S and live R cells	R cells transformed by genetic material from S cells; heat-stable factor may not be protein
Avery	S and R strains of bacteria	Purify chemicals from heat-killed S cells; test for transforming live R cells	Transforming agent was DNA
Hershey & Chase	Bacteriophage (phage T2) and the bacterium, *E. coli.*	Label DNA, protein of phage; infect *E. coli*; blend and centrifuge so bacteria in bottom, viral parts in top.	DNA of phage injected into host; proteins stay outside; DNA directed host to replicate phage—DNA hereditary material
Chargaff	DNA from several organsisms	Paper chromatography to separate A, T, G, C bases; compare base ratios	DNA species-specific—ratio of bases different between species; but A=T and G=C
Franklin	DNA	X-ray crystallography of DNA molecules	DNA is in shape of helix
Watson & Crick	DNA	Information from X-ray photo; wire scale models	Double helix with rungs of A-T and G-C, sides of phosphate-sugar chain
Meselson & Stahl	*E. coli.*	*E. coli* labeled with heavy isotope of N; grow one generation with light N; centrifuge to find density	Semiconservative replication of DNA

2.

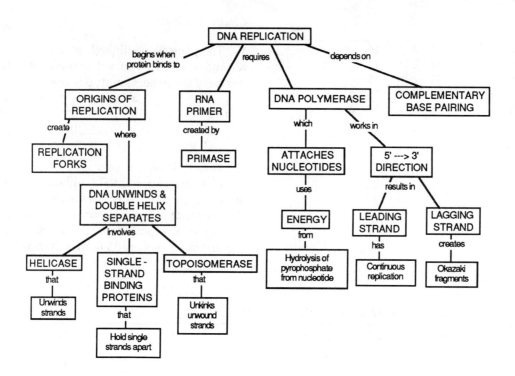

Answers to Test Your Knowledge

1. c	11. c	21. h
2. a	12. d	22. g
3. c	13. b	23. d to c
4. b	14. c	24. a
5. d	15. a	25. e
6. b	16. d	26. a
7. a	17. d	27. b
8. b	18. d	28. c
9. a	19. b	

CHAPTER 16: FROM GENE TO PROTEIN

Suggested Answers to Structure Your Knowledge

1.

	TRANSCRIPTION	TRANSLATION
Template	DNA	mRNA
Location	nucleus	ribosomes
Molecules involved	ribonucleotides, DNA and enzymes	amino acids, attached to tRNA, mRNA, ribosomes, enzymes, ATP, GTP and others
Enzymes needed	RNA polymerase, and processing enzymes	aminoacyl-tRNA synthetase, peptidyl transferase, and others
Control—start & stop	promoter sequence on DNA, terminator sequence	initiation factors, AUG codon, release factor
Product	mRNA	protein
Processing involved	RNA splitting: introns removed, Poly A and cap added	spontaneous folding, coiling; disulfide bridges; removal of some amino acids; cleaving, or quarternary structure
Energy source	ribonucleoside triphosphate	ATP and GTP

2. The genetic code includes the sequence of nucleotides on DNA that is transcribed into the codons found on mRNA and translated into their corresponding amino acids. There are 64 possible codons created from the four nucleotides used in the triplet code (4^3). Codons are needed only for 20 amino acids and stop codons. Redundancy of the code refers to the fact that several triplets may code for the same amino acid. Often these triplets only differ in the third nucleotide. The wobble hypothesis explains the fact that there are only about 40 different tRNA molecules that pair with the 64 possible codons. The third nucleotide of many tRNAs can form H bonds with more than one base. Because of the redundancy of the genetic code, these wobble tRNAs still place the correct amino acid in position. The genetic code was thought to be universal—each codon coded for exactly the same amino acid no matter which organism it was found in. Recently, some exceptions to this universality have been found in a few ciliates.

3.

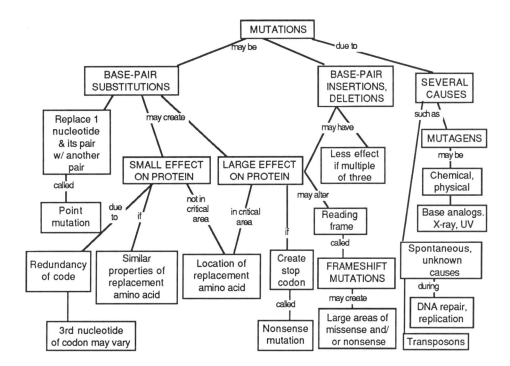

Answers to Test Your Knowledge

Multiple Choice:

1. d	6. d	11. b	16. a
2. a	7. a	12. d	17. a
3. c	8. a	13. c	18. c
4. d	9. d	14. d	19. c
5. b	10. a	15. c	20. b

Fill in the Blanks:

DNA	1 TAC	4 CAG	TTC	10 ATC
mRNA	2 AUG	5 GUC	7 AAG	UAG
Anti-codon	3 UAC	CAG	8 UUC	11 none—release factor binds
Amino acid	methionine	6 valine	9 lysine	12 STOP

CHAPTER 17: THE GENETICS OF VIRUSES AND BACTERIA

Suggested Answers to Structure Your Knowledge

1.

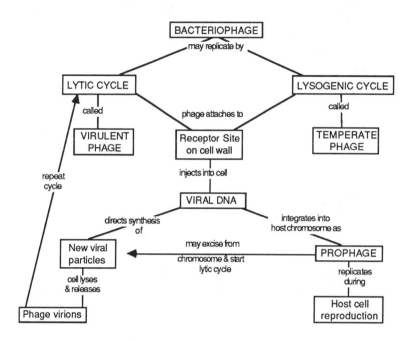

2.

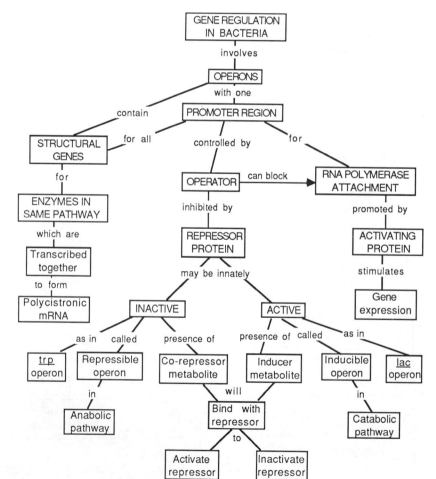

Answers to Test Your Knowledge

Matching:

1. C   6. M
2. I   7. D
3. H   8. J
4. B   9. K
5. A   10. G

Matching:

1. C   5. F
2. D   6. B
3. G   7. A
4. E   8. H

Multiple Choice:

1. b   6. c   11. b   16. d
2. d   7. d   12. c   17. b
3. a   8. b   13. d   18. c
4. b   9. c   14. a   19. a
5. a   10. d   15. c   20. d

CHAPTER 18: GENE EXPRESSION AND DEVELOPMENT IN EUKARYOTES

Suggested Answers to Structure Your Knowledge

1.

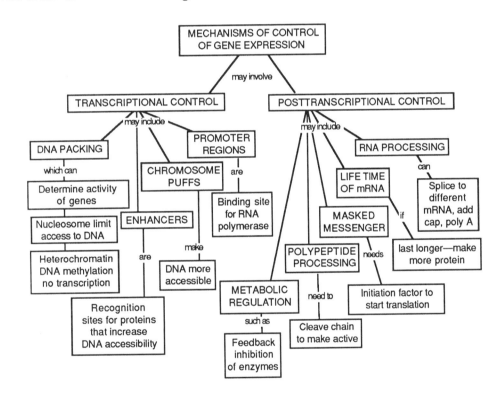

2.

ORGANIZATION OF EUKARYOTIC DNA				
PHYSICAL STRUCTURE	REGULATORY SEQUENCES	STRUCTURAL SEQUENCES	GENES	
			Coded sequences	Noncoded sequences
DNA packing	Recognition sequences (for genes in same pathway)	Highly repetitive sequences (satellite DNA at centromere)	EXONS	INTRONS
Nucleosome			Multigene families (copies of same or simlar genes; tandem arrays)	Pseudogenes (genelike sequences in tandem arrays)
30-nm chromatin fiber	Promoters (bind RNA polymerase)			
looped domain	Enhancers (increase transcription)		Gene amplification (many copies of same gene)	Gene diminution (chromosomes lost from cell)
hetero-chromatin			Development control genes (proto-oncogenes, homeotic genes)	

3. Development involves the complex processes through which a fertilized egg grows into a multicellular organism consisting of differentiated cells organized into tissues and organs. The key to this process is the selective expression of genes from the complete genome present in every cell. The zygote is said to be totipotent, capable of forming all the specialized cells of the adult organism. Determination is the progressive restriction of the developmental potential as the fates of the cells become more limited during development. Even though the cells are still genomically equivalent, long-term changes in gene expression may have irreversibly turned off some genes. Differentiation is the specialization of cells by selective gene expression by which a cell produces its characteristic proteins and develops specialized structures and functions.

Answers to Test Your Knowledge

1. b	6. c	11. c	16. d
2. c	7. a	12. c	17. c
3. d	8. b	13. d	18. a
4. b	9. d	14. c	19. b
5. a	10. a	15. d	20. c

CHAPTER 19: RECOMBINANT DNA TECHNOLOGY

Suggested Answers to Structure Your Knowledge

1.

Technique or Tool	Brief Description of Technique or Tool	Use in Recombinant DNA Technology
Restriction enzymes	Bacterial enzyme that cuts foreign DNA, leaving "sticky ends" that can base-pair with other restriction fragments	Form restriction fragments that can be used to sequence DNA, and to make recombinant DNA
DNA ligase	Enzyme that forms phosphodiester bonds in DNA molecule	Seal sticky ends of restriction fragments, make recombinant DNA
Gel electrophoresis	Apply mixture of macromolecules to slab of gel under electric field, molecules separate by movement due to size and charge	Separate restriction fragments. Bands permit identification or sequencing of DNA or removal of fragments for gene library.
Reverse transcriptase	Retrovirus enzyme that produces DNA from mRNA	Create copy DNA (cDNA) from mRNA, gives smaller gene (RNA processed--no introns) more easily translated by bacteria
Labeled probe	Radioactively labeled single-stranded DNA or mRNA that will base-pair with complementary sequences of DNA	Locate gene in clone of bacteria, identify similar nucleic acid sequences, make cytological map of genes
Gene cloning	Reproduce large quantities of a gene in phage, bacteria, or yeasts; use bacteria or phage as vectors	Produce many copies of a gene for basic research or for large-scale production of gene product

2. Basic Research:

(1) In the Sanger method, techniques using restriction enzymes and gel electrophoresis are used to determine nucleotide sequences in DNA. (2) Radioactively labeled probes are used to map genes on chromosomes. (3) Molecules that control metabolism and development are often present in very small quantities. With cDNA, their genes can be cloned and the molecules produced in large enough quantities to study. (4) cDNA can be used as a probe to locate genes and study their noncoded regions such as introns and regulator sequences. (5) Using probes to locate genes with similar sequences, the evolutionary relationships between genes and between organisms can be studied.

Agricultural Applications:

(1) Vaccines, hormones, and growth factors, which will improve the health or productivity of livestock, can be produced using recombinant DNA technology.

(2) The genome of agricultural plants and animals may eventually be directly altered and improved. (3) Improvements in the nitrogen-fixing capacity of bacteria or plants are being developed. (4) Soil bacteria that produce pesticides are being produced. (5) Plant varieties that have genes for resistance to adverse conditions, diseases, and herbicides are being developed.

Medical Applications:

(1) Vaccines are being developed, either through the production of virus proteins to determine if these subunits produce immune responses, or in the development of attenuated viruses. (2) Diagnosis of genetic diseases is possible by probing for a potentially defective gene with a normal gene. (3) Treatment of genetic disorders may be possible through the replacement of a defective gene with a normal one. (4) Insulin, human growth factor, and several products of the immune system have been produced by biotechnology.

Answers to Test Your Knowledge

Multiple Choice:

1. c	5. a	9. a
2. b	6. c	10. b
3. d	7. d	11. c
4. a	8. a	12. d

Unit IV

Evolution

CHAPTER 20

DESCENT WITH MODIFICATION: A DARWINIAN VIEW OF LIFE

FRAMEWORK

This chapter describes Darwin's formulation of *evolution*—descent with modification from a *common ancestor* by the mechanism of *natural selection*, which results in the evolution of species adapted to their environments. The scientific and philosophical climate of Darwin's day was quite inhospitable to the implications of evolution, but most biologists accepted the theory of evolution quite rapidly. Acceptance of natural selection as the mechanism of evolution occurred later, with the incorporation of genetic principles into the modern synthesis of evolution in the 1940s. Evidence for evolution is drawn from biogeography, the fossil record, taxonomy, comparative anatomy, comparative embryology, and molecular biology.

CHAPTER SUMMARY

Evolution refers to all the changes that have occurred in the history of life. Darwin presented the first convincing case for evolution in his book *The Origin of Species*, published in 1859. Darwin made two major claims: that species had not been specially created in their present forms but had evolved from ancestral species, and that *natural selection* provided the mechanism for evolution.

Pre-Darwinian Views

Darwin's theory was truly radical, for it challenged both the prevailing scientific views and the world view that had been held for centuries in Western culture.

The Greek philosopher Plato believed in two worlds, an ideal and eternal real world and the illusory world perceived by the senses. The variations in plant and animal populations were simply imperfect representatives of ideal forms. Plato's philosophy of ideal forms is known as idealism or essentialism.

Aristotle, noting the range of organisms from simple to complex, believed that all living forms could be arranged on a scale of nature of increasing complexity. Each group of organisms was fixed, permanent, and did not evolve. This view of life prevailed for over 2000 years.

The Judeo–Christian account of creation, the special design of each species during the week in which the Creator formed the universe, embedded the idea of the fixity of species in Western thought. Biology during Darwin's time was dominated by natural theology, the study of nature to reveal the Creator's plan.

One of the goals of natural theology was to classify species to reveal the rungs on the scale of life that God had created. Carl Linnaeus, working in the eighteenth century, developed the binomial system for naming organisms according to their genus and species and a hierarchy of classifications for grouping species. *Taxonomy*, the branch of biology that names and classifies organisms, was begun by Linnaeus.

Fossils are remnants or impressions of organisms laid down in rock—usually *sedimentary rocks*, such as sandstone and shale, that form through the compression of layers of sand and mud into superimposed layers called strata. Fossils reveal a succession of *flora* and *fauna* (plant and animal life).

Cuvier, the father of *paleontology*, the study of fossils, believed that the history of life was recorded in series of strata containing fossils. Cuvier observed the differences between older fossils and modern life forms, and the occurrences of extinctions, but he maintained that these changes were the result of catastrophic events such as floods or drought, and not indicative of evolution. This view of the history of the earth is called *catastrophism*.

Gradualism, the idea that profound change is the cumulative result of slow but continuous processes, was proposed by Hutton in 1795 to explain the geological state of Earth. Lyell, the leading geologist of Darwin's time, extended gradualism to a theory of *uniformitarianism*, that held that the uniform rates and effects of geological processes cause their effects to balance out through time.

Darwin took two ideas from the observations of Hutton and Lyell: that Earth must be very old if geological change is slow and gradual, and that very slow and subtle processes occurring over long periods of time could effect substantial change.

Although several naturalists suggested that life had evolved along with Earth, Lamarck, in 1809, was the first to publish a theory of evolution that explained how life evolves. Lamarck believed that evolution was driven by the tendency toward greater complexity, which he equated with perfection. As organisms evolved, they became better adapted to their environments, and if the environment changed, new adaptations evolved. Lamarck explained the mechanism of evolution with two principles: use or disuse of body parts leading to their development or deterioration, and the inheritance of acquired characteristics. In the creationist–essentialist climate of the time, Lamarck's views were dismissed and vilified. Lamarck's theory presented many key evolutionary ideas: that evolution is the best explanation for the fossil record and the current diversity of life, that Earth is very old, and that adaptation to the environment is the main product of evolution.

The Evolution of Charles Darwin

Darwin was 22 years old when he sailed from England as the naturalist on the *Beagle*. He spent the voyage collecting thousands of specimens of the fauna and flora of South America, observing the various adaptations of organisms living in very diverse habitats, and making note of the geographic distribution of species. He was particularly struck by the uniqueness of the fauna of the Galapagos Islands.

Through his experiences on the *Beagle* and his reading of Lyell's *Principles of Geology*, Darwin came to doubt the Church's orthodox view that Earth was static and had been created only a few thousand years ago. Upon learning that the 14 types of finches he had collected on the Galapagos were indeed separate species, Darwin began, in 1837, the first of several notebooks on the origin of species. He began to perceive that the origin of new species was closely related to the process of adaptation to different environments.

In 1844, Darwin wrote a long essay on the origin of species and natural selection but was reluctant to in-

troduce his theory publicly. In 1858, Wallace sent Darwin a manuscript in which he had developed a theory of natural selection identical to Darwin's. Lyell presented Wallace's paper and extracts of Darwin's unpublished essay of 1844 jointly to the Linnean Society on July 1, 1858, and Darwin quickly finished and published *The Origin of Species* the next year. Within a decade, Darwin's book and its defenders had convinced the majority of biologists that evolution was the best explanation for the forms of life on earth.

The Concepts of Darwinism

Darwin's concept of *descent with modification* included the notion that all organisms were related through descent from some unknown ancient prototype, and had developed increasing modifications as they adapted to various habitats. Darwin's view of the history of life is analogous to a tree with common ancestors at the fork of each new branching and modern species at the tips of the living twigs. Most branches are dead ends; about 99% of all species that have lived are extinct.

Darwin's book focused on how populations of individual species become adapted to their environments through natural selection. The concept of *natural selection* includes several premises: (1) there is variation between individuals, (2) these variations can be inherited, (3) individuals produce more offspring than the environment can support, and (4) individuals whose inherited characters best fit them to the environment are likely to leave more offspring. Natural selection is the differential success in reproduction among individuals within a population that tends to adapt it to its environment.

Variability arises through chance, but natural selection is not a chance phenomenon. The environment sets definite criteria for reproductive success. The excessive production of offspring sets up the struggle for life; only a small proportion can live to leave offspring of their own. Darwin's ideas were influenced by the writing of Malthus on the potential for the human population to grow much faster than the supply of food and resources.

Darwin found evidence in the *artificial selection* used in the breeding of domesticated plants and animals that selection among the variations present in a population can lead to substantial changes. He reasoned that, if artificial selection could create such differences in food crops and domestic animals in short periods of time, then natural selection, working over thousands or millions of generations, could gradually create the modifications essential for the present day diversity of life. Gradualism is basic to the Darwinian view of evolution.

Natural selection results in the evolution of *populations*, groups of interbreeding individuals of the same species. Individuals are acted on by natural selection, but evolution is measued only as change in a population over time. Natural selection affects only those traits that are heritable. Acquired, nongenetic characteristics cannot evolve. Natural selection is a local and temporal phenomenon, depending on the specific environmental factors present in a region at a given time.

The best-known example of natural selection in action is the English peppered moth, *Biston betularia*. Before the Industrial Revolution, the light-colored morph of this moth was the one most commonly collected, and dark morphs were rare. As industrial pollution darkened the tree trunks in the late 1800s, the light morphs began to decrease in number and the dark forms to increase. This change in the color of moths in polluted areas is known as industrial melanism.

Predation by birds on these moths has been observed and shown experimentally to increase when the moths contrast with their background. Predation served as the natural selection agent that caused the shift in the composition of peppered moth populations in industrial regions in a short period of time.

Evidence for Evolution

Darwin backed his theory with several lines of evidence, including biogeography, the fossil record, taxonomy, and comparative anatomy.

The geographic distribution of species, or biogeography, first suggested common descent to Darwin. Islands have many species that are endemic (native, found nowhere else) but are related to species on the nearest island or mainland. Widely separated islands having similar environments are more likely to have species related to the nearest mainland than to each other.

The succession of fossil forms supports the existence of the major branches of descent that were established with evidence from anatomy and other sources. Darwin was troubled by the fragmentary fossil record that did not contain many transitional fossils linking modern species to their ancestral forms. The record is still fragmentary, although many key links have been found, including *Archeopteryx*, a fossil that links birds to their reptilian ancestors.

The taxonomy developed by Linnaeus provided a hierarchical organization of groups that suggested the branching on the tree of life. The anatomical similarities among species grouped in the same taxonomic category provide evidence of common descent. The same skeletal elements make up the forelimbs of all mammals regardless of function or external shape. These forelimbs are *homologous structures*, similar because of their common ancestry. Comparative anatomy illustrates that evolution is a remodeling process in which ancestral structures are modified for new functions.

Vestigial organs are rudimentary structures, of little or no value to the organism, that are historical remnants of ancestral structures. The skeletons of some snakes retain vestigial pelvic and leg bones, once used in their walking ancestors.

Comparative embryology shows that closely related organisms have similar stages in their embryonic development. Early embryos of vertebrates look very similar and pass through a stage in which they have gill slits on the sides of their throats. These slits develop into gills in fishes, but into different structures in other vertebrates, such as eustachian tubes in humans.

In the late nineteenth century, many embryologists adopted the view that "ontogeny recapitulates phylogeny," stating that the embryonic development (*ontogeny*) of an organism is a replay of its evolutionary history (*phylogeny*). Although the recapitulation theory is an overstatement, ontogeny does provide clues to phylogeny. Early embryological development may provide evidence of the homology among structures that are very different in the adult form.

An organism's DNA reflects its ancestry; closely related species should have a larger proportion of DNA and proteins in common than do more distantly related species. Molecular biologists have measured the degree of similarity in DNA sequences and amino acid sequences in proteins and have found that the closer two species are taxonomically, the greater are the DNA and protein similarities.

Darwin's boldest speculation, that all forms of life descended from the earliest organisms and are thus related, is supported through molecular biology. For example, all aerobic species, even though they may be as unrelated as grass and fungi and butterflies, have the respiratory protein, cytochrome *c* (with variations in the amino acid sequences, of course, but with the same essential structure and function). The universal genetic code, evidently passed along through all branches of evolution, is also important evidence that all life is related.

The Modern Synthesis

Most biologists rapidly accepted evolution but not Darwin's suggestion of natural selection as the mechanism of evolution. Without a theory of genetics, the perpetuation of parental traits in offspring along with the existence of variations could not be explained. Although Mendel was a contemporary of Darwin, his contribution to the theory of inheritance and natural selection was not recognized. The genetic basis of variation and natural selection was not established until the late 1930s.

In the early 1940s, a comprehensive theory of evolution, known as the *modern synthesis*, or neo-Darwinism, was developed from the application of genetics to Darwin's theory. This theory includes the importance of populations as the units of evolution, the essential role of natural selection, and the gradualness of evolution.

Darwin provided a strong scientific basis to the study of the diversity and the unity of life. Today, nearly all biologists accept the theory of evolution, although debate still continues on models of how evolution occurs. The rate of evolution and the role of mechanisms other than natural selection are areas in which evolutionists, such as Steven Stanley, are challenging the modern synthesis.

STRUCTURE YOUR KNOWLEDGE

1. Briefly state the main components of Darwin's theory of evolution.
2. Darwin and present-day biologists have drawn on six sources of evidence for evolution. Briefly describe the contributions from each of these areas.

TEST YOUR KNOWLEDGE

MATCHING: *Match the theory or philosophy and its proponent(s) with the following descriptions.*

A. catastrophism	a. Aristotle
B. early theory of mechanism of evolution	b. Cuvier
	c. evolutionists of this century
C. essentialism	
D. gradualism	d. Hutton
E. modern synthesis	e. Lamarck
F. natural theology	f. Linnaeus
G. ontogeny recapitulates phylogeny	g. some embryologists
H. scale of nature	h. Plato
I. uniformitarianism	i. Wallace

1. __ __   discovery of the Creator's plan through the study of His works
2. __ __   history of Earth marked by floods or droughts that resulted in extinctions
3. __ __   replication of ancestry of organism during its development
4. __ __   inheritance of acquired characteristics
5. __ __   profound change is the cumulative product of slow but continuous processes
6. __ __   fixed species on a continuum from simple to complex

7. __ __   evolutionary theory with genetic explanation of natural selection
8. __ __   ideal world with perfect forms of which world of senses is imperfect representation

MULTIPLE CHOICE: *Choose the one best answer.*

1. The classification of organisms into hierarchical groups is called
   a. the scale of nature.
   b. taxonomy.
   c. natural theology.
   d. ontogeny.

2. The study of fossils is called
   a. phylogeny.
   b. gradualism.
   c. paleontology.
   d. biogeography.

3. The ideas of Hutton and Lyell that Darwin incorporated into his theory
   a. concerned the age of Earth and gradual changes.
   b. concerned the observation of extinctions in the fossil record.
   c. were that evolution is driven by adaptation to the environment.
   d. involved the classification of organisms into a hierarchy of groups.

4. Which of the following is not a premise of natural selection?
   a. There is variation among individuals.
   b. The diploid character of organisms allows variability to be maintained in the population.
   c. The variations among individuals may be inherited.
   d. Differential success in reproduction is determined by fitness for the environment.

5. Artificial selection
   a. was used by Darwin as evidence for changes possible with natural selection.
   b. involves the artificial insemination of females of a species.
   c. resulted in the change in the English peppered moth population.
   d. All of the above are correct.

6. In the case of the English peppered moth,
   a. bird predation was the agent of artificial selection.
   b. the soot incorporated into the moths resulted in industrial melanism.

c. natural selection was shown to affect a population's phenotypic makeup in a short span of time.

d. the light morphs were better adapted to pollution-free air.

7. The biogeographic distribution of species
   a. provides evidence for evolution.
   b. provides evidence for natural selection.
   c. shows that many endemic island species are related to species on the nearest mainland.
   d. All of the above are correct.

8. The gill slits of reptiles and birds
   a. are vestigial structures.
   b. provide support for the fact that ontogeny recapitulates phylogeny.
   c. are homologous structures.
   d. provide evidence for the degeneration of unused body parts.

9. Similar embryological development
   a. indicates that organisms have evolved from a common ancestor.
   b. can help to identify homologous structures that look different in the adult form.
   c. is evident in most early vertebrate embryos.
   d. All of the above are correct.

10. The best evidence that all of life has descended from a common ancestry comes from
    a. comparative anatomy.
    b. comparative embryology.
    c. molecular biology.
    d. the fossil record.

CHAPTER 21

HOW POPULATIONS EVOLVE

FRAMEWORK

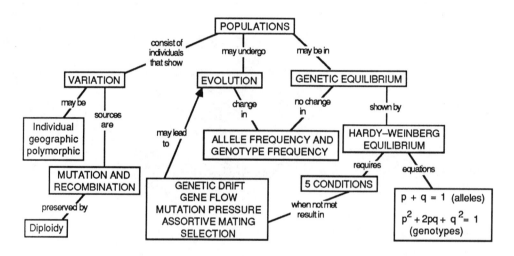

CHAPTER SUMMARY

Although individuals are selected for or against by natural selection, it is populations that actually evolve. The application of genetics to the theory of natural selection has been the most important post-Darwinian advance in evolutionary biology.

The Genetics of Populations

A *species* is a group of populations that have the ability to interbreed in nature. Within the geographic range of a species, populations of individuals of that species may be clustered together or spread out and isolated.

The *gene pool* is the term for all the genes present in a population at any given time. It contains two alleles for each gene locus for each of the individuals in the population of a diploid species. If all individuals are homozygous for the same allele, the allele is said to be

fixed in the gene pool. More often, two or more alleles are present, each at a certain frequency in the gene pool. *Microevolution* is a change in the relative frequencies of alleles in a population over time.

In the absence of selection and other agents of change, the allele frequencies within a population will remain constant from one generation to the next. This stasis is formulated as the *Hardy–Weinberg equilibrium*. The allele frequency within a population determines the proportion of gametes that will contain that allele. The random combination of gametes will yield offspring with genotypes that reflect and reconstitute the allele frequencies.

With the Hardy–Weinberg equation, the frequencies of alleles within a population can be estimated from the genotype frequencies, and vice versa. The letters p and q represent the proportions of the two alleles within a population (more than two alleles can be represented as r, s, etc.). The combined frequencies of the alleles must equal 100% of the alleles present for that gene: $p + q = 1$.

The frequencies of the genotypes in the offspring reflect the frequencies of the alleles and the probability of each combination. According to the rule of multiplication, the probability that two gametes containing the same allele will come together in a zygote is equal to $(p \times p)$ or p^2, or $(q \times q)$ or q^2. For a p and q allele to combine, the p allele could come from one parent or the other. The frequency of a heterozygous offspring reflects these two possibilities and is equal to 2pq. The sum of the frequencies of all possible genotypes in the population adds up to one: $p^2 + 2pq + q^2 = 1$.

Allele frequencies can be determined from genotype frequencies. If the frequencies of the homozygous genotype (p^2) and the heterozygous genotype (2pq) are known, then the frequency of the p allele can be determined. All the gametes from the homozygotes and one-half the gametes from the heterozygotes will contain p. The frequency of p in the gene pool will equal the frequency of homozygous and one-half the frequency of heterozygous genotypes. If the frequency of homozygous recessive individuals is known (q^2), then the frequency of q is equal to the square root of q^2 (assuming the population is in Hardy–Weinberg equilibrium for that gene).

As an example, if p = frequency of allele A = 0.7, and q = frequency of allele a = 0.3, then, according to $p^2 + 2pq + q^2 = 1$, AA = 0.49, Aa = 0.42, and aa = 0.09. To determine allele frequencies, p = frequency AA + 1/2 frequency Aa, or 0.7, and q = frequency aa + 1/2 frequency Aa, or 0.3.

The Hardy–Weinberg equilibrium is maintained only if all the following five conditions are met: (1) a large population, (2) no migration into or out of the population, (3) no net changes in the gene pool due to mutation, (4) random mating, and (5) equal reproductive success of all genotypes. The Hardy–Weinberg predictions can serve as a baseline for comparison with actual populations, in which these five conditions are almost never met and gene pools are changing.

Causes of Microevolution

Five potential agents of microevolution arise from the five conditions of Hardy–Weinberg equilibrium. Natural selection (a result of condition 5) tends to increase the fitness of a population to its environment; the other four agents are chance events and usually nonadaptive.

If a population is small, the random drawing of alleles to form the next generation may not represent the allele frequencies in the gene pool due to *sampling error*, the relatively large influence of each event in a small sample. *Genetic drift* is change in the gene pool of a small population due to chance and is not related to the fitness of individuals. It may play a

major role in the microevolution of populations of less than 100 individuals.

The *bottleneck effect* occurs when some disaster reduces the population size dramatically, and the few surviving individuals are unlikely to represent the original genetic makeup of the population. Genetic drift will remain a factor in the population until it is large enough for chance events to be less significant. A bottleneck usually reduces variability because some alleles are lost from the gene pool.

The genetic drift found when a few individuals colonize a new area is known as the *founder effect*. The small sample size represented by the few colonists is unlikely to be representative of the parent population, and genetic drift will continue to affect the new population until it is large enough that sampling errors will not be a factor in determining allele frequencies.

The migration of individuals among populations may result in the gain or loss of alleles. This phenomenon is known as *gene flow*. The differences in allele frequencies among populations, which may have developed by natural selection or genetic drift, tend to be reduced by gene flow.

The altering of allele frequencies due to *mutation pressure* is probably of little importance in microevolution due to the very low mutation rates for most gene loci. Mutation is the original source of genetic variation, however; as such, it is central to evolution.

Assortive mating, the nonrandom choice of genotypes for mates, may be due to geographic proximity or preference for like individuals. In a large population, assortive mating does not change the allele frequencies in the gene pool, but may change the ratio of genotypes. Assortive mating also fosters inbreeding and may increase the numbers of gene loci that are homozygous.

For the Hardy–Weinberg equilibrium to be maintained, there must be no differential success in reproduction; there must not be a genotype/fitness correlation. This condition is probably never met due to selection; some individuals are more successful in producing offspring and passing their alleles to the next generation than are others. Selection is likely to be adaptive; favorable genotypes are maintained in a population. With environmental changes, selection will favor genotypes adapted to the new conditions.

Genetic Basis of Variation

Individual variation, the slight differences among individuals as a result of their unique genomes, is the raw material for natural selection.

Geographic variations are regional differences that occur within a species that exists over a wide range. These variations are greatest among populations, but may also exist within a population whose range includes a variety of habitats. If an environmental parameter changes gradually across a distance, there may be variations within a species, called a *cline*, that parallel the environmental gradient.

Polymorphism is a situation in which two or more distinct forms, or *morphs*, coexist in a population. In *balanced polymorphism*, the relative frequencies of the two morphs do not change much over many generations. Each morph may have an adaptive advantage in a particular season or habitat.

The extent of genetic variation is evident in the molecular differences found by using biochemical methods such as electrophoresis to compare the protein products of gene loci within individuals in a population.

All new alleles originate by mutations, most of which occur in somatic cells and cannot be passed on to the next generation. Mutations that alter a protein enough to affect its function are more often harmful than beneficial. Random changes in a genome that has been refined over thousands of generations are not likely to improve the genetic makeup of an individual. Rarely, however, a mutation may result in an individual who is better adapted to the environment; or a mutation already present in the population may be selected for when the environment changes.

Chromosomal mutations also are most often deleterious. Occasionally, a translocation may bring alleles together that are beneficial in combination. A *supergene* is a cluster of genes on a chromosome that have a cooperative function. An inversion of a supergene may preserve it by preventing crossing over with its noninverted homologous section.

Duplications that do not upset the genetic balance within cells may provide an expanded genome with superfluous loci that could eventually take on new functions by mutation.

Most of the genetic variation present in a population is due to the unique recombinations of existing alleles that each individual inherits from the gene pool. Crossing over and independent assortment produce gametes with a great deal of genetic variation, and each zygote has a unique assortment of genes from two parents. Animals and higher plants depend on sexual recombination for the genetic variation needed for adaptation.

Mutation can be a source of genetic variation, however, for bacteria and microorganisms with very short generation times. A new beneficial mutation can increase in frequency rapidly in a bacterial population that is growing by the asexual expansion of clones.

Natural selection selects for favorable genotypes and tends to eliminate others, setting a trend toward genetic uniformity. The diploid character of most eukaryotes maintains genetic variation by the hiding of alleles from selection when present with a dominant allele in the heterozygous form, perpetuating a huge pool of alleles that could be selected for should the environment change.

Heterozygote advantage, when individuals heterozygous for a certain gene have a selective advantage, tends to maintain both alleles at relatively high frequencies. In the case of sickle-cell anemia in countries with malaria, a lethal recessive allele is maintained in a population due to the malaria-resistant advantage of the heterozygote.

Hybrid vigor may be seen in plants when two highly inbred varieties are crossbred. The crossbreeding may segregate harmful recessive alleles that were homozygous in the inbred varieties, and produce a heterozygote advantage at other loci.

Balanced polymorphism may be maintained in a population when diverse individuals have increased reproductive success. In the land snail, *Cepaea nemoralis*, each of the distinctively colored morphs is camouflaged for a particular section of the habitat of the population. The reduced predation of these morphs in the various habitats maintains genetic variability within the population. The polymorphism of female African swallowtail butterflies that resemble several different species of noxious butterflies increases the effectiveness of the mimicry.

Some of the diversity seen in populations may be called *neutral variation*, genetic variations that do not result in a selective advantage for some individuals over others. Many of the mutations affecting the beta chain of human hemoglobin do not appear to confer a selective advantage or disadvantage. It is impossible to estimate how much variation is neutral, since it cannot be shown that an allele brings no benefits at all to an organism.

Adaptive Evolution

Adaptive evolution is a combination of the chance occurrence of new genetic variations by mutation and sexual recombination, and the selection of those chance variations that fit organisms to their environments.

Darwinian fitness is measured by the relative contribution of an individual to the next generation's gene pool. Success is determined, not simply by survival, but also by the number of offspring produced, called reproductive success.

The fitness of specific genotypes can sometimes be assigned relative values. The most fecund variants are said to have a fitness of 1.0, whereas the fitness of another genotype is measured as the percentage of offspring it produces compared with the most fit variant. The *selection coefficient* is the difference between the two fitness values, a measure of the selection against the inferior variant. A lethal gene in a homozygous state would have a coefficient of 1.0.

The rate at which a deleterious allele declines in a population depends both on the selection coefficient and on whether the allele is dominant or recessive. Harmful recessives are rarely eliminated due to heterozygote protection; they can occur within the heterozygote without being selected against. Likewise, beneficial recessives increase slowly because they must occur as a homozygote to be selected for. Dominant alleles can be selected for and against much more rapidly.

Selection acts on phenotype, the physical traits of an organisms, and indirectly adapts a population to its environment by selecting for and maintaining favorable genotypes in the gene pool.

Genes may have pleiotropic effects, some of which may be positive while others may be negative. Some traits are polygenic, influenced by several genes. Selection for or against alleles that are pleiotropic or involved in polygenic traits is not straightforward.

Selection acts on an organism that is an integrated composite of many phenotypic features. The fitness of any allele depends on the entire genetic context of the individual. Genes that have related functions make up a *coadapted gene complex*. The integrated development of an organ requires coadaptation of alleles at many gene loci.

Under the influence of the environment, genotypes determine phenotypes within a range of possibilities, known as the *norm of reaction*. The norm of reaction can be very specifically determined, as in blood groups, or quite broadly variable, as in mental abilities and behavioral patterns.

Three common adaptive trends can result from natural selection, seen especially for quantitative traits determined by many gene loci. *Stabilizing selection* acts against extreme phenotypes and favors intermediate variants. *Directional selection* occurs most frequently during periods of environmental change. It selects for individuals that deviate from the average and shifts the frequency curve for some phenotypic trait. *Disruptive selection* occurs when the environment favors individuals on both extremes of a phenotypic range. Balanced polymorphism may be an example of disruptive selection.

Sexual dimorphism is the distinction between males and females on the basis of secondary sexual characteristics such as differences in size, plumage, or antlers. In vertebrates, the male is usually the showier sex. Sexual selection is the selection for traits that may not be adaptive to the environment, but do enhance reproductive success by increasing the individual's success in attracting a mate.

There are at least four reasons why evolution does not produce perfect organisms. First, each species has evolved from a long line of ancestral forms, many of whose structures have been co-opted for new situations. Second, adaptations are often compromises between the needs to do several different things, such as swim and walk, be fast and strong. Third, the evolution that occurs as the result of chance events is not adaptive. Fourth, natural selection can act on only those variations that are available; new genes do not arise when needed.

Natural selection is an agent of change and also an agent of the status quo. Stabilizing selection probably acts most of the time, resisting maladaptive change. Evolution occurs in spurts, when a population is faced with a major change in environment or genome. The population either becomes extinct or evolves.

STRUCTURE YOUR KNOWLEDGE

1. For the Hardy–Weinberg equation, explain what p and q are, and how they can be used to predict the frequencies of the genotypes of the next generation. Make sure you understand both how to use the Hardy–Weinberg equation and why it works.

2. Create a concept map that organizes your understanding of the possible causes of microevolution.

3. Natural selection tends to work toward genetic unity; the genotypes that are most fit produce the most offspring and increase the frequency of adaptive alleles in the population. Yet there remains a great deal of variability within the populations of a species. Describe some of the factors that contribute to this genetic variability.

TEST YOUR KNOWLEDGE

MULTIPLE CHOICE: *Choose the one best answer.*

1. A gene pool consists of
   a. all the genes that are present in a given species.
   b. the total of all the alleles present in a population at a given time.
   c. all the genes within a cline.
   d. the frequencies for the alleles of a gene within a population.

2. Assuming the conditions of the Hardy–Weinberg equilibrium are met,
   a. the gene pool should remain constant from one generation to the next.
   b. the genotype frequencies for a population should remain constant.
   c. the allele frequencies within a population should not change over time.
   d. All of the above should be true.

3. If a population has the following genotype frequencies, what are the allele frequencies? AA = 0.42, Aa = 0.46, aa = 0.12.
   a. A = 0.42, a = 0.12
   b. A = 0.88, a = 0.12
   c. A = 0.65, a = 0.35
   d. A = 0.6, a = 0.4

4. In a population with two alleles, B and b, the allele frequency of B is 0.8. What would be the frequency of heterozygotes if the population is in Hardy–Weinberg equilibrium?
   a. 0.8
   b. 0.16
   c. 0.32
   d. 0.64

5. In a population that is in Hardy–Weinberg equilibrium, 16% of the population shows a recessive trait. What percent shows the dominant trait?
   a. 84%
   b. 36%
   c. 48%
   d. 72%

6. Genetic drift is likely to be seen in
   a. a population that has a high migration rate.
   b. a population that has a high mutation rate.
   c. a population in which there is assortive mating.
   d. a population that is very small.

7. The bottleneck effect usually results in
   a. reduced genetic variability.
   b. many alleles that are fixed in a population.
   c. a small population that is then subject to genetic drift.
   d. all of the above.

8. Gene flow usually results in
   a. populations that are better adapted to the environment.
   b. an increase in sampling error in the formation of the next generation.
   c. a gain or loss of alleles caused by migration.
   d. nonassortive matings.

9. The existence of two distinctly colored forms in a species is known as
   a. geographic variation.
   b. sexual selection.
   c. heterozygote advantage.
   d. polymorphism.

10. Assortive mating may result in
    a. a change in allele frequency.
    b. an increase in the gene loci that are homozygous.

c. sexual selection.
d. neutral variations.

11. Variations in a species existing over a large temperature range may
    a. result in stabilizing selection.
    b. be a cline if the variations parallel the temperature change.
    c. result in directional selection, favoring the milder climate.
    d. increase gene flow.

12. Mutations are rarely the cause of microevolution because
    a. they are most often harmful and do not get passed on.
    b. even if they are not harmful, they may be masked in the diploid condition and thus not be able to be selected for.
    c. they occur very rarely.
    d. of all of the above.

13. The rate at which a harmful allele declines in a population depends on
    a. whether the allele is dominant or recessive.
    b. the selection coefficient for the genotypes in which it occurs.
    c. the frequency of the allele in the population.
    d. all of the above.

14. The phenotype that a genotype determines may vary due to
    a. the occurrence of pleiotropic genes.
    b. the effect of the environment.
    c. a very narrow norm of reaction.
    d. coadapted gene complexes.

15. Disruptive selection may result in
    a. balanced polymorphism.
    b. one phenotype gradually disappearing from the population.
    c. sexual dimorphism.
    d. a high selection coefficient.

16. Sexual selection will
    a. result in individuals better adapted to the environment.
    b. increase assortive mating.
    c. select for traits that enhance an individual's chances of mating.
    d. result in stabilizing selection.

17. Which of these is not a reason why evolution does not produce perfect organisms?
    a. Much of evolution may occur as the result of chance.
    b. Adaptations are often compromises between different needs.

c. Selection works on the structures available at the time.

d. Unless conditions are always changing, equilibrium sets in and evolution does not occur.

18. The greatest source of genetic variation in most populations is from
    a. mutations.
    b. recombination.
    c. selection.
    d. polymorphism.

19. A supergene is a
    a. coadapted gene complex located close together on a chromosome.
    b. pleiotropic gene.
    c. polygenic complex.
    d. gene found on a polytene chromosome.

20. A population of a plant lives in an area which is becoming more arid. The average surface area of leaves has been decreasing over the generations. This is an example of
    a. stabilizing selection.
    b. directional selection.
    c. disruptive selection.
    d. heterozygote superiority.

CHAPTER 22

THE ORIGIN OF SPECIES

FRAMEWORK

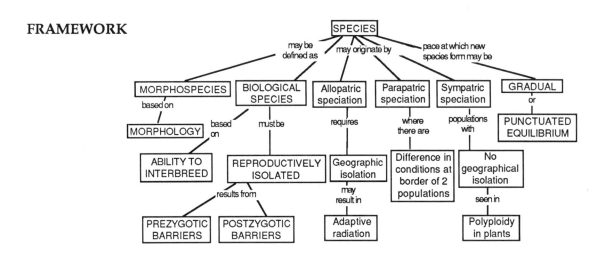

CHAPTER SUMMARY

The origin of species, or *speciation*, is the basis of evolution and biological diversity. *Anagenesis*, or *phyletic evolution*, involves the transformation of an entire population into a different enough form that it is renamed a new species. In *cladogenesis*, or branching evolution, a new species arises from a parent species that continues to exist. Cladogenesis is both the more common pattern of evolution and the process that increases biological diversity. Evolutionary theory attempts to determine the mechanisms by which new species originate.

Defining Species

Taxonomists often find that their scientific classification of local species corresponds to the folk taxonomy of the region. *Species* (a Latin word for "kind" or "appearance") are discrete, identifiable units. Groups that are

determined on the basis of morphology or anatomical features are known as *morphospecies*. The concept of *biological species*, developed by Mayr in 1942, goes beyond the physical differences among species and considers reproductive isolation to be the basis for separating species.

A biological species consists of a population or group of populations of individuals that have the potential to interbreed with each other in nature. A biological species is the largest unit in which gene flow is possible and that is genetically isolated from other populations.

The biological species concept does not work for species that are completely asexual, such as prokaryotes, some protoctists, and fungi. It is also impossible to group extinct species on the criterion of interbreeding. Even in some sexual, geographically neighboring species, it may be difficult to apply the biological species concept. Phenotypically distinct populations that are separated geographically but presumably capable of interbreeding are sometimes called *subspecies*. The gene flow among subspecies may be so circuitous or slight that it is difficult to decide whether some subspecies should be designated as separate species.

Reproductive Isolating Mechanisms

Any mechanism that prevents two species from producing fertile hybrids is a reproductive isolating mechanism that serves to preserve the genetic integrity of a species. *Prezygotic barriers* function before the formation of a zygote by preventing the mating between species or the successful formation of a zygote. Should a hybrid zygote form, *postzygotic barriers* prevent it from developing into a fertile adult.

Prezygotic barriers include *ecological isolation*, in which two species may live in the same area but occupy different habitats, and *temporal isolation*, in which two species breed at different times. *Behavioral isolation* includes courtship rituals and physical or chemical signals that attract mates of the proper species. *Pheromones* are volatile chemical compounds that may serve as sexual attractants.

Anatomical incompatibility may provide a *mechanical isolation* for closely related species. Mechanical barriers also may be present in flowering plants in which the floral anatomy is adapted to a specific pollinator. Should these other prezygotic barriers fail, *gametic isolation* will usually prevent interbreeding between species because the gametes may not fuse to form a zygote. Gamete recognition may depend on molecular receptors on the egg cell that are specific for complementary molecules on sperm cells. Similar molecular recognition mechanisms are involved in pollen discrimination within flowers.

Postzygotic barriers may include *hybrid inviability*, in which a hybrid zygote fails to develop due to genetic incompatibility, or *hybrid sterility*, in which a viable hybrid individual is sterile, often due to the inability to produce normal gametes in meiosis. Hybrid breakdown is a postzygotic barrier in which the hybrids are viable and fertile, but their offspring are defective or sterile.

When fertile hybrids occasionally successfully mate with one of their parent species, genes may pass between species in a process called *introgression*. This small amount of gene transplantation may increase the reservoir of genetic variation present in a species.

Mechanisms of Speciation

Allopatric speciation occurs when the gene pool of a population is segregated geographically from other populations and follows its own evolutionary course of selection, genetic drift, and mutation. *Sympatric* specia-

tion may occur when a radical change in the genome of a subpopulation results in its reproductive isolation in the midst of the parent population. If gene flow is very slight at the boundary between two populations, *parapatric* speciation may occur when the gene pools of both populations diverge without the dilution of genes from their neighbors.

Geological changes can cause sympatric populations to become allopatric. A separated population is more likely to undergo speciation if it is small; the evolutionary inertia of a large gene pool slows the process of microevolutionary changes in gene frequencies. The effects of both chance events and selection pressures are amplified in a small gene pool.

The geographic isolation of a small population at the fringe of the parent population's range may result in speciation because of four factors. First, the gene pool of the peripheral isolate probably represents the extremes of any clines in the population, and thus may initially differ from the parent population. If the isolated population is small, the founder effect also may have produced a population whose gene pool differs initially from the parent pool. Second, genetic drift in the small gene pool may cause phenotypic divergence from the parent population by chance. Third, natural selection can change a small gene pool rapidly. The genetic contribution of favored individuals will be proportionally great in a small gene pool. Fourth, selection pressures are likely to be different, and possibly more severe, on the fringe of a population's range, pushing the peripheral isolate's gene pool in a different direction than that of the parent population. These factors, however, do not guarantee speciation. Most pioneer populations probably become extinct before they change enough to become new species.

When allopatric populations become sympatric again, they may have developed an intrinsic reproductive isolation mechanism that prevents them from successfully mating; they may not be reproductively isolated and may continue as one species; or the result of their coming together may push them to become more different and to complete speciation. Individuals of the two populations that are most different will compete the least with the other population and may have a reproductive advantage. Natural selection for individuals with differing requirements for food, habitats, or other resources results in the divergence of two populations in a phenomenon called *character displacement*.

Allopatric speciation may occur on island chains, where small founding populations may diverge in isolation. A single ancestral species of finch probably gave rise to the 13 species of finches now found on the Galapagos. *Adaptive radiation* results in the formation of numerous species from an ancestral species introduced to a new and diverse habitat.

Parapatric populations occupy separate ranges that

abut along a common border. This border may represent a region of change for some important environmental factor, such that the gene pools of the two populations have different selective pressures. The two gene pools may diverge enough that the limited breeding at the border may not maintain reproductive compatability. Hybrids produced at the border may be less well adapted to conditions on either side, and postzygotic barriers may facilitate speciation. Parapatric speciation is theoretically possible, although no specific instances can be cited.

In sympatric speciation, reproductive isolation is achieved without geographic isolation. New plant species can result from the production and fusion of mutant gametes that have formed by nondisjunction. In *autopolyploidy*, a single species doubles its chromosome number to the tetraploid state through the fusion of diploid gametes. Tetraploids can fertilize themselves or mate with other tetraploids but cannot mate with diploids from the parent population. This postzygotic barrier results in reproductive isolation in just one generation. De Vries has documented this form of speciation in the evening primrose.

Polyploid species can also arise from *allopolyploidy*. A hybrid plant may be sterile due to difficulties in the meiotic production of gametes. Should the hybrid undergo nondisjunction, the *4N* chromosome number will then be able to produce gametes with complete sets of chromosomes, able to fuse with gametes from the same plant or other hybrid tetraploids, creating a new species that is reproductively isolated from both parent species.

Speciation of polyploids has been frequent and important in plant evolution. Plant geneticists are now hybridizing plants and using chemicals to induce nondisjunction in order to create new, specially adapted polyploid species.

Gradual and Punctuated Interpretations of Speciation

The traditional evolutionary concept of the origin of species involves the gradual divergence of species, with each new species evolving continuously over long spans of time. The fossil record, however, provides few cases of gradually transitioning forms. Rather, new forms appear rather suddenly, persist unchanged for a long time, and then disappear. The fossil evidence for evolution seems to support the theory of *punctuated equilibrium*, in which long periods of stasis are punctuated by episodes of relatively rapid change and speciation.

Speciation both by polyploidy in plants and by small allopatric populations may occur fairly rapidly. In geological time, a few thousand years for a species to

evolve is small compared to the millions of years a successful species may remain in existence.

The degree to which a species changes after its origin may be very little if it remains in a stable environment. Once selection has resulted in a complex of coadapted genes, new changes in the genome will tend to be disruptive. Some gradualists maintain, however, that fossils report stasis only in external anatomy and that changes in internal anatomy, physiology, and behavior go unrecorded.

Natural selection is the most important mechanism underlying the modifications a species acquires. The act of speciation, however, often may be a matter of chance events, random mutations, or genetic drift in a small population. According to this view, reproductive isolation may occur through random changes, and then natural selection acts on the isolated gene pool to create adaptive changes that may allow the new species to survive.

STRUCTURE YOUR KNOWLEDGE

1. How are speciation and microevolution different?
2. What is probably the key event in the origin of a species? How might this event occur?
3. Compare the gradual and punctuated equilibrium theories of evolution. Which theory seems to have the most evidence supporting it? Describe this evidence.

TEST YOUR KNOWLEDGE

MULTIPLE CHOICE: *Choose the one best answer.*

1. Most new species probably have arisen by
   a. anagenesis.
   b. cladogenesis.
   c. phyletic evolution.
   d. both a and c are correct.

2. Which of the following is not a type of *intrinsic* reproductive isolation?
   a. geographical isolation
   b. behavioral isolation
   c. mechanical isolation
   d. gametic isolation

3. The individuals placed in a morphospecies and a biological species may differ because
   a. organisms that look different may be able to interbreed.
   b. organisms that do not breed in nature may do so in captivity.

c. hybrids formed from the mating of two morphospecies are sterile.

d. anatomical features do not provide a reliable basis for grouping organisms.

4. The largest unit in which gene flow is possible is
   a. a population.
   b. a species.
   c. a genus.
   d. a subspecies.

5. Subspecies are usually
   a. separated geographically.
   b. phenotypically distinct.
   c. assumed to be capable of interbreeding.
   d. all of the above.

6. The reproductive barrier that maintains the species boundary between horses and donkeys is
   a. mechanical isolation.
   b. gametic isolation.
   c. hybrid inviability.
   d. hybrid sterility.

7. Introgression occurs when
   a. a hybrid successfully breeds with an individual of a parent species.
   b. hybrids successfully breed with each other.
   c. individuals from two species successfully breed.
   d. all of the above take place.

8. Gametic isolation is a
   a. prezygotic barrier.
   b. postzygotic barrier.
   c. mechanical isolating mechanism.
   d. hybrid breakdown isolating mechanism.

9. Speciation is most likely to occur in
   a. large sympatric populations with a lot of genetic variability.
   b. small parapatric populations.
   c. small allopatric populations.
   d. a large parent population separated from a small peripheral isolate.

10. Due to character displacement, two closely related species will be most different when their ranges are
    a. allopatric.
    b. sympatric.
    c. parapatric.
    d. very small.

11. Adaptive radiation may occur
    a. when a small population is introduced to a new and diverse habitat.
    b. when numerous invasions and allopatric speciations take place on an island chain.
    c. when character displacement takes place as once allopatric populations become sympatric again.
    d. through all of the above.

12. When a species doubles its chromosome number due to the fusion of diploid gametes, it is called
    a. allopolyploidy.
    b. autopolyploidy.
    c. introgression.
    d. nondisjunction.

13. According to advocates of the punctuated equilibrium theory,
    a. natural selection is unimportant as a mechanism of evolution.
    b. given enough time, most existing species will branch gradually into new species.
    c. a new species accumulates most of its unique features as it comes into existence and changes little for most of its duration.
    d. evolution is punctuated by times of mass extinctions and the evolution of new species.

14. The act of speciation depends on
    a. natural selection.
    b. allopatric populations.
    c. chance events and reproductive isolation.
    d. changing environments.

15. The long periods of stasis found in many species may be due to
    a. evolutionary inertia of a large gene pool.
    b. the existence of a large number of coadapted genes.
    c. relatively stable environments.
    d. all of the above.

16. For which of the following is the biological species concept least appropriate?
    a. plants
    b. prokaryotes
    c. two subspecies that are geographically isolated
    d. field biologists

CHAPTER 23

MACROEVOLUTION

FRAMEWORK

This chapter considers the major events and evolutionary trends that have led to the biological diversity of today. A goal of *systematics* is to determine the phylogenetic history of species. The branch of systematics called *taxonomy* names and classifies species according to their presumed evolutionary relationships. *Phylogenetic trees* are constructed from evidence gathered from the *fossil record* and from anatomical and molecular *homologies*.

Mechanisms for macroevolution include the gradual modification of *preadapted structures* for new functions, alterations in *regulatory genes* that result in major morphological changes, evolutionary trends resulting from *species selection*, and *adaptive radiations* as new *adaptive zones* appear when evolutionary novelties develop or mass *extinctions* occur. *Continental drift* and other major geological events have shaped the direction of macroevolution. The relationship between *microevolution* and *macroevolution* and the role of *natural selection* and *chance* in macroevolution are controversial topics in contemporary evolutionary theory.

CHAPTER SUMMARY

The term *macroevolution* refers to major events and evolutionary trends in the history of life. When biologists study macroevolution, they consider the origin of the major taxonomic groups, the novel biological designs associated with these taxa, and the mechanisms that may have produced these evolutionary developments.

The Record of the Rocks

Paleobiologists reconstruct evolutionary history by studying the succession of organisms found in the fossil record. Paleontologists focus on animal fossils; paleobotanists study the fossils of plants.

Fossils are preserved impressions or remnants of past organisms. Sometimes enough organic material is retained in the fossil that biochemical analysis and electron microscopic study of cells can be done. Hard parts of animals, such as bones, teeth, or shells, may remain as fossils. Petrification, the replacement of tissues with dissolved minerals, may turn the fossil to stone. Molds of organisms, left when they were covered by mud or sand, are a common type of fossil.

Sedimentary rocks, formed from the compression of deposits of mud or sand, are the richest source of fossils. Layers, or strata, of rock form during the varying periods of sedimentation within bodies of water. The order in which fossils appear in the strata of sedimentary rocks indicates the relative ages of the fossils.

Index fossils are used to correlate strata from different locations. Gaps in the sequence of fossils may occur at a location due to changes in sea level or erosion. Index fossils, such as shells of widespread animals, allow geologists to develop a composite picture of a consistent sequence of geological periods. There are four geological eras, the Precambrian, Paleozoic, Mesozoic, and Cenozoic, that are delineated by major transitions in the fossils found in the rocks. The eras are subdivided into epochs associated with particular evolutionary developments.

Radioactive dating is used to determine the age of rocks and fossils. During an organism's lifetime, it accumulates radioactive isotopes of certain elements in proportions equal to the relative abundances of the isotopes in the environment. After the organism dies, the isotopes begin to decay at a fixed rate, known as its *half-life*, or the number of years it takes for one-half of the radioactive isotopes present in a specimen to decay. Carbon-14 dating is used for determining the age of relatively young fossils; potassium-40, with a half-life of 1.3 billion years, can be used to date rocks hundreds of millions of years old.

During an organism's life, only *l* amino acids are synthesized, but following death, *l* amino acids are slowly converted to *d* amino acids. The proportion of *l* and *d* forms of amino acids can be used to date some fossils.

The formation of a fossil is an unlikely occurrence. The incompleteness of the fossil record is understandable considering that a large number of species that lived probably left no fossils, most fossils that form are destroyed, and only a fraction of existing fossils have been found. Although more transitional fossils have been found since Darwin's day, there is still a scarcity of fossils that show a progression of changes from ancestral to present-day forms. Some paleontologists interpret the discontinuities in the fossil record as evidence for the punctuated equilibrium model of evolution.

Systematics: Tracing Phylogeny

Phylogeny is the evolutionary history of a species. A phylogenetic tree is a diagram of the proposed evolutionary relationships of various groups. *Systematics* is the branch of biology concerned with the diversity of life and its phylogenetic history.

Taxonomy involves the identification and classification of species. Linnaeus developed a system of taxonomy that assigned to each species a two-part Latin name—a *binomial* consisting of a genus and species name. Both words of the binomial are italicized, and the first letter of the name of the genus is capitalized. Linnaeus also developed a hierarchy of classifications that organizes similar groups into more general categories proceeding from species to *genus, family, order, class, phylum,* and finally to *kingdom.*

Taxonomy describes diagnostic characteristics for closely related organisms and assigns names to new species. Species are ordered into the *taxa,* or units, of the increasingly broader taxonomic categories. The names for all taxa at the genus level and higher are capitalized.

Species exist in nature as a biological identifiable group, connected by interbreeding and reproductively isolated from other organisms. The assignment of taxa to the next category is often a subjective enterprise, depending on the distinctions the taxonomist deems important. A goal of classification is to reflect the evolutionary relationships of species. Each taxon should be monophyletic, meaning that it should contain only species that are derived from a single ancestor. A polyphyletic taxon would include groups having different ancestry.

Species are generally classified into higher taxa based on similarities in morphology and other characteristics. *Homology* is likeness based on a shared ancestry; *analogy* is similarity due to convergent evolution. In *convergent evolution,* unrelated species develop similar features because they have similar ecological roles and natural selection has chosen for analogous adaptations.

Phylogenetic trees are built on homologous similarities. In general, the more homology between two species, the more closely they are related. The issue can be confused when adaptive radiation results in large differences and convergence creates misleadingly similiar structures. If two similar structures are fairly complex, they are less likely to have had separate origins.

A powerful taxonomic tool is the comparison of sequences of amino acids in proteins and nucleotides in DNA. The protein cytochrome *c* is found in all aerobic organisms. Comparisons of the amino acid sequences of this protein from many species show that differences increase as the groups are more taxonomically distant. Phylogeny based on cytochrome *c* is consistent with evidence from comparative anatomy and the fossil record.

DNA–DNA hybridization is a technique that compares the genomes of two species. DNA is extracted and heated until it "melts" into separate strands. Single-stranded DNA from two species is mixed and base pairing occurs between complementary sequences. The hybrid DNA is reheated, and the temperature required for the strands to separate is indicative of the extent of the base pairing and thus of the homology between DNA sequences of the species. Evidence from this technique generally supports phylogenetic relationships established by other methods.

Some proteins seem to change or evolve at consistent rates. Thus, the number of amino acid substitutions in homologous proteins is proportional to the time elapsed from when two species branched off from a common ancestor. DNA–DNA hybridization also can be used to date phylogenetic branchings. Dates obtained by this method are generally consistent with the fossil record and may be more closely correlated with the time species have been separated than are morphological differences.

The consistent rate of protein change and the rate of DNA divergence indicate that the accumulation of neutral mutations may change the genome as a whole more than do changes brought about by selection and adaptation. Disagreement about the extent of neutral mutations has caused some evolutionists to claim that molecular clocks can determine only the sequence of branches in phylogeny and not the actual dates for the origin of taxa.

Phylogenetic trees indicate the relative time of origin of different taxa and the degree of divergence or difference that develops between branches. *Phenetics* is a school of taxonomy that determines taxa strictly on the basis of measurable similarities and differences determined for as many anatomical characteristics as possible.

Cladistics is a taxonomic school that classifies organisms according to the order in time that clades, or branches, arise. Phylogeny is diagrammed on a cladogram, a series of dichotomous forks, each of which is defined by novel homologies for the species on that branch. A *shared primitive character* is common for all the species to be grouped, and *shared derived characters* are identified as homologies that evolved after a branch point and apply to all species on that branch.

In the cladistic approach, in which the timing of branch points is central, birds are closer relatives of crocodiles than crocodiles are of lizards and snakes. The branch point between birds and crocodiles is more recent, and they share derived characters not found in snakes and lizards. According to cladistics, the class *Aves* does not exist because birds are found within the clade of reptiles. The degree of morphological difference between branches is not a consideration in cladistics; only the time and sequence of evolutionary origin are noted.

Classical evolutionary taxonomy considers both the homology of structures and the sequence of branching. In cases of conflict between these two characteristics, a subjective decision is made. According to classical taxonomy, even though crocodiles may share a closer phylogenetic branch with birds than with snakes and lizards, they are grouped with reptiles on the basis of homologies.

The rate of morphological change is not constant, so that phenetics and cladistics will never agree. Taxonomic differences arise due to the need for compromise between a classification system that accurately reflects genealogies and one that provides a useful filing system for organizing the diversity of species.

Mechanisms of Macroevolution

Evolutionary novelties that define higher taxa may evolve by the gradual refinement of existing structures for new functions. *Preadaptation* is the term for structures that evolve in one context and are co-opted for another function. Feathers and wings may have developed as insulation and prey-capturing structures that were then shaped into structures adapted for flight. Along with the Darwinian tradition of gradualism, preadaptation provides a mechanism by which novel designs arise slowly through many small changes in structures that are then adapted for another function.

In regulatory genes, slight changes in their function may create major changes in development and morphology. Since each regulatory gene may influence hundreds of structural genes, evolutionary novelties could arise fairly rapidly due to a change in a regulatory gene.

Allometric growth involves different rates of growth in various parts of the body. These differences result in the final shape of the organism. A minor genetic alteration that affects allometric growth could have a major effect.

Novel organisms also could be produced by genetic changes that affect the timing of development. *Paedomorphosis* is the retention in the adult of juvenile traits of ancestral organisms. The continuation of growth of the human brain for several years longer than the chimpanzee brain is a retention of a juvenile trait that has greatly influenced human characteristics.

Evolutionary trends rarely occur in the fossil record as a sequence of gradually changing intermediate forms. *Equus* is not the direct result of trends of increasing size, reduction in the number of toes, and changes in dentition proceeding from the dog-sized ancestor *Hyracotherium* to the modern horse of today. Many different species related to the ancestral form appeared in the fossil record, remained fairly unchanged, and then become extinct.

Branching evolution can result in a trend even though some of the new species go against the trend. According to Stanley's model of *species selection*, an evolutionary trend is analogous to a trend in a population produced by natural selection when the best-adapted individuals are most successful reproductively. The species that live the longest before extinction and speciate most (produce the most offspring) determine the direction of the trend.

Biogeography, the geographic distribution of species, is correlated with the geological history of Earth. The continents rest on great plates of crust and upper mantle that float on the plastic zone of the mantle. These plates shift and move, creating earthquakes, volcanoes, and mountains in regions where plates abut, and affecting the geographical relationships of populations.

Large-scale continental drift brought all the land masses together into a supercontinent named *Pangaea* about 250 million years ago, near the end of the Paleozoic era. This tremendous change undoubtedly had a great environmental influence as shorelines were eliminated, the oceans got deeper, large shallow coastal areas were drained, and the climate of the land mass changed. Species that had evolved in isolation came together and competed, many species became extinct, and new opportunities for remaining species became available.

About 180 million years ago, during the early Mesozoic era, Pangaea broke up and the continents drifted apart, creating a huge geographic isolation event. The current biogeography reflects this separation.

Major adaptive radiations have occurred during the early history of some taxa when the evolution of a novel characteristic opened a new *adaptive zone*, or way of life with unexploited opportunities. Between the Precambrian and Paleozoic eras, around 570 million years ago, a large increase in the diversity of sea animals took place. Nearly all the animal phyla that exist today evolved during the first 10 to 20 million years of the Cambrian, the first period of the Paleozoic era. The origin of shells and skeletons in a few taxa opened an adaptive zone by making more complex designs possible.

An empty adaptive zone can be exploited only if appropriate evolutionary novelties arise; and novelties that do arise cannot enable organisms to move into adaptive zones that do not exist or are filled. Mammals existed at least 75 million years before their first major adaptive radiation, which occurred in the early Cenozoic era and has been linked to the ecological void left by the extinction of the dinosaurs. Mass extinctions are often followed by new adaptive radiations.

A species may become extinct due to a change in its physical or biological environment. An evolutionary change in one species may affect other species in the community. Extinction, inevitable in a changing world, usually occurs at a rate of between 2.0 and 4.6 families per million years. During periods of major environmental change, mass extinctions may occur.

There have been at last five mass extinctions in the history of Earth. The Permian extinctions, which define the boundary between the Paleozoic and Mesozoic eras (about 225 million years ago), claimed over 90% of the species of marine animals. This extinction occurred around the time that Pangaea formed and may be related to the changes associated with that event.

The Cretaceous extinction, which marks the boundary between the Mesozoic and Cenozoic eras about 65 million years ago, claimed over one-half the marine species, many families of terrestrial plants and animals, and nearly all species of dinosaurs. The climate was cooling during that time, and shallow seas were receding from continental lowlands.

There is also evidence that an asteroid or comet collided with the earth during the Cretaceous extinctions. Separating Mesozoic and Cenozoic sediments is a thin layer of clay enriched in iridium, an element rare on Earth but common in meteorites. Walter and Luis Alvarez have suggested that this layer is the fallout from a huge cloud of dust that was created when an asteroid hit the earth. This cloud, similar to what is projected in a nuclear winter, would have blocked light and severely affected weather for several months.

The linking of the evidence for an asteroid and the mass extinctions occurring at this time does not indicate cause and effect. Many scientists, including Stanley, believe that changes in climate due to continental drift and other processes were sufficient to account for the mass extinctions that took place.

Is a New Synthesis Necessary?

The modern synthesis, which has dominated evolutionary theory for 50 years, combines contributions from paleontology, systematics, and population genetics, along with recent discoveries in molecular biology, to reaffirm the Darwinian view of life. This paradigm maintains that the gradual accumulations of many small changes occurring over vast periods of time can result in large-scale evolutionary changes. Microevolution, changes in gene frequencies in populations, is sufficient to explain most macroevolution, and natural selection is seen as the major cause of evolution at all levels.

A current evolutionary debate concerns both the rates of evolution (gradualism versus punctuated equilibrium) and the relative contribution of microevolution to macroevolution. Most evolutionists agree that natural selection is the mechanism of adaptation that fine-tunes a population to its environment. But some scientists favor a hierarchical theory, which maintains that events that lead to speciation and episodes of macroevolution may have little to do with adaptation.

According to the hierarchical theory, the beginnings of most new species result from chance geographic isolation, genetic accidents, and genetic drift in small, isolated populations. Macroevolution related to continental drift and mass extinctions is believed to have affected biological diversity as much as has the gradual adaptation that results from selection acting on populations' gene pools. And most evolutionary trends occur not by phyletic transitions caused by microevolution, but by species selection—the differential survival and speciation of separate species that change little after they come into existence.

The modern synthesis does not claim that evolution is always gradual or that chance events do not change gene pools. The major difference between the modern synthesis and the hierarchical theory is the relative importance of the different evolutionary mechanisms of natural selection and chance events, and the connection between microevolution and macroevolution.

STRUCTURE YOUR KNOWLEDGE

1. Organize a concept map that describes the major activities and objectives of taxonomy.

2. Compare microevolution and macroevolution. To what extent are these two related?

TEST YOUR KNOWLEDGE

MULTIPLE CHOICE: *Choose the one best answer.*

1. The richest source of fossils is found
   a. where roadways cut through rock layers.
   b. along gorges.
   c. within sedimentary rock strata.
   d. encased in volcanic rocks.

2. Which of the following is least likely to leave a fossil?
   a. a soft-bodied land organism, such as a slug
   b. a marine organism with a shell, such as a mussel
   c. a vascular plant embedded in layers of mud
   d. a freshwater snake

3. Index fossils are fossils of
   a. unique organisms that are used to determine the relative rates of evolution in different areas.
   b. widespread organisms that allow geologists to correlate strata of rocks from different locations.
   c. transitional forms that link major evolutionary groups.
   d. extinct organisms that mark the separation of different eras.

4. The half-life of Carbon-14 is 5600 years. A fossil that has one-eighth the normal proportion of C-14 to C-12 is probably
   a. 2800 years old.
   b. 11,200 years old.
   c. 16,800 years old.
   d. 22,400 years old.

5. Which of the following is not a reason for an incomplete fossil record?
   a. Fossils form only when organisms are buried in sand or mud.
   b. Erosion and other processes may destroy fossils.
   c. The large majority of fossils are not found.
   d. Some of the transitional forms expected by evolutionists may not exist due to a punctuated equilibrium pace of evolution.

6. The relatively youngest fossils
   a. will have the greatest amount of *d* amino acids.
   b. will be found in the highest strata.
   c. will have the greatest amount of DNA–DNA hybridization.
   d. all of the above.

7. In the binomial, *Homo Sapiens,*
   a. *Homo* is the name of the genus.
   b. *Sapiens* should not be capitalized.
   c. the words are italicized because they are Latin.
   d. all of the above are correct.

8. Related families are grouped into the next highest taxon called a
   a. class.
   b. order.
   c. phylum.
   d. genus.

9. The only biologically identifiable group in nature is the
   a. kingdom.
   b. phylum.
   c. genus.
   d. species.

10. When two structures are homologous,
    a. they indicate that the species have evolved from a common ancestor.
    b. they share common embryological development patterns.
    c. they still may be adapted for different functions in the two species.
    d. all of the above are correct.

11. DNA–DNA hybridization compares the genomes of two species by
    a. determining the temperature at which hybrid DNA separates.
    b. determining the extent of base pairing between single-stranded DNA of two species.
    c. establishing the degree of DNA homology and thus the degree of relatedness between two species.
    d. all of the above.

12. Convergent evolution may result
    a. when species have similar ecological roles.
    b. when homologous structures are adapted for different functions.
    c. from adaptive radiation.
    d. all of the above may be correct.

13. Which of the following is not a means of dating evolutionary branch points?
    a. the fossil record
    b. phenetics
    c. the number of amino acid substitutions in homologous proteins
    d. DNA–DNA hybridization

14. The development of feathers in the ancestors of birds is an example of
    a. convergent evolution.
    b. divergent evolution.

14. The development of feathers in the ancestors of birds is an example of
    a. convergent evolution.
    b. divergent evolution.
    c. preadaptation.
    d. adaptive radiation.

15. Shared derived characters
    a. are used to characterize all the species on a branch of a cladogram.
    b. are common for all groups on a major line of a cladogram.
    c. are homologous structures that develop during adaptive radiation.
    d. are used to indicate the degree of morphological differences between clades.

16. Evolutionary trends in the fossil record may be the result of
    a. adaptive radiation.
    b. mass extinctions.
    c. changing environments.
    d. species selection.

17. Continental drift may explain
    a. mass extinctions at the end of the Paleozoic era.
    b. the biogeography of species seen today.
    c. the formation of volcanoes, earthquakes, and mountains.
    d. all of the above.

18. Allometric growth
    a. is the uneven growth of different body parts.
    b. involves paedomorphosis.
    c. results in an evolutionary trend of increasing body size.
    d. results in the phyletic evolution of a change in body size.

19. The differences between the modern synthesis and hierarchical theory of evolution
    a. concern the tempo or pace of evolution.
    b. concern the role of microevolution in macroevolution.
    c. concern the role of chance and natural selection in evolution.
    d. concern all of the above.

20. Punctuated equilibrium is a part of the hierarchical theory because
    a. it emphasizes the role of chance in speciation events before natural selection has a chance to develop adaptations.
    b. it emphasizes phyletic evolution of preadapted structures.
    c. it emphasizes continental drift and mass extinctions as major factors of macroevolution.
    d. it recognizes the central role of natural selection in microevolution.

TRUE OR FALSE: *Indicate T or F, and then correct the false statements.*

1. ____ A monophyletic taxon includes only species that share a common ancestor.

2. ____ The more the sequences of amino acids in homologous proteins vary, the more recently the two species have diverged.

3. ____ Phylogenetic trees determined on the basis of similar structures may be inaccurate when adaptive radiations have created large differences or convergent evolution has created misleading similarities.

4. ____ Phenetics is the school of taxonomy that is concerned with the order in which new groups branch over time on the phylogenetic tree.

5. ____ The retention in the adult of juvenile traits of ancestors is called paedomorphosis.

6. ____ A slight change in the function of a regulator gene may result in a major morphological change because there are so many regulator genes present in a genome.

7. ____ An adaptive zone is the region in which a species can successfully live.

8. ____ The layer of clay enriched in iridium between the sediments of the Mesozoic and Cenozoic eras is indicative of a nuclear winter.

9. ____ According to species selection, the best-adapted species will be selected for by the environment.

10. ____ According to the modern synthesis, natural selection is seen as the major cause of evolution at all levels.

ANSWER SECTION

CHAPTER 20: DESCENT WITH MODIFICATION

Suggested Answers to Structure Your Knowledge

1. The two major components of Darwin's evolutionary theory are that all of life has descended from a common ancestral form and that the modification of that form has been the result of natural selection of the offspring best adapted to the environment.

2. (a) The distribution of species geographically (biogeography) supports evolution because of the similarities among species in close proximity, and the differences in species that are separated by great distances, even if their habitats are similar. (b) The fossil record documents the changes that have occurred in species throughout time, and occasionally provides transitional forms that link different phylogenetic groups. (c) The taxonomic groupings of species into a hierarchy of related groups points to the evolution of species from common ancestors. (d) Comparative anatomy illustrates the relatedness of groups based on their similar structures. Homologous structures and vestigial structures provide evidence of evolution. (e) Comparative embryology illustrates the commonalities of developmental patterns in related groups. It also helps to identify homology of structures that may appear different in adult form. (f) The similarities in the nucleotide sequences of DNA and protein products among related species indicate descent from a common ancestor.

Answers to Test Your Knowledge

Matching:

1. F f	5. D d
2. A b	6. H a
3. G g	7. E c
4. B e	8. C h

Multiple Choice:

1. b	6. c
2. c	7. d
3. a	8. c
4. b	9. d
5. a	10. c

CHAPTER 21: HOW POPULATIONS EVOLVE

Suggested Answers to Structure Your Knowledge

1. In the Hardy–Weinberg equation, p and q refer to the frequencies of two alleles (*A* and *a*) in the gene pool. The frequency or proportion of gametes containing the A allele is equal to p, and the gametes containing the a allele is equal to q. The proportion of offspring resulting from gametes containing two A alleles is (p x p) or p^2. Likewise, aa offspring will equal q^2. Aa individuals can be formed with the A allele coming from the egg or sperm while the a allele is contributed by the sperm or egg. Thus the frequency of Aa offspring is equal to 2pq.

2.

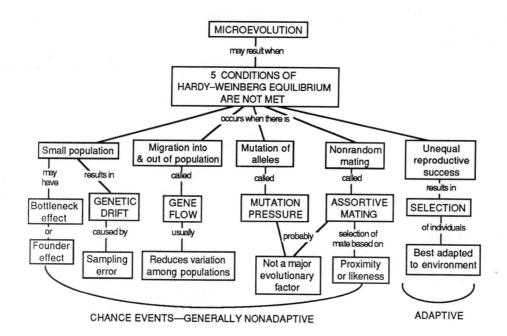

3. Genetic variation is retained within a population by three general mechanisms: diploidy, heterozygote advantage, and balanced polymorphism. Diploidy masks recessive alleles from selection when they occur in the heterozygote. Thus, less adaptive or even harmful alleles are maintained in the gene pool and are available should selection pressures change. In situations in which there is heterozygote advantage or balanced polymorphism, two or more alleles are simultaneously selected for and retained within the gene pool.

Answers to Test Your Knowledge

1. b	6. d	11. b	16. c
2. d	7. d	12. d	17. d
3. c	8. c	13. d	18. b
4. c	9. d	14. b	19. a
5. a	10. b	15. a	20. b

CHAPTER 22: THE ORIGIN OF SPECIES

Suggested Answers to Structure Your Knowledge

1. Speciation is the process by which a new, reproductively isolated species evolves from its predecessor. It is part of the process of evolution and the increase in biological diversity. Microevolution is the process by which changes occur within the gene pool of a population, as a result of either chance events or natural selection. If the makeup of the gene pool changes enough, microevolution may lead to speciation.

2. Reproductive isolation is probably the key event in the origin of a species. Reproductive isolating mechanisms may develop as the result of chance events, and then natural selection may work on the newly isolated gene pool to develop a population adapted for its environment. These chance events may occur because of the geographic separation of a small population (in which genetic drift may play a large role) or chance mutations (as in polyploidy in plants).

3. In the gradualism theory of evolution, small changes accumulate within populations as a result of chance

events and natural selection, and these may lead (particularly as environmental conditions change) to the gradual evolution of new life forms. The punctuated equilibrium theory relies on the fossil evidence that evolution occurs in spurts of relatively rapid change inserted within large periods of stasis. The fossil record seems to indicate that new species appear fairly rapidly and then remain relatively unchanged during their duration on earth. The lack of transitional forms, intermediate between species or between higher taxonomic groups, is further evidence for the punctuated pace of evolution.

Answers to Test Your Knowledge

1. b	5. d	9. c	13. c
2. a	6. d	10. b	14. c
3. a	7. a	11. d	15. d
4. b	8. a	12. b	16. b

CHAPTER 23: MACROEVOLUTION

Suggested Answers to Structure Your Knowledge

1.

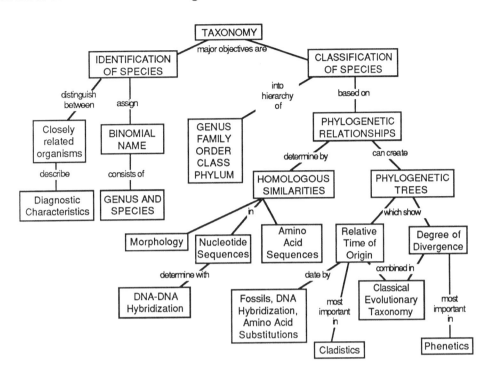

2. Microevolution consists of the changes in gene frequencies that occur within a population over the generations due both to chance events, such as genetic drift, gene flow, and mutation pressure, and to differential reproductive success as a result of natural selection. New species may form as the result of microevolution, or of geographical separations or mutations. Macroevolution includes the major events and trends that have occurred in the history of life as taxa higher than species evolve. Macroevolution can result from the gradual modification of preadapted structures for new functions through natural selection (as in microevolution), or more likely, from chance–related mechanisms such as major geological changes resulting in geographic rearrangements of species, alterations in regulatory genes causing relatively rapid morphological changes, and species selection. Microevolution and macroevolution both involve natural selection and the element of chance. Natural selection appears to play a larger role in microevolution, whereas chance may play the largest role in macroevolution.

Answers to Test Your Knowledge

Multiple Choice:

1. c	6. b	11. d	16. d
2. a	7. d	12. a	17. d
3. b	8. b	13. b	18. a
4. c	9. d	14. c	19. d
5. a	10. d	15. a	20. a

True or False:

1. True
2. False, change the *more* they vary to *less*
3. True
4. False, change *phenetics* to *cladistics*
5. True
6. False, change to: because *regulator genes control so many structural genes.*
7. False, change to: is *an unexploited niche or way of life*
8. False, change to: indicative of *an asteroid having crashed into the earth.*
9. False, the species that exist the longest and speciate the most often will determine the trend of evolution.
10. True

The History of Life

CHAPTER 24

EARLY EARTH AND THE ORIGIN OF LIFE

FRAMEWORK

This chapter describes the formation of Earth and a scenario for the chemical evolution of life between 4.1 and 3.5 billion years ago. Conditions on the primitive Earth are thought to have favored the spontaneous formation of organic monomers, the linking of these monomers into polymers, the grouping of aggregates of organic molecules into droplets called protobionts, which became capable of metabolism and reproduction, and the development of self–replicating genetic information capable of directing metabolism and reproduction.

From this proposed beginning, an incredible biological diversity has evolved. This heterogeneous group of organisms is now classified into five kingdoms: Monera, Protoctist, Plantae, Fungi, and Animalia.

CHAPTER SUMMARY

Life on Earth has been evolving for 3.5 billion years. Geological events have altered the course of biological evolution, and life has changed Earth. The fossil record documents, however incompletely, the development of much of the diversity of life, but the origin of life on the young planet Earth remains unrecorded and is a matter of speculation.

Formation of the Earth

Most astronomers believe that all matter was once concentrated in a giant mass that blew apart with a "big bang" 10 to 20 billion years ago. Our sun formed about 5 billion years ago when most of the swirling matter in the cloud of dust that formed our solar system condensed in the center. Earth and the rest of the planets formed about 4.6 billion years ago as gravity attracted the remaining dust and ice to a few kernels.

Geologists believe that Earth went through a molten period, when its components sorted into layers of different densities. Nickle and iron sank to the core, less dense material formed a mantle, and the least dense material solidified into a thin crust. The continents formed from plates of crust that float on the plastic mantle.

Scientists speculate that the early atmosphere of Earth consisted mostly of water vapor, carbon monoxide and dioxide (CO, CO_2), nitrogen (N_2), methane (CH_4), and ammonia (NH_3). The first seas were formed by torrential rains; and lightning, volcanism, and UV radiation were quite intense.

The Antiquity of Life

During Darwin's time, no fossils prior to the Cambrian period had been found. Now, Precambrian fossils provide evidence of animals dating back 700 million years, and a succession of microorganisms, most of which were prokaryotes, spanning nearly 3 billion years. Fossils that appear to be prokaryotes about the size of bacteria have been discovered in a rock formation called the Fig Tree Chert, which is 3.4 billion years old. Fossils resembling spherical and filamentous prokaryotes have been found in stromatolites in the Kromberg formation in southern Africa, which is 3.5 billion years old. *Stromatolites* are banded domes of sediment that form around the jellylike coats of motile microbes, which again and again migrate out of the layer of sediment and form a new one above it.

The Origin of Life

Between 4.1 billion years ago, when Earth's crust began to solidify, and 3.5 billion years ago, when records are found of stromatolites, life began. Most biologists believe that life developed on Earth from nonliving materials that became ordered into molecular aggregates capable of self-replication and metabolism.

According to one hypothesis, a chemical evolution in four stages produced the first organisms: (1) the abiotic (nonliving) synthesis and accumulation of small organic molecules such as amino acids and nucleotides; (2) the joining of organic monomers into polymers, including proteins and nucleic acids; (3) the aggregation of molecules into droplets, called *protobionts*, with chemical characteristics different from their surroundings; and (4) the origin of heredity.

Oparin and Haldane independently postulated that primitive Earth conditions, in particular the reducing atmosphere, favored the synthesis of organic compounds from inorganic precursors available in the atmosphere and seas. Energy sources for this synthesis could have been provided by lightning or the intense UV radiation that penetrated the early atmosphere.

In 1953, Miller and Urey tested the Oparin–Haldane hypothesis with an apparatus that simulated early Earth conditions. After a week of applying sparks (lightning) to a warmed flask of water (the primeval sea) in an "atmosphere" of H_2O, H_2, CH_4, and NH_3, cooled by a condensor that created rain, Miller and Urey found a variety of amino acids and other organic compounds in the flask.

Laboratory replications of the early Earth have been able to produce all 20 amino acids, several sugars, lipids, the purine and pyrimidine bases of DNA and RNA, and even ATP. The abiotic synthesis and accumulation of organic monomers could have been a natural process on primeval Earth.

The abiotic synthesis of polymers may have occurred when organic monomers splashed onto hot rocks. Fox has created polypeptides he calls *proteinoids* in the laboratory by dripping dilute solutions of organic monomers onto hot sand or rock.

Clay may have been an important substratum for polymerization because of its ability to concentrate amino acids and other organic monomers when they bind to charged sites on the clay particles. Metal atoms, such as iron and zinc, present at some of the binding sites may function as catalysts facilitating the dehydration reactions that link monomers.

Protobionts, aggregates of abiotically produced organic molecules that exhibit some of the properties associated with life, probably preceded living cells. Protobionts may have developed such capabilities as metabolism, excitability, self-replication, and ability to maintain an internal chemical environment different from the surroundings.

Laboratory experiments indicate that protobionts could have formed spontaneously. Proteinoids, when mixed with cool water, self-assemble into tiny droplets called *microspheres*, which are coated by a selectively permeable membrane across which a membrane potential may develop. Protobionts can discharge this voltage in nervelike fashion.

When the organic ingredients include certain lipids, droplets called *liposomes*, surrounded by a lipid bilayer, may spontaneously form. Oparin has made protobionts he calls *coacervates*, which are colloidal droplets that form when a solution of polypeptides, nucleic acids, and polysaccharides is shaken. If enzymes are included, the coacervates are capable of catalyzing reactions and releasing the products.

The last step necessary for the evolution of life was the origin of genetic information. Successful protobionts would need not only to grow and divide, but also to develop a mechanism for replicating their successful characteristics, for creating instructions for making their key molecules. The hereditary pathway of DNA—>RNA—>protein was too complex to evolve all at once. One hypothesis maintains that the first genes were short strands of RNA, shown in the laboratory to be capable of self-replication without the aid of enzymes. Natural selection also has been observed operating on populations of RNA molecules in test tubes, in which the molecules that are more stable or facile at self-replication successfully compete for monomers and become more common.

RNA-directed protein synthesis may have begun with the weak binding of specific amino acids to bases along the RNA molecules. Perhaps with a catalyst such as zinc, the amino acids could link together to form a short polypeptide. This polypeptide may have then behaved as an enzyme to help the RNA molecule replicate.

When these early RNA and polypeptide molecules became packaged into protobionts, molecular cooperation could become more efficient due to the concentration of components and the potential for the protobiont to evolve as a unit. If primitive enzymes that developed the ability to extract energy from an organic fuel were contained within a membrane, that energy would be made available for the other reactions within the protobiont.

This four-step scenario results in a hypothetical antecedent of a cell, in which an aggregate of molecules selectively incorporates monomers from its surroundings and uses enzymes produced by genes to make polymers and to carry out chemical reactions. The protobiont could grow and split, and distribute copies of its genes to its offspring. Laboratory simulations cannot establish that this sequence actually occurred in the evolution of life, but they have shown that some of the key steps are possible.

In another scenario, Cairns-Smith has proposed that inorganic crystals in the form of layered clays were the first systems to reproduce in the early chemical evolution of Earth. Different types of layered clay that self-assembled by the crystallization of dissolved minerals would be able to grow and split and grow again according to the template of the crystal. The first organic molecules to appear may have helped the clay replicate. Successful molecular cooperatives would have begun to replicate their organic molecules along with the clay crystal. Eventually, the organic genes may have continued to develop while the clay genes became subordinate and then disappeared.

Via whatever route, at some point membrane-bound compartments capable of metabolism and genetic replication evolved past the gray border separating them from true cells; by 3.5 billion years ago, prokaryotes were already flourishing.

The Kingdoms of Life

The kingdom is the most inclusive taxonomic category. Intuitively and historically, humans have divided the diversity of life into two kingdoms—plants and animals. The study of microbial life often confused the two-group classification, as researchers found organisms such as *Euglena* that moved, photosynthesized, and ingested food. The plant kingdom was stretched to make room for bacteria, unicellular eukaryotes with chloroplasts, and fungi, whereas unicellular creatures that move and ingest food were placed in the animal kingdom.

Whittaker's proposal for a five-kingdom system, made in 1969, has slowly gained favor among biologists. The five kingdoms are Monera, Protoctist (as proposed by Margulis), Plantae, Fungi, and Animalia. The prokaryotes are set apart from the eukaryotes and placed in the kingdom Monera. The kingdoms Plantae, Fungi, and Animalia are multicellular eukaryotes, defined by characteristics of structure, life cycle, and modes of nutrition. Plants are autotrophic organisms; animals are heterotrophic organisms that ingest and digest their food; and fungi are absorptive heterotrophic organisms. Protoctists are unicellular organisms, or simple multicellular organisms that descended from them.

STRUCTURE YOUR KNOWLEDGE

1. What do scientists think the primitive Earth was like? How could life possibly evolve in such an inhospitable environment?

2. Describe the significance of the four steps that are proposed as a possible route for the chemical evolution of protobionts.

TEST YOUR KNOWLEDGE

MATCHING: *Match the scientist with the theory or experiment.*

1. _____ produced organic compounds in a lab apparatus

2. _____ proposed clay genes as the first replicating molecules

3. _____ developed the five-kingdom system

4. _____ claimed that conditions of primitive Earth favored synthesis of organic compounds

5. _____ created polypeptides abiotically by dripping organic molecules onto hot sand

6. _____ produced protobionts called coacervates in the laboratory

A. Fox

B. Oparin and Haldane

C. Oparin

D. Cairns-Smith

E. Whittaker

F. Miller and Urey

MULTIPLE CHOICE: *Choose the one best answer.*

1. The "big bang" theory has to do with the formation of
   a. the expanding universe.
   b. the solar system.
   c. the sun.
   d. Earth and the other planets.

2. The primitive atmosphere of Earth may have favored the synthesis of organic molecules because
   a. it was highly oxidative.
   b. it was reducing and had energy sources in the form of lightning and UV radiation.
   c. it had a great deal of methane and organic fuels.
   d. it had plenty of water vapor, lightning, and UV radiation.

3. Life is thought to have begun
   a. 700 million years ago.
   b. 3 billion years ago.
   c. between 3.5 and 4.1 billion years ago.
   d. 5 billion years ago.

4. Stromatolites are
   a. the earliest fossils.
   b. similar to the layered mats formed by cyanobacteria.

c. banded domes of sediment formed around motile microbes.

d. all of the above.

5. Proteinoids are
   a. droplets of molecules covered with a selectively permeable membrane.
   b. colloidal droplets formed from polypeptides, nucleic acids, and polysaccharides.
   c. polypeptides formed in the laboratory by dripping organic monomers onto hot rocks.
   d. organic molecules associated with clay crystals.

6. The first genes may have developed when
   a. protobionts grew and divided.
   b. short RNA strands self-replicated.
   c. DNA molecules produced RNA molecules.
   d. zinc or other metals catalyzed the formation of polypeptides.

7. Protobionts are thought to be
   a. the ancestral antecedents of the cell.
   b. aggregates of molecules that maintain an environment different from their surroundings.

c. droplets that developed the properties of metabolism, excitability, and self-replication.

d. all of the above.

8. Margulis is associated with
   a. the clay gene hypothesis.
   b. the production of liposomes in the laboratory.
   c. the five-kingdom classification system.
   d. the protobiont hypothesis.

9. Microspheres
   a. were the first protobionts.
   b. form when proteinoids are mixed with cool water and may develop a charge across their membranes.
   c. are bounded by a lipid bilayer.
   d. involve all of the above.

10. In the two-kingdom system of classification,
    a. prokaryotes were placed in the plant kingdom.
    b. only multicellular organisms were considered animals.
    c. the animal kingdom included all heterotrophs.
    d. All of the above are correct.

CHAPTER 25

PROKARYOTES AND THE ORIGINS OF METABOLIC DIVERSITY

FRAMEWORK

This chapter describes the bacteria and cyanobacteria of the kingdom Monera. It reviews the morphology of the prokaryotic cell and presents an overview of the groups archaebacteria and eubacteria. Every mode of nutrition and most metabolic pathways that are found today evolved within the prokaryotes. Prokaryotes exist in all conceivable habitats and in various symbiotic relationships with other organisms. These most numerous and diverse of all organisms are essential to the chemical cycles necessary to maintain life on Earth.

CHAPTER SUMMARY

The history of prokaryotes spans at least 3.5 billion years. The kingdom Monera, containing *bacteria* and *cyanobacteria* (formerly blue-green algae), is characterized by the prokaryotic cell. Prokaryotes evolved alone for the first 2 billion years of life, and still outnumber all eukaryotes combined. They flourish in all habitats, including ones that are too harsh for any other forms of life. The collective effect of these microscopic organisms on Earth is huge.

Prokaryotic Form and Function

Prokaryotes exist primarily as single cells, although a few species may form aggregates of cells that stick together after dividing, and some cyanobacteria exhibit a simple multicellular form. The three most common shapes of prokaryotes are spherical (cocci), rod (bacilli) and spiral (spirilla). Prokaryotic cells are usually 1 to 10 μm in diameter, one-tenth the size of eukaryotic cells.

The haploid genome of prokaryotes contains only 1/1000 as much DNA as a eukaryote. The circular double-stranded DNA chromosome has very little associated protein and is concentrated in a *nucleoid region*. Smaller rings of DNA, called plasmids, may carry genes for antibiotic resistance, metabolism of unusual nutrients, or other functions. They replicate independently of the main chromosome and may be transferred between bacteria during conjugation.

Prokaryotic ribosomes are smaller than eukaryotic ribosomes and differ in their protein and RNA content. Because of these differences, selective antibiotics can bind to prokaryotic ribosomes and block protein synthesis without inhibiting eukaryotic ribosomes. Extensive internal membrane systems are not found, although there may be some infoldings of the plasma membrane that function in respiration, and thylakoids in cyanobacteria that function in photosynthesis.

Nearly all prokaryotes have cell walls composed of *peptidoglycan*, a matrix composed of polymers of sugars cross-linked by short polypeptides. Many antibiotics, such as penicillin, inhibit the cross-links in peptidoglycan and prevent wall formation. The *gram stain* is an important tool for identifying bacteria as *gram-positive* (bacteria with simpler walls containing a larger amount of peptidoglycan) or *gram-negative* (bacteria with more complex walls and an outer membrane). Pathogenic gram-negative bacteria are often more harmful because the lipopolysaccharides of their walls may be toxic, and their outer membrane protects them from the defenses of their hosts and from antibiotics.

Many prokaryotes secrete a sticky *capsule* outside the cell wall that serves as protection and as glue for adhering to a substratum or each other. Bacteria may also attach by means of surface appendages called *pilli*. Pilli may be specialized for transfer of DNA during conjugation.

Some motile bacteria secrete slime and glide, whereas others are equipped with flagella, either scattered over the cell surface or concentrated at one or both ends of the cell. Many motile bacteria exhibit *taxis*, an oriented movement in response to chemical, magnetic, or light stimuli. Alternating runs and tumbles result in the movement of a bacterium in a beneficial direction along a chemical gradient.

Prokaryotes reproduce by binary fission, producing a colony of progeny. The geometric growth of bacterial colonies usually stops at some point due to the exhaustion of nutrients or the toxic accumulation of wastes.

Mutations are the major source of genetic variation in prokaryotes. Because generation times are less than a few hours, a favorable mutation can be passed rapidly to a large number of progeny. Despite the lack of meiosis and sexual cycles, some genetic recombination occurs by conjugation and viral transduction.

Nutrition refers to how an organism obtains both energy and the carbon it uses for synthesizing organic compounds. Prokaryotes may be *phototrophs*, using light for energy, or *chemotrophs*, obtaining energy from chemicals. If CO_2 is the only carbon source, organisms are called *autotrophs*; when organic nutrients are needed, organisms are called *heterotrophs*.

Thus, there are four major nutritional categories of bacteria: (1) *photoautotrophs*—cyanobacteria and other photosynthetic prokaryotes—which use light energy and CO_2 to synthesize organic compounds; (2) *photoheterotrophs*—which use light to generate ATP but obtain carbon in organic form; (3) *chemoautotrophs*—which obtain energy by oxidizing inorganic substances (such as H_2S, NH_3, or Fe^{++}) and need only CO_2 as a carbon source; and (4) *chemoheterotrophs*—which use organic molecules as both an energy and carbon source.

The majority of bacteria are chemoheterotrophs, including *saprophytes*, which decompose and absorb nutrients from dead organic matter, and *parasites*, which absorb nutrients from living hosts. There is such diversity of chemoheterotrophs that almost any organic molecule can serve as food for some species. Synthetic organic compounds that cannot be broken down by any bacteria are called *nonbiodegradable*.

Obligate aerobes need oxygen for cellular respiration; *facultative anaerobes* can use oxygen but also can grow in anaerobic conditions using fermentation; *obligate anaerobes* cannot use annot use oxygen and are poisoned by it.

Diversity of Prokaryotes

Molecular systematics, using comparisons of amino-acid sequences and base sequences in DNA and RNA, provide the best approach to the classification of this immense and heterogeneous kingdom that diversified so long ago. These comparisons show that the prokaryotes split into at least two divergent lineages very early in the history of life: the *archaebacteria* and the *eubacteria*.

The archaebacteria include only a few surviving genera that inhabit extreme environments, perhaps resembling habitats on early Earth. Their cell walls lack peptidoglycan, and their plasma membranes contain a unique lipid composition. The archaebacteria presently consist of three subgroups: the methanogens, the extreme halophiles, and the thermoacidophiles.

Methanogens have a unique energy metabolism in which H_2 is used to reduce CO_2 to methane (CH_4). These strict anaerin swamps and marshes—their presence there is indicated by the bubbling out of marsh gas or methane. They are important decomposers in sewage treatment, and have been used to convert manure to methane on some farms. Methanogens in the guts of cattle and other herbivores contribute to the animals' nutrition; in the human large intestine, they cause flatulence.

The extreme halophiles (salt lovers) live in saline places such as the Great Salt Lake and the Dead Sea. They use a simple mechanism of photophosphorylation in which bacteriorhodopsin, the pink pigment built into the plasma membrane, absorbs light and uses the energy to pump H^+ out of the cell. The proton gradient drives the synthesis of ATP.

Thermoacidophiles need an environment that is both hot and acidic. Optimal conditions are $60°$ to $80°$ C with a pH between 2 and 4. They may be found oxidizing sulfur for energy in the hot sulfur springs in Yellowstone National Park, or at the openings of deep-sea vents in the ocein the ocean floor.

Within the group eubacteria, the cyanobacteria represent a photoautotrophic mode of nutrition. Commonly known as blue-green algae, these prokaryotes have a plantlike mechanism of photosynthesis in which they use chlorophyll *a* and two photosystems to transfer electrons from water to $NADP^+$, releasing O_2. The components of the light reaction are built into thylakoid membranes. Accessory pigments, called phycobilins, are held in knobs, called phycobilisomes, located on the thylakoid surface.

Some filamentous cyanobacteria are capable of *nitrogen fixation*, the assimilation of atmospheric nitrogen into organic compounds. *Heterocysts* are specialized cells with enzymes that reduce N_2 to NH_3 (ammonia). Some nitrogen-fixing cyanobacteria exhibit true multicellularity in which the heterocysts fix nitrogen for all the cells in the filament and the vegetative cells provide organic substances to the heterocysts.

Other phototrophs include green and purple sulfur bacteria. They have only one photosystem and reduce $NADP^+$ with electrons from H_2S rather than from H_2O. These photoautotrophic bacteria commonly are found in anaerobic habitats in muddy sediments at the bottom of bodies of water. Their pigments can absorb infrared light that passes through algae living near the surface.

The genus *Pseudomonas*, found in almost all aquatic and soil habitats, includes the most versatile chemoheterotrophs. rotrophs. Species of this group are capable of utilizing a large range of organic nutrients, including some pesticides and synthetic compounds, and even some drug and antiseptic solutions.

Spirochaetes are spiral-shaped bacteria that have an undulating movement caused by the action of strands of flagellin spiraled between the plasma membrane and flexible wall. *Treponema pallidum*, the pathogen that causes syphilis, is a member of this group.

Endospore-forming bacteria can survive ds of harsh conditions, including desiccation, heat, cold, and exposure to most poisons, when they are protected within their resistant, thick-walled endospores. Microbiologists use an autoclave to kill endospores and sterilize laboratory media and glassware. Canned foods must be heated to temperatures high enough to kill the endospores of bacteria, including those of *Clostridium botulinum*, the bacterium that produces the lethal botulism toxin.

Enteric bacteria inhabit the intestinal tracts of animals. Many are harmless permanent residents, but a few may cause diseases such as typhoid fever and salmonella poisoning.

Rickettsias are small parasitic bacteria, including ones that cause Rocky Mountain spotted fever and typhus. Chlamydias are similar to rickettsias. Nongonococcal urethritis (NGU), a sexually transmitted disease, usually is caused by a chlamydia.

Mycoplasmas may be the smallest of all cells. They are the only prokaryotes that lack cell walls. Most live as parasites, and one species causes a form of pneumonia in humans.

Actinomycetes form branching colonies that resemble fungi and may produce resistant spores at the tips of filaments. Tuberculosis and leprosy are caused by actinomycetes, although most species are free-living soil organisms. Cultured species of *Streptomyces* are used to produce commercial antibiotics.

Myxobacteria form gliding colonies that move through the soil on a secreted slime. Under adverse conditions, the cells congregate to form an erect fruiting body in which durable spores are formed.

The Importance of Prokaryotes

Prokaryotes are indispensable in the *chemical cycles* through which chemical elements are recycled between the biological and physical worlds. Bacteria, along with fungi, are *decomposers* of dead organisms and of the waste products of living ones, and thus return carbon, nitrogen, and other elements to the environment for assimilation into new living forms. Autotrophic bacteria bring carbon from CO_2 into the food chain, and nitrogen-fixing bacteria—the only organisms capable of fixing atmospheric nitrogen—supply plants with the nitrogen they need to make proteins.

Symbiosis is an ecological relationship involving direct contact between organisms of two different species. Usually the smaller organism, or *symbiont*, lives within or on the larger organism, the *host*. In *mutualism*, both host and symbiont benefit. *Nodules* on the roots of legumes house symbiotic bacteria that fix nitrogen for the host. In *commensalism*, the symbiont benefits and the host is neither harmed nor helped. Many of the bacteria found in the human body are commensal, although some may be mutualistic. In *parasitism*, the symbiont, now called a parasite, benefits at the expense of the host. Disease-causing bacteria can be considered parasitic.

About one-half of all human diseases are caused by pathogenic bacteria that manage to invade the body, resist internal defenses, and grow enough to harm the host. *Opportunistic* bacteria are normal inhabitants of the human body that cause illness when the body's defenses are weakened.

Koch, the first person to link particular diseases to specific bacteria, suggested four criteria for establishing this connection. Called *Koch's postulates*, they include the following: (1) find the same pathogen in each diseased individual studied; (2) isolate and grow the pathogen in a pure culture; (3) induce the disease in experimental animals with the cultured pathogen; (4) isolate the same pathogen from the experimental animal after it develops the disease.

Pathogenic bacteria commonly cause disease by producing toxins. *Exotoxins*, among the most potent poisons known, are proteins secreted by the bacterium that can induce symptoms without the bacterium being present. *Endotoxins*, whi, which are components of the outer membrane of certain gram-negative bacteria, all produce fever and aches in the host.

In the past century, improved hygiene, sanitation, and the development of antibiotics have decreased the incidence and severity of bacterial disease, reduced infant mortality, and extended life expectancy in developed countries.

The diverse metabolic capabilities of prokaryotes have been put to work by humans to digest organic wastes, produce chemical products, make vitamins and antibiotics, and produce food products such as yogurt and cheese. Research using *Escherichia coli* and other bacteria has expanded our understanding of molecular biology, and recombinant DNA techniques using bacteria are used to develop both basic knowledge and practical applications.

The Origins of Metabolic Diversity

All forms of nutrition and nearly all metabolic pathways evolved in prokaryotes before there were eukaryotes. Geological evidence about conditions of early Earth, combined with molecular systematics and the energy metabolism found in existing groups, indicates a possible sequence of events in metabolic evolution.

The first prokaryotes probably were chemoheterotrophs that absorbed abiotically synthesized organic compounds. ATP must have become an important energy molecule early in life's history, based on its universal role in energy exchange in all modern organisms. Cells that could regenerate a supply of free ATP using other organic nutrients would have had an advantage. Selection may have resulted in the gradual evolution of glycolysis, the metabolic pathway that breaks down organic molecules and generates ATP by substrate phosphorylation. As the only metabolic pathway common to all modern organisms, glycolysis must have developed relatively early in metabolic evolution.

Glycolysis and fermentation, the transfer of electrons extracted during glycolysis to other organic molecules, do not require molecular oxygen. The archaebacteria and other obligate anaerobes that use fermentation are believed to have forms of nutrition most like that of the original prokaryotes that developed on anaerobic Earth.

The excretion of organic acids formed by fermentation into the surroundings would have acidified the environments of early prokaryotes. Transmembrane proton pumps, driven by ATP, may have developed to help regulate internal pH. An electron transport chain that coupled the oxidation of organic acids to the transport of H^+ out of the cell would have conserved ATP. Highly effective electron transport systems may have transferred an excess of H^+, which could then diffuse back and reverse the proton pump, thus generating ATP. This type of anaerobic respiration persists in some modern bacteria. The basic chemiosmotic mechanism that uses proton gradients to transfer energy from redox reactions to ATP synthesis is common to all cells.

The development of pigments and photosystems, co-opting components of the electron-transport chains, may have allowed some prokaryotes to use light energy to drive electrons from H_2S to $NADP^+$ and thus generate reducing power to fix CO_2 and make their own organic molecules. The nutrition of the anaerobic green and purple sulfur bacteria probably is most like that of the early photosynthetic prokaryotes.

The first cyanobacteria evolved mechanisms for using water as a source of electrons and hydrogen for fixing CO_2 and began to release O_2 as a byproduct of their photosynthesis. Cyanobacteria evolved at least 2.5 billion years ago, building the fossil stromatolites that have been found all over the world. Banded iron formations, rich in iron oxide, are found in marine sediments from that same time. The oxygen evolved by the cyanobacteria probably reacted with dissolved iron ions, which precipitated out as iron oxide and prevented the accumulation of free O_2 for a few hundred million years. When the dissolved iron was finally exhausted, oxygen began to accumulate in the seas and eventually to bubble out into the atmosphere. Oxidized iron is found in terrestrial rocks from about 2 billion years ago.

The change to a more oxidizing atmosphere probably caused the extinction of many bacteria, while others survived in anaerobic habitats where they are still found today. Some bacteria evolved antioxidant mechanisms that protected them from rising oxygen levels. Some photosynthetic prokaryotes, however, began using the oxidizing power of O_2 to pull electrons from organic molecules down existing transport chains, leading to the development of aerobic respiration by the co-opting of transport chains from photosynthesis. Several bacterial lines gave up photosynthesis, and their electron transport chains became adapted to function exclusively in aerobic respiration.

Thus, on ancient Earth, every type of nutrition and energy metabolism evolved within the prokaryotes.

STRUCTURE YOUR KNOWLEDGE

1. Fill in the following chart with a brief description of the characteristics of prokaryotic cells.

CHARACTERISTICS OF THE PROKARYOTIC CELL

Characteristic	Description
Cell shape	
Genome	
Membranes	
Structures of the cell surface	
Forms of motility	
Reproduction and growth	

2. Create a concept map that organizes the various modes of nutrition (with their associated terminology) found in the kingdom Monera.

3. List the five or six stages that may have occurred in the early evolution of nutrition and metabolism in the prokaryotes.

TEST YOUR KNOWLEDGE

MULTIPLE CHOICE: *Choose the one best answer.*

1. A prokaryotic genome is different from a eukaryotic genome because
   a. it has only one-half as much DNA as an eukaryote does.
   b. it consists of a single-stranded circular chromosome.
   c. it has less protein associated with its DNA and is not enclosed in a nuclear envelope.
   d. of all of the above.

2. Which of the following is not true of plasmids?
   a. They replicate independently of the main chromosome.
   b. They may carry genes for antibiotic resistance or special enzymes.
   c. They are essential for the existence of bacterial cells.
   d. They may be transferred between bacteria during conjugation.

3. Gram-negative bacteria
   a. have peptidoglycan in their cell walls.
   b. have an outer membrane around their cell walls.
   c. may be more pathogenic than gram-positive bacteria because of the lipopolysaccharides of their walls.
   d. All of the above are correct.

4. Some selective antibiotics function by
   a. binding to prokaryotic ribosomes and blocking protein synthesis.
   b. dissolving the prokaryotic cell wall.
   c. preventing plasmid replication.
   d. doing any one of the above.

5. Many prokaryotes secrete a sticky capsule outside the cell wall that
   a. allows them to glide through a slime layer.
   b. serves as protection and glue for adherence.
   c. reacts with the gram stain.
   d. is used for attaching cells during conjugation.

6. Chemoautotrophs
   a. are photosynthetic.
   b. use organic molecules for an energy and carbon source.
   c. oxidize inorganic substances for energy and use CO_2 as a carbon source.
   d. use light to generate ATP but need organic molecules for a carbon source.

7. The major source of genetic variation in prokaryotes is
   a. binary fission.
   b. mutation.
   c. conjugation.
   d. plasmid exchange.

8. A symbiotic relationship in which the symbiont benefits and the host is neither harmed nor helped is called
   a. saprocism.
   b. opportunism.
   c. mutualism.
   d. commensalism.

9. According to Koch's postulates, a specific bacteria can be linked to a disease if
   a. the same pathogen is isolated from each diseased individual studied.
   b. the cultured pathogen causes the disease when introduced to experimental animals.
   c. the same pathogen can be isolated from the experimental animal that has developed the disease.
   d. all of the above are done.

10. The first prokaryotes were probably
    a. cyanobacteria.
    b. chemoheterotrophs that used abiotically made organic compounds.
    c. anaerobic photosynthetic organisms.
    d. green and purple sulfur bacteria.

11. Banded iron formations in marine sediments indicate that
    a. early cyanobacteria probably were producing oxygen when the sediment was deposited.
    b. the early atmosphere was very reducing.
    c. purple sulfur bacteria were the dominant life form in the seas at that time.
    d. the pH value of the early seas was quite low due to the excretion of organic acids from fermentation by early prokaryotes.

12. The first transmembrane proton pumps may have developed to
    a. produce ATP by the creation of a proton gradient.
    b. pass electrons from H_2S to $NADP^+$ in a primitive photosynthesis.
    c. regulate internal pH.
    d. function in anaerobic respiration.

13. Glycolysis
    a. is the only metabolic pathway common to all modern organisms.
    b. breaks down organic molecules and generates ATP by substrate phosphorylation.
    c. may have been the first metabolic pathway to evolve.
    d. all of the above are correct.

14. The first electron transport chains may have
    a. coupled the oxidation of organic acids to the transfer of H^+ out of the cell, thus saving the cell ATP.
    b. transferred electrons to $NADP^+$ to create the reducing power needed to fix CO_2 in an early form of photosynthesis.
    c. eventually developed into a chemiosmotic mechanism of using proton gradients to produce ATP.
    d. done both a and c.

15. Aerobic respiration may have originated when
    a. some bacteria evolved antioxidant mechanisms to protect them from rising oxygen levels.
    b. photosynthetic bacteria produced an excess of organic molecules.
    c. some bacteria used modified electron-transport chains from photosynthesis to pass electrons from organic molecules to O_2.
    d. some bacteria gave up photosynthesis and reverted to chemoheterotrophic nutrition.

FILL IN THE BLANKS

1. _____ the name for rod-shaped bacteria

2. _____ region in which the prokaryotic chromosome is found

3. _____ important method for identifying bacteria.

4. _____ surface appendages of bacteria used for adherence to a substrate.

5. _____ an oriented movement in response to light or chemical stimuli

6. _____ type of nutrition in which organic molecules are used for energy and carbon source

7. _____ type of nutrition of cyanobacteria

8. _____ organism that can use oxygen or grow anaerobically

9. _____ organisms that decompose and absorb nutrients from dead organic matter

10. _____ bacterial group that lacks peptidoglycan in the cell walls and is found in harsh habitats

11. _____ proteins that are secreted by bacteria and are potent poisons

12. _____ thickenings on roots that house nitrogen-fixing bacteria

MATCHING: *Match the bacterial group with its characteristics.*

1. ___ grow in hot and acidic habitats

2. ___ genus of versatile chemoheterotrophs found in aquatic and soil habitats

3. ___ bacteria found in animal intestinal tracts

4. ___ small parasitic bacteria; one causes Rocky Mountain spotted fever

5. ___ spiral-shaped bacteria with undulating movement; one causes syphilis

6. ___ photosynthetic bacteria that use chlorophyll *a* and phycobilin pigments

7. ___ gliding colonies that erect fruiting bodies

8. ___ anaerobes that produce marsh gas

A. cyanobacteria
B. endospore-forming bacteria
C. actinomycetes
D. mycoplasmas
E. *Pseudomonas*
F. myxobacteria
G. methanogens
H. spirochaetes
I. rickettsias
J. mycoplasmas
K. green and purple sulfur bacteria
L. enteric bacteria
M. halophiles

CHAPTER 26

PROTOCTISTS AND THE ORIGIN OF EUKARYOTES

FRAMEWORK

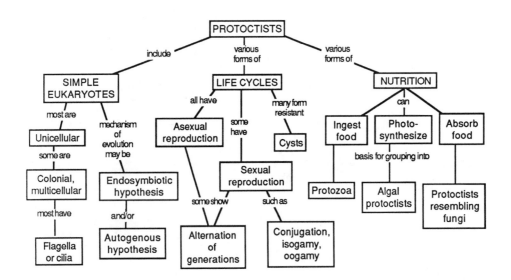

CHAPTER SUMMARY

The kingdom Protoctista includes eukaryotic unicellular (with a few colonial and multicellular) organisms. Ancient protoctists, probably the first eukaryotes to evolve from prokaryotes, were the ancestors of modern protoctists, plants, fungi, and animals. The eukaryotic characteristics of a true nucleus, mitochondria, chloroplasts, ER, Golgi bodies, "9 + 2" flagella and cilia, mitosis, and meiosis evolved in the protoctists.

Characteristics of Protoctists

Protoctists abound almost anywhere there is water—they occupy freshwater, marine, and moist terrestrial habitats, or live symbiotically within their hosts' bodies. They form an important part of *plankton*, the drifting community found in the oceans, ponds, and lakes. Some forms attach to or creep along the substrate.

Nearly all protoctists are aerobic and use mitochondria for their respiration. Their nutrition can be photoautotrophic, heterotrophic, or both. Most protoctists have flagella or cilia at some point during their lifetime. These structures are called *undulipodia* by Margulis to distinguish them from prokaryotic flagella.

Mitosis occurs in most phyla of protoctists, but the process may have variations. All protoctists can reproduce asexually; some use a form of sexual reproduction. Many protoctists form resistant *cysts* to withstand harsh conditions.

These organisms are exceedingly complex at the cellular level. A single cell must be capable of performing all the basic functions that can be shared among the specialized cells of plants and animals.

The Boundaries of the Kingdom Protoctista

Whittaker originally assigned unicellular eukaryotes to the kingdom Protista. Because multicellularity evolved several times within the protoctists, giving rise to plants, animals, fungi, and some organisms that lack the distinctive traits of those three groups, there is disagreement over which groups to include in this kingdom. Margulis and others have advocated the use of the name *Protoctista* and the inclusion of some phyla of multicellular organisms that seem to be closely related to unicellular eukaryotes. The classification scheme used by this book moves the multicellular algae and the funguslike *slime molds* into the kingdom Protoctista.

The morphology and life styles of the protoctists are diverse. An informal scheme for surveying the diversity of protoctists (but one that does not reflect evolutionary relationships) involves the use of three general categories: (1) protozoa (animallike), (2) algal protoctists (plantlike), and (3) protoctists resembling fungi.

Protozoa

Protoctists that ingest food are informally grouped as protozoans. Six phyla are briefly described in the text.

1. *Rhizopoda*: The *amoebas* are naked or shelled, simple protoctists that move and feed with *pseudopodia*. The cytoskeleton functions in the formation of these cellular extensions. Nontypical forms of mitosis occur in asexual reproduction; meiosis and sexual reproduction do not occur. Most amoebas are found freeliving in freshwater, marine, or soil habitats. Some are parasites, such as *Entamoeba histolytica*, which causes amoebic dysentery in humans.

2. *Actinopoda*: The slender projections called axopods, which radiate from these protoctists, help these planktonic organisms remain buoyant and capture microscopic food organisms. *Heliozoa* are freshwater actinopods. *Radiolarians* are primarily marine and may have delicate silica shells.

3. *Foraminifera*: *Forams* are marine organisms known for their porous shells hardened with calcium carbonate. Cytoplasmic strands, extending through the pores, function in swimming, shell formation, and feeding. Some forams derive food from symbiotic algae. Foram fossils are useful for dating sedimentary rocks.

4. *Sporozoa*: These parasites of animals have complex life cycles that often include several host species. *Sporozoa* disseminate by tiny infectious cells called spores or sporozoites. Malaria, caused by the sporozoan *Plasmodium* that is carried by the anopheles mosquito, affects over 200 million people per year.

5. *Zoomastigina*: These protozoa are characterized by whiplike flagella used for motility. *Zoomastigotes* are heterotrophic organisms that may be freeliving or symbiotic. Mutualistic flagellates that live in the gut of termites digest the cellulose eaten by the host. One of the parasitic zoomastigotes pathogenic to humans is *Trypanosoma gambiense*, which causes African sleeping sickness.

6. *Ciliophora*: These protoctists are characterized by the cilia they use to move and feed. Most *ciliates* are complex unicellular organisms found in freshwater. Various arrangements of cilia are used for swimming, "walking," or sweeping food particles into the cell "mouth." Ciliates have two types of nuclei: a large *macronucleus* that controls everyday functions of the cell and splits during binary fission (asexual reproduction), and many small *micronuclei*, one of which will be involved in the sexual process called *conjugation*. Micronuclei are exchanged between partners; syngamy occurs; the new micronucleus divides repeatedly; and one of the daughter micronuclei replicates its genome many times and develops into a new macronucleus.

Algal Protoctists

The protoctists of the algal phyla are mainly photosynthetic. Except for cyanobacteria (blue-green algae), all organisms generically called *algae* are included in the kingdom Protoctista. Some biologists place the green, red, and brown algae in the plant kingdom.

All photosynthetic protoctists have chlorophyll *a*, as do the cyanobacteria and plants. The accessory pigments vary among the algal protoctists and are used, along with cell-wall chemistry, stored food products, and chloroplast structure, as classification characteristics. The text describes seven of the algal phyla.

1. *Dinoflagellata*: *Dinoflagellates* are abundant components of the phytoplankton that forms the basis of most marine food chains. Blooms, or episodes of explosive population growth, of dinoflagellates are responsible for the often harmful red tides. The photosynthesis of symbiotic dinoflagellates, living in small coral organisms, is the main food source for coral reef communities. Some heterotrophic dinoflagellates may be parasitic or carnivorous.

Dinoflagellates have brownish plastids containing chlorophylls *a* and *c* and a mixture of pigments, including peridinin, which is unique to this phylum. Food is stored as starch. The beating of two flagella in perpendicular grooves in the "armor," or hardened plates, of the cell produces a characteristic spinning movement. The division of the dinoflagellate nucleus during asexual reproduction is unusual. Sexual reproduction does not occur.

2. *Chrysophyta*: The color of *golden algae* results from yellow and brown carotenoid pigments, present in addition to chlorophylls *a* and *c*. Carbohydrates are stored in the form of laminarin. There are many colonial forms of these flagellated, freshwater

plankton. Golden algae may form resistant cysts to survive freezing and desiccation.

3. *Bacillariophyta*: *Diatoms* are common unicellular plankton with unique glasslike shells. Massive quantities of fossilized shells make up diatomaceous earth. Although diatoms have the same pigments as golden algae, most *phycologists* (specialists in algae) classify them in a separate phylum. Food reserves are stored in the form of an oil.

4. *Euglenophyta*: The best-known member of this phylum is *Euglena*, a common green flagellate found in pond water. Euglenophytes have chlorophyll *b* (also found in the green algae and in plants) in addition to *a*, and store the polysaccharide paramylon. *Euglena* photosynthesizes in the light, but can ingest food particles by phagocytosis in the dark. These organisms once again point out that there is no sharp distinction between the artificial groups of protozoa and unicellular algae.

5. *Chlorophyta*: The chloroplasts of *green algae* resemble those of plants, and botanists believe that plants evolved from chlorophytes. Most species of green algae live in freshwater, although many are marine. Various unicellular forms may be planktonic, inhabitants of damp soil, or symbionts in invertebrates or in fungi. Associations between green algae and fungi are called *lichens*. Colonial species may be filamentous. Some multicellular forms are large and complex and resemble true plants, but they are more closely related to other members of the green algae. The *thallus*, or body of an alga, lacks true roots, stems, and leaves.

Three separate evolutionary trends within the green algae have produced the volvicine series of increasingly complex flagellated colonies; the siphonous series of coenocytes, masses of cytoplasm containing many nuclei; and the tetrasporine line with definite multicellularity, as seen in *Ulva*.

The life cycles of most green algae include sexual (with biflagellated gametes) and asexual stages. *Conjugating algae*, such as *Spirogyra*, produce amoeboid gametes.

In the life cycle of the unicellular flagellated *Chlamydomonas*, the haploid mature organism reproduces asexually by dividing mitotically to form four zoospores, which grow into mature haploid cells. Under harsh conditions, the cell may produce many haploid gametes, which are released and pair off with similar-looking gametes (*isogamy*) of the opposite mating strain. The fusion of the isogametes results in a diploid zygote, which secretes a protective coat. Following dormancy, the zygote undergoes meiosis, forming four haploid cells that grow into mature individuals. Some species of green algae exhibit *oogamy*, in which a flagellated sperm fertilizes a nonmotile egg.

Some multicellular green algae have an *alternation of generations*, in which a haploid multicellular individual (called the *gametophyte*) produces by mitosis haploid gametes that fuse to form a diploid zygote. The zygote develops into the multicellular *sporophyte*, which produces haploid cells called *zoospores* by meiosis. Zoospores grow into the haploid gametophyte generation. In *Ulva*, the two generations are isomorphic.

6. *Phaeophyta*: The mostly marine *brown algae* are the largest and most complex protoctists. They contain the pigment fucoxanthin in addition to chlorophylls *a* and *c*, and produce laminarin, the same storage compound produced by the golden algae. The brown and golden algae are closely related.

Most common seaweeds are brown algae (or red algae). The gelatinous material, algin, is found in the cells walls and protects the thallus. The giant seaweeds known as kelps are brown algae that fasten to the sea floor by holdfasts. The long leaflike blades are supported by gas-filled bladders and have tubular cells for transporting photosynthetic products.

In most brown algae, there is an alternation of generations. The gametophyte and sporophyte may be *heteromorphic*, as in *Laminaria*.

7. *Rhodophyta*: The color of *red algae* is due to the accessory pigment phycoerythrin, which can absorb the wavelengths of light that reach fairly deep in the oceans. Red algae that live in shallow water may be more green because less phycoerythrin is present to mask the green color of chlorophyll. Floridean starch is the storage product. Red algae are multicellular, often with filamentous, delicately branched thalli. The base of the thallus may be differentiated as a holdfast. The *coralline algae* are red algae with walls encrusted with calcium carbonate.

All red algae reproduce sexually, and they often have an alternation of generations. Unlike other algal protoctists, they have no flagellated cells in their life cycles.

Protoctists Resembling Fungi (Slime Molds)

Although the slime molds resemble fungi in appearance and lifestyle, their cellular organization, reproduction, and life cycles are different enough to suggest that they are most closely related to protoctists. General consensus places the following two phyla with the protoctists, although they are still studied by *mycologists* (biologists who study fungi).

1. *Myxomycota*: Slime molds are heterotrophic, engulfing food particles by phagocytosis as they grow through leaf litter or rotting logs. A coenocytic amoeboid mass called a *plasmodium* is the feeding stage. A sexual stage develops erect stalks that produce resistant spores by meiosis. When the spores germinate, they fuse to form diploid zygotes in which the nucleus repeatedly divides without cytoplasmic divisions.

2. *Arasiomycota*: *Cellular slime molds* consist of solitary cells during the feeding stage of the life cycle. In the

absence of food, the cells congregate into a mass resembling a plasmodial slime mold, except that it is not coenocytic. Also, since the cellular slime molds are haploid, the fruiting bodies produce resistant spores asexually by mitosis.

The Origin of Eukaryotes

The oldest reputed fossils of eukaryotes, found in rocks 1.5 billion years old, are *architarchs*, Precambrian structures that resemble ruptured coats of cysts similar to those of algal protoctists. The cyanobacteria probably had produced enough oxygen to accumulate in the atmosphere during the 2 billion years of prokaryotic history predating that time. Almost all eukaryotes require oxygen.

There are two major theories on the origin of the eukaryotic cell: the *autogenous hypothesis* and the *endosymbiotic hypothesis*. In the first model, eukaryotic cells are believed to have evolved by the specialization of internal membranes derived from invaginations of the plasma membrane.

According to the endosymbiotic model, developed by Margulis, eukaryotic cells evolved from symbiotic combinations of prokaryotic cells. Chloroplasts are thought to be descendants of photysynthetic prokaryotes that became endosymbionts within larger cells. Mitochondria may have developed from aerobic heterotrophic bacteria that were endosymbionts.

Chloroplasts appear to have had at least three separate origins, as evidenced by three lineages seen in algal protoctists. The red line of the chloroplasts of red algae shows similarities to the thylakoid arrangement and pigment composition of cyanobacteria. The chloroplasts of green algae, the green line, are more similar to a eubacterium named *Prochloron*, which is the only known prokaryote with chlorophyll *b* as well as *a*. The thylakoids of the brown line generally are in stacks of three, and chlorophyll *c* and fucoxanthin are accessory pigments. A modern prokaryote that resembles a possible ancestor of the prokaryotic symbiont of the brown plastids has not been found.

Various similarities, including size, membrane enzymes and transport systems, and the process of division, exist between eubacteria and the chloroplasts and mitochondria of eukaryotes. Chloroplasts and mitochondria contain circular DNA molecules not associated with proteins, the same as in prokaryotes. They also contain the ribosomes and equipment needed to transcribe and translate their DNA into proteins. The ribosomes of chloroplasts and mitochondria resemble prokaryotic ribosomes. These similarities are cited as evidence of the endosymbiotic hypothesis. Molecular systematics also suggests eubacterial origins for chloroplasts and mitochondria.

Even though chloroplasts and mitochondria are not genetically autonomous, the billions of years of co-evolution, and the existence of transposons (jumping genes) may have allowed the host cell to develop nuclear control over its symbionts.

The autogenous and endosymbiotic models need not be mutually exclusive. The chloroplasts and mitochondria may have originated as endosymbionts, whereas the components of the endomembrane system may have developed as modifications of the plasma membrane.

The Origins of Multicellularity

Multicellularity evolved in the kingdom Protoctista many times, probably by two mechanisms: colonial connection as aggregations of cells became more interdependent, and membranous partitioning of coenocytic masses. Multicellular algae, plants, and fungi may have descended from colonial protoctists, whereas most animals may have had their origin by the second mechanism.

The larger size possible with aggregations of cells may have increased the speed of motile organisms, and improved the ability of sessile organisms to resist being dislodged. The division of labor is probably the most important characteristic of truly multicellular organisms.

STRUCTURE YOUR KNOWLEDGE

1. (a) What are the criteria or characteristics of the organisms placed in the kingdom Protoctista? (b) Why is there a controversy over which groups belong in this kingdom?
2. Describe the two hypotheses for the origin of eukaryotic cells. What evidence is given to support the endosymbiotic model?

TEST YOUR KNOWLEDGE

MATCHING: *Match the protoctistan phyla with their descriptions.*

1. _____ Naked and shelled amoebas
2. _____ Parasites with complex life cycles, spread by sporozoites; one causes malaria
3. _____ Plasmodial slime molds, coenocytic

A. Acrasiomycota
B. Actinopoda
C. Bacillariophyta
D. Chlorophyta
E. Chrysophyta
F. Ciliophora

4. ____ Marine, with calcareous porous shells

5. ____ Green algae, probable
ancestors of plants

6. ____ Brown algae, large
and complex, seaweeds

7. ____ Golden algae, flagellated, freshwater plankton

8. ____ Motile organisms
with whiplike flagella,
heterotrophic, free–living
or symbiotic

9. ____ Complex, unicellular
organisms, move with cilia,
macronucleus and
micronuclei

10. ____ Cellular slime molds,
unicellular feeding stage,
aggregate to form asexual
fruiting body

11. ____ Red algae, pigment
phycoerythrin

12. ____ Heliozoa, radiolarians, siliceous skeletons

13. ____ Whirling movement
caused by two flagella in
grooves of shell, cause red
tides

14. ____ Diatoms, two-piece
shells of silica

15. ____ Photosynthetic or
heterotrophic flagellates

G. Dinoflagellata

H. Euglenophyta

I. Foraminifera

J. Myxomycota

K. Phaeophyta

L. Rhizopoda

M. Rhodophyta

N. Sporozoa

O. Zoomastigina

MULTIPLE CHOICE: *Choose the one best answer.*

1. The term *undulipodia* is used to refer to
   a. protoctists that use an amoeboid type
      of movement.
   b. the drifting community of organisms found
      in aquatic environments.
   c. the eukaryotic 9 + 2 form of flagella
      and cilia.
   d. the multicellular forms included in the
      kingdom Protoctista.

2. The slime molds and multicellular algae are
   included in the kingdom Protoctista because
   a. they appear to be more closely related to
      unicellular eukaryotes.
   b. they lack important characteristics of the
      fungi and plants.

   c. the classification system is trying to reflect
      phylogenetic relationships.
   d. of all of the above.

3. The category of protozoa includes the
   protoctists that
   a. are motile.
   b. ingest their food.
   c. do not have chlorophyll.
   d. are all of the above.

4. Genetic variation is generated in the ciliates when
   a. a micronucleus replicates its genome many
      times and becomes a macronucleus.
   b. isogametes, in the form of zoospores, fuse.
   c. micronuclei are exchanged in conjugation.
   d. oogamous sexual reproduction occurs.

5. Phytoplankton
   a. form the basis of most marine food chains.
   b. include the radiolarians and forams.
   c. are the main food source for coral
      reef communities.
   d. All of the above are correct.

6. Phycologists are
   a. specialists in the study of algae.
   b. specialists in the study of fungi.
   c. protoctistan taxonomists.
   d. plant biologists.

7. Plants are believed to have evolved from
   a. the euglenophytes, because they have
      chlorophyll *b*.
   b. the chrysophyta, because they have similar
      storage products.
   c. the chlorophyta, because of similarities in
      chloroplasts and pigments composition.
   d. the volvicine series of increasing
      multicellularity.

8. Which of the following protoctist phyla is
   incorrectly paired with its accessory pigments?
   a. Rhodophyta—phycoerythrin
   b. Phaeophyta—fucoxanthin and chlorophyll *c*
   c. Dinoflagellata—fucoxanthin and chlorophyll *c*
   d. Euglenophyta—chlorophyll *b*

9. Architarchs are
   a. the oldest reputed eukaryotic fossils.
   b. fossils that resemble the ruptured coats
      of cyanobacteria.
   c. the fossilized remains of diatoms.
   d. symbiotic combinations of prokaryotic cells.

10. Which of the following is not evidence that chloroplasts and mitochondria originated as prokaryotic endosymbionts of larger cells?
    a. They contain circular DNA molecules not associated with proteins.
    b. Their membranes have enzymes and transport systems that resemble those found on the plasma membranes of prokaryotes.
    c. They contain their own genome and produce all their own proteins.
    d. Their ribosomal RNA is more similar to that of eubacteria than to the RNA of ribosomes in the eukaryotic cytoplasm.

11. The endosymbiotic hypothesis maintains that chloroplasts may have had at least three different origins because
    a. the phyla of algal protoctists can be divided into three groups based on photosynthetic pigments and chloroplast structure.
    b. the chloroplasts of green plants show three distinct morphologies.
    c. the pigment composition and thylakoid arrangement of cyanobacteria occur in three different arrangements.
    d. molecular systematics link the red line, green line, and brown line to three different eubacteria.

12. Multicellularity probably evolved
    a. when colonies of cells became more and more interdependent.
    b. by cellular separation within a coenocyte.
    c. many times in the kingdom Protoctista.
    d. All of the above are correct.

CHAPTER 27

PLANTS AND THE
COLONIZATION OF LAND

FRAMEWORK

This chapter details the evolution of plants and their adaptations to the terrestrial habitat. The kingdom Plantae includes multicellular, photoautotrophic eukaryotes that develop from embryos retained in a protective jacket of cells. Most plants exhibit an alternation of generations, in which the sporophyte is the most conspicuous stage in all divisions except the bryophytes. The development of a cuticle and jacketed reproductive organs, vascular tissue, seeds and pollen, and flowers are linked to the major periods of plant evolution in which mosses, ferns, conifers, and flowering plants appeared and radiated.

CHAPTER SUMMARY

For the first 3 billion years, life was confined to the seas, ponds, lakes, and streams. Plants began the movement of life onto land about 400 million years ago, and the evolutionary history of the plant kingdom involves the increasing adaptation to changing terrestrial conditions.

Introduction to the Plant Kingdom

Plants are multicellular, photoautotrophic eukaryotes. Nearly all plants are terrestrial, although some have returned to water during their evolution. Adaptations to life on land include a waxy *cuticle* that prevents desiccation and *stomata* or pores to allow for the exchange of gases through the cuticle. Plant chloroplasts contain chlorophylls *a* and *b* and a variety of carotenoids. The cell walls of plants consist of cellulose, and starch is the food storage compound.

Nearly all plants reproduce sexually; most also can propagate asexually. Gametes are produced and the egg

is fertilized and develops into an embryo within a protective jacket of sterile cells. Margulis defined plants as those eukaryotes that develop from an embryo supported by maternal tissue. The life cycle of plants involves the alternation of generations between the haploid gametophyte and the diploid sporophyte. In all extant plants, these generations are *heteromorphic*. In all but the bryophytes, the sporophyte is the more conspicuous stage. There has been an evolutionary trend toward the reduction of the haploid generation, the dominance of the diploid, and the replacement of flagellated sperm by pollen.

Three major periods of plant evolution are linked to the evolution of structures that opened new adaptive zones on land. The first period involved the origin of plants, probably from green algae, over 400 million years ago. Two distinct groups of early plants developed, both showing the terrestrial adaptations of a cuticle and jacketed reproductive organs. The first group included the ancestors of all plants except the bryophytes. Within these *vascular* plants evolved *vascular tissue*, specialized cells joined into tubes for transport throughout the plant. The second group, the *nonvascular plants*, were the ancestors of the mosses.

About 360 million years ago, the origin of the first vascular plants with *seeds*, or embryos enclosed with a store of food inside a resistant coat, marked the second major period of plant evolution. Early seed plants gave rise to *gymnosperms* (naked seed plants). At the same time, some seedless plants, such as the ferns, continued to evolve; they dominated the landscape along with gymnosperms for over 200 million years.

The emergence of flowering plants about 130 million years ago started the third major evolutionary episode. The majority of contemporary plants are *angiosperms*, or flowering plants, which bear seeds in protective chambers called ovaries.

Most botantist use the term "division" in place of "phylum" for the major taxonomic groups within the plant kingdom. This textbook recognizes ten divisions within the kingdom Plantae. Alternative classification schemes separate plants into two divisions: the Bryophyta (mosses) and the Tracheophyta (vascular plants).

The Move onto Land

Almost all botanists agree that plants evolved from green algae, most of which share the following characteristics with plants: (1) chlorophyll *a*, and the accessory pigments chlorophyll *b* and beta-carotene; (2) thylakoid membranes stacked into grana; (3) cell walls of cellulose; (4) starch as carbohydrate reserve; (5) a phragmoplast (in some green algae) that leads to cytoplasmic division in cytokinesis.

Fiiamentous chlorophytes that carpeted the fringes of bodies of water may have been the first green algae to colonize the land. Fluctuations in water levels may have resulted in the selection for species that could survive exposed periods. The development of waxy cuticles and jacketed reproductive organs opened the adaptive zone of land. Bryophytes and vascular plants either had separate origins from green algae or diverged early in their evolution from an algal ancestor.

The *division Bryophyta* includes three classes of *bryophytes*: mosses, liverworts, and hornworts. The gametophyte is the dominant generation. Gametes develop in reproductive organs called *gametangia*, which have jackets of sterile cells. Flagellated sperm are produced in *antheridia*. One egg is produced and fertilized within the *archegonium*, where the zygote develops into an embryo. Even with cuticles and protected embryos, mosses must inhabit damp, shady places because they have no vascular tissue to carry water from the soil to the aerial parts of the plant and their sperm need a watery film in which to swim to the archegonium.

Mosses grow as mats of many plants in a tight pack. Each gametophyte plant attaches to the soil with rootlike structures called rhizoids. Photosynthesis occurs in small leaflike appendages. The embryo grows into the sporophyte generation while still attached within the archegonium and dependent on the gametophyte for water and nutrients. At the tip of the sporophyte stalk, a *sporangium* produces haploid spores by meiosis. Germinating spores grow into a small, green, threadlike protonema from which the gametophyte plant develops.

Most *liverworts* have bodies that are divided into lobes. Their life cycle is similar to that of mosses, although they also reproduce asexually by the formation of bundles of cells called gemmae that grow in cups on the surface of the gametophyte. *Hornworts* resemble liverworts, but their sporophytes are elongated capsules that grow like horns from the flat gametophyte.

Mosses are a separate phylogenetic branch that are represented in the fossil record from about 350 million years ago. Vascular plants were already established on land by that time.

During the evolution of vascular plants, there was an increasing differentiation of an underground root system to absorb water and minerals and an aerial shoot system of stems and leaves to receive light for photosynthesis. *Lignin*, a hard material embedded in the cellulose of plant cell walls, was an important adaptation to provide support for the aerial parts of the plant. The vascular system of *xylem* and *phloem* provided the conducting vessels to connect roots and leaves. Empty xylem cells carry water and minerals up from the roots; their lignified walls also provide support. The living cells of phloem distribute organic nutrients throughout the plant.

Fossils of vascular plants are present in the sedimentary rocks of the late Silurian and early Devonian periods (around 400 million years ago). *Cooksonia* is the oldest plant that has been discovered; *Zosterophyllum* and *Rhynia* are characteristic genera of the diversity of early Devonian species that are the probable ancestors, respectively, of the division Lycophyta and the other divisions of vascular plants.

Seedless Vascular Plants

Four extant divisions have retained the primitive seedless condition.

Division Psilophyta *Psilotum*, known by the common name *whiskfern*, is one of two genera of relatively simple, primitive vascular plants in this division. True roots and leaves are absent. Emergences from the dichotomously branching stems lack vascular tissue. Sporangia develop in the emergences and produce haploid spores by meiosis. The tiny gametophyte lacks chlorophyll and depends on symbiotic fungi for its nutrition. Flagellated sperm reach the egg through the moist soil; and the zygote and young sporophyte begin development in the archegonium.

Division Lycophyta *Lycopods* were a major part of the landscape during the Carboniferous period. The division Lycophyta had split into two evolutionary lines: one that led to huge woody trees and a second consisting of lycopods that remained small and herbaceous. The giant lycopods thrived for millions of years, but became extinct when the Carboniferous swamps dried up about 280 million years ago. The small lycopods are represented today by about a thousand species, most in the genera *Lycopodium* and *Selaginella*, commonly called club mosses and ground pines. Many tropical species grow on trees as *epiphytes*.

In the club moss, the spores are borne on *sporophylls*, leaves specialized for reproduction. Spores germinate and grow into inconspicuous, subterranean gametophytes, which depend on symbiotic fungi for nutrition. These gametophytes produce both archegonia and antheridia. *Lycopodium* is said to be *homosporous* because it produces only one kind of spore, which produces bisexual gametophytes. *Selaginella* is *heterosporous*: it produces *megaspores* that develop into female gametophytes bearing archegonia, and *microspores* that develop into male gametophytes with antheridia.

Division Sphenophyta   This division produced giant plants during the Carboniferous period. Today, only about 20 species of the genus *Equisetum* are found. Commonly called horsetails, these homosporous plants exist in damp locations. The gametophyte is minute but freeliving and photosynthetic.

Division Pterophyta   Ferns were also found in the great forests of the Carboniferous era, and are the most numerous of seedless plants in the modern flora. The leaves of ferns, commonly called fronds, are much larger than those of lycopods. The leaves of lycopods probably evolved as emergences from the stem containing a single strand of vascular tissue, and are thus called *microphylls*. Fern leaves, called *megaphylls*, have branching systems of veins. One theory holds that megaphylls evolved by the formation of webbing between close-growing separate branches.

Some fern leaves are specialized sporophylls with sporangia, arranged into clusters called sori, on their undersides. Most ferns are homosporous, although the archegonia and antheridia on the small, photosynthetic gametophyte mature at different times, ensuring cross-fertilization. Sperm swim to fertilize the egg in the archegonia, and the young sporophyte grows out from the archegonium.

The seedless plants of the Carboniferous forests left behind extensive beds of coal. Dead plants did not completely decay in the stagnant swamp waters, and great accumulations of peat developed. When the swamps were later covered by the sea, marine sediments piled on top, generating heat and pressure that converted the peat to coal.

Gymnosperms

Three life-cycle modifications contributed to the success of seed plants on land: (1) the further-reduced gametophyte generation was retained within the sporophyte plant; (2) pollination replaced the swimming of sperm to egg; (3) the seed, with embryo surrounded by a food supply within a seed coat, maintained dormancy through harsh conditions and functioned in dispersal. The seed replaced the spore as the agent of dispersal.

Of the two groups of seed plants, gymnosperms appear earlier in the fossil record and are considered more primitive. Three of the four divisions are relatively small: Cycadophyta, Ginkgophyta, and Gnetophyta.

Division Coniferophyta   The name *conifer* comes from the reproductive structure, the cone. Most conifers are evergreens, with needle-shaped leaves covered with a thick cuticle. Coniferous trees are among the tallest, largest, and oldest living organisms.

The pine tree is a heterosporous sporophyte. Pollen cones consist of many tiny sporophylls or scales, each of which bear two sporangia. Spores, produced on separate cones, develop into male and female gametophytes. Meiosis gives rise to microspores that develop into pollen grains, the immature male gametophyte consisting of two haploid cells. Scales (actually modified branches) of the ovulate cone hold two ovules. An ovule contains a sporangium, called a nucellus, within a protective integument. One of the cells of the nucellus undergoes meiosis, but only one of the resulting cells develops into a spore. The spore undergoes repeated divisions to produce a female gametophyte in which a few archegonia develop.

Pollination occurs when a pollen grain is drawn through the micropyle, an opening in the integument around the nucellus. A pollen tube grows and digests its way through the nucellus. After 1 year, the male and female gametophyte mature: the female gametophyte develops its two or three egg cells, and one of the two nuclei of the pollen grain divides to form two sperm nuclei.

Fertilization occurs when the two sperm nuclei enter an egg cell. One nucleus disintegrates; the other joins with the egg nucleus to produce a zygote. The zygote develops into a sporophyte embryo with a rudimentary root and several embryonic leaves, or cotyledons. The female gametophyte tissue nourishes the embryo. A seed coat develops from the integuments of the ovule that were part of the parent sporophyte generation. The production and maturation of seeds takes three years.

Gymnosperms probably developed from a group of plants that had evolved seeds by the end of the Devonian period. The divisions of gymnosperms arose during the Carboniferous and early Permian period. As the Permian progressed, continental interiors became warmer and drier. The lycopods, horsetails, and ferns that dominated the Carboniferous swamps were replaced by conifers, which appeared in the late Carboniferous, and the cycads. The end of the Permian marked the boundary between the Paleozoic and Mesozoic eras. The Mesozoic is sometimes called the Age of Dinosaurs. The vegetation supporting these giant reptiles consisted of conifers and great palmlike cycads. When the climate became cooler at the end of the Mesozoic, the dinosaurs and many plants became extinct, but some gymnosperms, particularly conifers, persisted.

Angiosperms (Flowering Plants)

Angiosperms are the most diverse and widespread of modern flora. The *division Anthophyta* is divided into two classes: Monocotyledones and Dicotyledones.

Most angiosperms rely on insects or other animals for transferring pollen from male to female structure, an evolutionary advance over the wind pollination of gymnosperms. Gymnosperms use *tracheids* for water transport. These relatively primitive, tapered cells also function in mechanical support. In angiosperms, shorter, wider *vessel elements* are arranged end to end, thus forming more specialized tubes for water transport. *Fibers* have thick lignified walls that add support to the xylem of angiosperms.

The *flower*, the reproductive structure of an angiosperm, is a compressed shoot with four whorls of modified leaves: *sepals, petals, stamens,* and *carpels*. Sepals are usually green, whereas petals are brightly colored in most flowers pollinated by insects. A stamen consists of a stalk, called a *filament*, and an *anther*, where pollen is produced. The carpel has a sticky *stigma*, which receives pollen, and a *style*, which leads to the ovary. The ovary contains ovules that develop into seeds.

Four evolutionary trends in angiosperms include: (1) reduction of the number of floral parts, (2) fusion of floral parts (compound carpels are common, *pistil* refers to single or fused carpels), (3) change from radial to bilateral symmetry, and (4) lowering of the position of the ovary to below petals and sepals for better protection.

A *fruit* is a ripened ovary, which functions in the protection and dispersal of seeds. Simple fruits develop from a single ovary; aggregate fruits come from several ovaries of the same flower; and multiple fruits, such as a pineapple, develop from several separate flowers. Fruits may be modified in various ways to disperse seeds.

Angiosperms are heterosporous. Immature male gametophytes, consisting of two haploid nuclei, are *pollen grains*, which develop within the anthers. *Ovules*

contain the female gametophyte, consisting of an embryo sac with eight haploid nuclei.

Most flowers have some mechanism to ensure *cross-pollination*, the transfer of pollen from one plant to another. A pollen grain germinates on the stigma and extends a pollen tube down the style to the ovule, where it releases two sperm nuclei into the embryo sac. In a process called *double fertilization*, one sperm unites with the egg to form the zygote, and the other sperm nucleus fuses with two nuclei in the center of the embryo sac. The *endosperm*, which develops from the triploid (3N) nucleus, serves as a food reserve for the embryo. The embryo consists of a rudimentary root and one (monocots) or two (dicots) seed leaves, called *cotyledons*. The seed is a mature ovule, containing an embryo, endosperm, and seed coat derived from the integuments.

Angiosperms suddenly appear in the fossil record about 120 million years ago in the early Cretaceous period, with no transitional links to ancestors. By the end of the Cretaceous, 65 million years ago, angiosperms had radiated and become the dominant plants, as they are today. Nearly all paleobotanists agree that the angiosperms must have evolved from some group of gymnosperms, although the lack of transitional forms obscures the angiosperms' ancestry.

Angiosperms rose to prominence during the Cretaceous, another period of climatic change and extinctions. Dinosaurs, many cycads, and conifers disappeared and were replaced by mammals and flowering plants. The end of the Cretaceous marks the boundary between the Mesozoic and Cenozoic eras.

Animals influence the evolution of plants and vice versa, a condition known as *co-evolution*. Predation of animals on plants may have provided the selective pressure for plants to keep their spores and vulnerable gametophytes attached to themselves. As flowers and fruits evolved, some predators became beneficial as pollinators and seed dispersers. Co-evolution of angiosperms and their pollinators is largely responsible for the diversity of flowers. The color, fragrance, and shape of flowers usually is matched to the pollinator's sense of sight and smell, or its particular morphology. As seeds mature, the fruit softens and increases in sugar content, attracting bird and mammal seed dispersers.

Nearly all our fruit and vegetable crops are angiosperms. Grains are grass fruits; their endosperm is the main food source for most people and their domesticated animals. As humans domesticated plants, they intervened in plant evolution by selective breeding, and developed cultures centered on the care of plants. Agriculture is a unique case of co-evolution between plants and animals.

STRUCTURE YOUR KNOWLEDGE

1. The evolution of plants shows a trend of increasing adaptations to a terrestrial habitat. In the following table, list as many plant characteristics that contribute to survival on land as you can, and check which of the major plant groups have those characteristics.

Plant Groups and Their Common Names / Characteristics	Bryophyta	Lycophyta Sphenophyta	Pterophyta	Conifero-phyta	Anthophyta
Cuticle, stomata	X	X	X	X	X

2. The following table will help you to organize plant evolution within a geological time frame. Plot the major events (origin, dominance, and decline) for vascular plants, seed plants, bryophytes, lycopods, sphenopsids, ferns, gymnosperms, conifers, cycads, angiosperms, and the animal groups of amphibians, dinosaurs, mammals, birds, and insects.

Era	Period	Age (millions of years ago)	Major Plant Events	Animal Events
CENOZOIC	Quaternary			
CENOZOIC	Tertiary	65		
Climatic Change--Cooler				
MESOZOIC	Cretaceous	135		
MESOZOIC	Jurassic	195		
MESOZOIC	Triassic	240		
Climatic Change--Warmer and Drier				
PALEOZOIC	Permian	285		
PALEOZOIC	Carboniferous	375		
PALEOZOIC	Devonian	400		
PALEOZOIC	Silurian	450		

TEST YOUR KNOWLEDGE

MULTIPLE CHOICE: *Choose the one best answer.*

1. Which of the following is not evidence that plants evolved from green algae?
   a. Both groups use chlorophyll *a* and the same accessory pigments.
   b. Both groups have cell walls of cellulose and starch as a storage product.
   c. Both groups have members that live on land.
   d. Their chloroplasts are similar.

2. Plants are thought to have originated from chlorophytes
   a. over 400 million years ago.
   b. during the Carboniferous period.
   c. when the climate became warmer and drier at the end of the Paleozoic era.
   d. as the ancient atmosphere became oxygen-rich.

3. Bryophytes must inhabit damp places because
   a. they do not have woody tissue.
   b. their sporophyte generation is dependent on the gametophyte.
   c. their sperm must swim to reach the egg.
   d. of all of the above.

4. Which of the following lack true roots and leaves?
   a. bryophytes
   b. bryophytes and whiskferns (*Psilotum*)
   c. bryophytes, whiskferns, and lycopods
   d. bryophytes, whiskferns, lycopods, and pterophytes

5. Megaphylls
   a. are gametophyte plants that develop from megaspores.
   b. are leaves specialized for reproduction.
   c. are leaves with branching vascular systems.
   d. are large leaves.

6. The great coal beds were created by
   a. the undecomposed bodies of gymnosperms and dinosaurs.
   b. the compression of great accumulations of peat formed from the anaerobic decomposition of mats of bryophytes.
   c. the remains of the seedless plants of the Carboniferous forests.
   d. the compressed remains of cycads and gymnosperms.

7. Pollination in a pine occurs when
   a. a pollen grain is drawn through the micropyle.
   b. a sperm nucleus fuses with an egg in an archegonium.
   c. two sperm nuclei enter the nucellus.
   d. pollen grains reach the ovulate cone.

8. Which of the following incorrectly pairs a sporophyte embryo with its food source?
   a. pine embryo—female gametophyte tissue in nucellus
   b. corn embryo—3N endosperm tissue in seed
   c. moss embryo—archegonium and gametophyte plant
   d. fern embryo—female sporophyte surrounding archegonium

9. A difference between seedless vascular plants and the plants with seeds is that
   a. the gametophyte generation is conspicuous in the seedless plants, whereas the sporophyte is dominant in the seeded plants.
   b. the spore is the agent of dispersal in the first, whereas the seed functions in dispersal in the second.
   c. the gametophyte is photoautotrophic in the seedless plants but is dependent on the sporophyte generation in the other group.
   d. the embryo is unprotected in the seedless plants but is retained within the female reproductive structure in the other group.

10. As a result of co-evolution,
    a. two organisms mutually influence each other's evolution.
    b. a flower may have nectar guides that lead bees to its nectaries.
    c. the red color of ripe fruits may attract birds or mammals that disperse the seeds.
    d. all of the above apply.

11. A fruit is
    a. a ripened ovary.
    b. a thickened carpel.
    c. an enlarged ovule.
    d. a ripe flower.

12. Which of the following does not function as support of angiosperms?
    a. lignin
    b. fibers
    c. xylem tracheids
    d. styles

TRUE OR FALSE: *Indicate T or F, and then correct the false statements.*

1. _____ All photoautotrophic, multicellular eukaryotes are plants.

2. _____ Heteromorphic plants produce male and female gametophytes.

3. _____ Bryophytes are terrestrial nonvascular plants.

4. _____ The club mosses and horsetails are naked seed plants.

5. _____ A sporangium produces spores, no matter what group it is found in.

6. _____ A seed consists of an embryo, nutritive material, and a protective coat.

7. _____ Tracheids are wide, specialized cells arranged end to end for water transport; they are found in angiosperms.

8. _____ A stamen consists of a filament and anther in which pollen is produced.

9. _____ The female gametophyte in angiosperms consists of haploid cells in which a few archegonia develop.

10. _____ The male gametophyte in angiosperms consists of a pollen grain.

FRAMEWORK

This chapter describes the morphology, life cycles, and ecological and economic importance of the kingdom Fungi. The six divisions of fungi are determined on the basis of variations in sexual reproduction. The lichens are symbiotic complexes of fungi and algae. The role of fungi as decomposers and their association with plant roots in mycorrhizae have great ecological significance.

CHAPTER SUMMARY

Characteristics of Fungi

Fungi are eukaryotic, mostly multicellular, heterotrophs. They acquire their nutrients by *absorption* of the small organic molecules formed when they secrete enzymes into their surrounding food media. Fungi may be saprophytes that feed on nonliving organic material as *decomposers*; parasites that absorb nutrients from living hosts; or mutualistic symbionts who feed on but also benefit their hosts.

Most fungi are composed of filaments called *hyphae*, which form a netlike mass called a *mycellium*. Hyphae may be divided into cells by *septa*, which usually have pores through which cell organelles and nutrients can pass. Cell walls are composed of chitin. Hyphae also may be *aseptate* and *coenocytic*, consisting of a continuous cytoplasmic mass with many nuclei.

The filamentous hyphae grow rapidly and provide a large surface area for absorption. Parasitic fungi may penetrate their host cells or tissues with modified hyphae called *haustoria*.

During mitosis, the spindle forms within the nucleus, which divides after the chromosomes separate. Nuclei of the mycelia are haploid except in the water molds. Some septate fungi are *dikaryons* with two nuclei per cell. If the nuclei came from different parents during sexual reproduction, the dikaryon is said to be *heterokaryotic*.

Huge quantities of asexual spores are produced when conditions are good. Sexual spores may be produced when environmental conditions are unfavorable. After a diploid zygote forms by syngamy, it undergoes meiosis to return to the haploid condition characteristic of most fungi.

Diversity of Fungi

Fungi can be classified into six divisions, based primarily on variations in sexual reproduction, and the lichens.

Division Oomycota The oomycetes include the water molds, white rusts, and downy mildews. They have coenocytic hyphae. Their cellulose cell walls, diploid predominance, and flagellated cells make them very different from typical fungi, and some classification systems include them with the protoctists. In the sexual reproduction of water molds, small sperm cells fertilize larger egg cells. Most water molds grow as saprophytes on dead algae and animals. White rusts and downy mildews are parasites of plants. They include the pathogens that cause downy mildew of grapes and late potato blight.

Division Chytridiomycota This diverse group includes some of the simplest fungi. Some unicellular forms have rootlike *rhizoids* that penetrate into their food source. Other chytrids may be multicellular with coenocytic hyphae. The cell walls are typically chitin, and the life cycle is predominantly haploid. In sexual reproduction, a sporangium releases zoospores. Some classification systems include absence of flagella at all stages of the life cycle as a diagnostic characteristic of fungi, and thus place Chytridiomycota and Oomycota with the protoctists. Some mycologists believe that chytrids may resemble the ancestors of more advanced fungi.

Division Zygomycota Zygomycetes are mostly terrestrial fungi that produce asexual spores in sporangia at the tips of aerial hyphae. Zygospores, formed in sexual reproduction, can remain dormant during harsh conditions.

Rhizopus stolonifer, the black bread mold, is a common zygomycete that spreads coenocytic, horizontal hyphae across its food substrate, produces rhizoids for anchorage and absorption, and erects hyphae with bulbous black sporangia containing hundreds of haploid spores. Sexual reproduction involves the conjugation of hyphal extensions from neighboring mycelia of opposite mating strains. The zygote develops a tough protective coat. *Pilobolus* is a dung fungus that can bend toward bright light to shoot its sporangia onto grass, which is eaten by grazing animals who scatter the spores in feces.

Division Ascomycota The ascomycetes, ranging from the unicellular yeasts to the cup fungi, produce their sexual spores in sacs called asci, which may be packed together to form ascocarps or "fruiting" structures. The hyphae of multicellular forms are septate. Ascomycetes produce chains of asexual spores, called *conidia*, at the tips of specialized hyphae.

Ascospores are haploid spores that are produced by sexual reproduction. In some ascomycetes, the eight ascospores formed from a zygote are lined up in order in the ascus, allowing geneticists to study the genetic recombination, crossing over, and independent assortment of chromosomes during meiosis.

Division Basidiomycota The basidiomycetes produce a club-shaped *basidium*, which represents a short diploid stage in the life cycles of the mushrooms, shelf fungi, puffballs, and stinkhorns. There is no asexual stage in the life cycle.

A large mass of heterokaryotic, septate hyphae extends under the ground from the reproductive structure, called a *basidiocarp*, of a mushroom. Syngamy and meiosis occur in the tip of hyphae lining the gills of a mushroom, forming basidia that produce four haploid basidiospores. Spores germinate to form monokaryotic hyphae that will fuse with hyphae of the opposite mating strain. A fairy ring of mushrooms may be produced at the perimeter of the expanding underground mycelium.

Division Deuteromycota The large group of fungi that seem to reproduce only asexually are lumped into the imperfect fungi. When a sexual stage is discovered in the life cycle of one of these, the fungus is reclassified into the appropriate group. Many deuteromycetes, including *Penicillium*, are probably related to the ascomycetes, as suggested by their formation of conidia.

Lichens Lichens are associations of millions of algal cells in a lattice of fungal hyphae. The fungus is usually an ascomycete, and the algae are usually unicellular or filamentous green algae or cyanobacteria. Lichens reproduce asexually either as fragments or by tiny airborn clumps called soredia. The algal and fungal components can also reproduce independently, sexually or asexually.

The symbiotic relationship between the algae and fungus may be mutual, or perhaps parasitic. The algae provide the fungus with food; it is debated whether the algae benefit from the fungus. These associations are given genus and species names, even though they are not single organisms. Naturalists categorize lichens according to their morphology, whether foliose (leafy), fruticose (shrubby), or crustose (crusty). Some vividly colored species are used to produce dyes.

Lichens can grow on bare rock and can withstand desiccation and great cold. In the tundra, herds of caribou and reindeer graze on lichens. Lichens cannot, however, tolerate air pollution, as they take up most of their minerals from the air in the form of compounds dissolved in rain.

Ecological and Commercial Importance of Fungi

Fungi and bacteria are the principal *decomposers* that make possible the essential recycling of chemical elements between living organisms and their abiotic surroundings.

Mycorrhizae are very common and important mutualistic associations of plant roots and fungi, in which the fungal hyphae provide minerals absorbed from the soil to the plant. Basidiomycetes are the fungi most often found in mycorrhizae.

Fungi have commercial uses as food (mushrooms, truffles, and flavoring in cheese), yeast (*Saccharomyces cerevisiae*) used in baking and brewing, and sources of antibiotics. In 1928, Fleming discovered that a strain of the fungus *Penicillium* killed bacteria, and Florey purified the antibiotic penicillin a decade later.

Fungi are also spoilers of food, clothing, and wood used in buildings and boats. Pathogenic fungi cause athlete's foot, ringworm, vaginal yeast infections, and lung infections.

Fungal diseases of plants include downy mildews, rusts, smuts, and blights. The Irish potato famine, which claimed the lives of a million people and caused the emigration of a million more, was caused by an oomycete. The Dutch elm disease is caused by an ascomycete. Another ascomycete forms purple structures called ergots on rye that can cause gangrene, spasms, burning sensations, temporary insanity, or death when accidentally milled into flour. Lysergic acid, the raw material of LSD (lysergic acid diethylamide), is one of the toxins in the ergots. Fungal infections of crops are treated and prevented by fungicides, and the breeding of resistant varieties of plants.

Evolution of Fungi

Fungi are not related to plants or animals; they appear to have had multiple origins from protoctists. The phylogeny of fungi is uncertain. Ascomycota may have evolved from Zygomycota and later given rise to Basidiomycota. The ancestors of Zygomycota may have been red algae, conjugating green algae, or perhaps Chytridiomycota.

All major groups of fungi had evolved by the end of the Carboniferous period, 300 million years ago. The oldest fossils of fungi date back 450–500 million years. The first vascular plant fossils have petrified mycorrhizae. Terrestrial communities seem to have always depended on fungi.

STRUCTURE YOUR KNOWLEDGE

1. Fill in the following table that summarizes the characteristics of the major groups of fungi.

Division	Common Members	Morphology	Asexual Reproduction	Sexual Reproduction	Miscellaneous Information
Oomycota					
Chitridio-mycota					
Zygomycota					
Ascomycota					
Basidio-mycota					
Deutero-mycota					
Lichens					

2. The kingdom Fungi contains members with saprophytic, parasitic, and mutualistic symbiotic modes of nutrition. How do these types of nutrition relate to the ecological and economic importance of this group?

TEST YOUR KNOWLEDGE

FILL IN THE BLANKS

1. _____ division between cells in fungal hyphae

2. _____ hyphal cells with two nuclei from different parents

3. _____ component of cell walls in most fungi

4. _____ asexual spores produced in chains at ends of hyphae

5. _____ club-shaped reproductive structure found in mushrooms

6. _____ sacs that contain sexual spores in cup fungi

7. _____ mutualistic associations between plant roots and fungi

8. _____ tough, protective zygote produced by zygomycetes

9. _____ type of hyphae with many nuclei

10. _____ hyphae of parasitic fungi that grows into cells or tissues

MULTIPLE CHOICE: *Choose the one best answer.*

1. Which of the following is not descriptive of the kingdom Fungi?
   a. absorptive form of nutrition
   b. alternation of generations, although haploid state is dominant
   c. eukaryotic heterotrophs
   d. most with body form of hyphae

2. The pores in septa
   a. are regions of high absorption into the mycelium.
   b. are the regions through which spores are expelled.
   c. provide a means for proteins and materials from other cells to move to the rapidly growing tips of hyphae.
   d. are involved in the unusual mitosis found in fungi.

3. A fungus that is both a parasite and a saprophyte is one that
   a. digests only the nonliving portions of its host's body.

b. lives off the sap within its host's body.
c. first lives as a parasite but then consumes the host after it dies.
d. lives as a mutualistic symbiont on its host.

4. The oomycetes differ from all the other fungi in that
   a. their hyphae are coenocytic.
   b. they have flagellated cells.
   c. they have cell walls of cellulose and are predominantly diploid.
   d. they have all the above differences.

5. The group that may resemble the ancestors of advanced fungi is the
   a. Oomycota.
   b. Chytridiomycota.
   c. Zygomycota.
   d. Ascomycota.

6. Sporangia on erect hyphae that produce asexual spores are found in the
   a. Basidiomycota.
   b. Ascomycota.
   c. Deuteromycota.
   d. Zygomycota.

7. In the Basidiomycota,
   a. monokaryotic hyphae fuse to grow into a heterokaryotic mycelium.
   b. spores line up in a sac in the order they were formed by meiosis.
   c. no sexual stage has been found.
   d. the vast majority of spores formed are asexual.

8. Lichens are symbiotic associations that
   a. usually involve an ascomycete and a green alga or cyanobacterium.
   b. can reproduce asexually by forming soredia.
   c. are able to colonize bare rocks and cold habitats.
   d. are all of the above.

9. The Deuteromycota
   a. have no known asexual stage.
   b. include *Penicillium* and yeasts.
   c. have many members that appear to be related to the ascomycetes because of their conidia.
   d. include fungi that reproduce by conjugation.

10. Fossil evidence indicates that
    a. all major fungal groups had evolved by the end of the Carboniferous.
    b. the Ascomycota evolved from the Zygomycota.
    c. mycorrhizae evolved from the first vascular plants.
    d. fungi evolved from protoctists.

CHAPTER 29

INVERTEBRATES AND THE ORIGIN OF ANIMAL DIVERSITY

FRAMEWORK

Animals are multicellular eukaryotic heterotrophs whose mode of nutrition is ingestion. Characteristics of animals include sexual reproduction; dominant diploid stage; specialized cells, tissues, and organs; embryonic development with a blastula stage; and the presence of muscles and nerves.

This chapter surveys the characteristics and representatives of the major animal phyla. Fossil evidence for the kingdom Animalia's origin and phylogeny is lacking. Comparative anatomy and embryology are used to construct phylogenetic trees. Phyla are organized into several broad groupings based on basic organization, symmetry, presence of a coelom, and embryology. The following sequence outlines these divisions.

are specialized in structure and function; in all but the simplest animals, cells are organized into tissues, tissues into organs, and, in the more complex animals, organs are grouped into systems. Animal cells lack walls. Desmosomes, gap junctions, and tight junctions may connect cells. Carbohydrates are stored as glycogen.

The diploid stage is dominant; gametes usually are the only haploid cells. Reproduction is primarily sexual, with a flagellated sperm fertilizing a larger, nonmotile egg. The zygote undergoes a series of divisions, called cleavage, and usually passes through a *blastula* (hollow ball of cells) stage during embryonic development. Some animal life cycles include a *larva*, a free-living, sexually immature form. *Metamorphosis* transforms a larva into a sexually mature adult.

Muscles and nerves are unique to animals and are

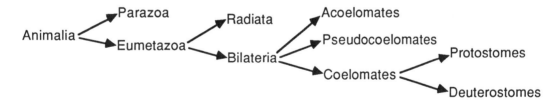

CHAPTER SUMMARY

The evolution of animal life in Precambrian seas opened a new adaptive zone of multicellular organisms that live by eating other organisms. Animals are grouped into about 30 phyla. All but one subphylum of these are *invertebrates*, animals that lack backbones.

Characteristics of Animals

Animals are multicellular eukaryotes that use *ingestion* as their mode of nutrition. The cells of an animal

necessary for their active behavior.

The greatest number of animal phyla are found in the stable environment of the oceans, where most invertebrates drift passively as plankton, live as sessile, attached animals, or burrow in the substrate. Most animals that live in freshwater environments are hypertonic to their surroundings and have osmoregulatory specializations. Terrestrial habitats have been exploited by only a few animal phyla, notably the vertebrates, arthropods, and ubiquitous nematodes.

Clues to Animal Phylogeny

Because of the lack of transitional fossils linking phyla to their predecessors, animal phylogeny depends on comparative anatomy and embryology. The animal kingdom appears to have had at least two separate origins from protoctists, leading to the sponges of the phylum Porifera, which are given a separate sub-kingdom, *Parazoa*, and to all other phyla, which are grouped into the subkingdom *Eumetozoa*.

The eumetozoans are divided into two branches on the basis of body symmetry. The branch *Radiata*, containing the jellyfishes and their relatives, have *radial symmetry*. The *Bilateria* include animals with *bilateral symmetry*, having distinct head and tail ends, and left and right sides. Bilateral symmetry is associated with *cephalization*, the concentration of sensory organs in the head end, which is an adaptation for movement. The radial symmetry of a sessile or planktonic animal allows it to sample its environment equally on all sides.

Early in the development of almost all animals in the branch Bilateria, the embryo develops three layers, called the *primary germ layers*. The *ectoderm* gives rise to the outer body covering and, in some phyla, to the central nervous system. The *endoderm* lines the primitive gut, or *archenteron*, and gives rise to the lining of the digestive tract and associated organs. The middle *mesoderm* forms the muscles and most other organs.

Animals of the Bilateria that have solid bodies are called *acoelomates*, and include the flatworms and a few other phyla. A fluid-filled cavity separating the digestive tract and the outer body wall is found in the other bilateria animals. In *pseudocoelomates*, including the rotifers and roundworms, the cavity is not completely lined by mesoderm. *Coelomates* have a true *coelom*, a body cavity completely lined by mesoderm. A fluid-filled coelom cushions internal organs, allows organs to grow and to move independently of the outer body wall, and also functions as a hydraulic skeleton in soft-bodied coelomates.

Differences in embryology and coelom formation divide the coelomates into two evolutionary lines. The *protostome* line includes annelids, mollusks, and arthropods. The *deuterostomes* include echinoderms and chordates. In protostomes, the coelom forms from splits within solid masses of mesoderm, called schizocoelous development. In deuterostomes, the mesoderm begins as lateral outpocketings from the archenteron, called enterocoelous development.

Protostomes have *spiral cleavage*, in which the planes of cell division are oblique to the polar axis of the egg and newly formed cells fit in the grooves between cells of adjacent tiers. The *determinate cleavage* of protostomes sets the developmental fate of each embryonic cell very early. Deuterostomes exhibit *radial cleavage*, in which parallel and perpendicular cleavage planes result in aligned tiers of cells in the dividing zygote. *Indeterminate cleavage* in deuterostomes means that early embryonic cells retain the capacity to develop into a complete embryo.

The *blastopore* is the opening of the developing archenteron. In protostomes, the blastopore develops into the mouth, and a second opening forms at the end of the archenteron to produce an anus. In the deuterostomes, the second opening develops into the mouth, whereas the blastopore becomes the anus.

Phylum Porifera (Sponges)

Sponges are sessile animals that are mostly marine. Water is drawn through pores in the wall of this saclike animal, into a central cavity, the *spongocoel*, from where it flows out through the *osculum*. Sponges are filter-feeders, collecting food particles from the water circulating through their bodies.

In the *mesoglea*, or gelatinous matrix, between the two layers of the sponge wall are *amoebocytes*. These cells carry food from the *choanocytes*, flagellated collar cells lining the inside cavity that phagocytize food particles, to the epidermal cells. Amoebocytes also form tough skeletal fibers, which may be spicules made of calcium carbonate or silicate, or flexible fibers composed of a protein called spongin.

Most sponges are *hermaphrodites*, producing both eggs and sperm that differentiate from amoebocytes. Sperm, released into the spongocoel and carried out through the osculum, cross-fertilize eggs retained in the mesoglea of neighboring sponges. Swimming larvae disperse through the osculum.

Sponges are capable of extensive *regeneration*, replacing damaged body parts and reproducing asexually from fragments broken off a parent sponge. Sponges lack organs and have relatively unspecialized cells. Ancestors of sponges may have been choanoflagellates, collared protoctists that resemble choanocytes and that may form colonies.

Radiata

The *phylum Cnidaria* includes hydras, jellyfishes, sea anemones and corals. These mostly marine animals have radial symmetry and a simple anatomy consisting of a sac with a central digestive cavity and a single opening serving as both mouth and anus. *Polyps* are sessile, cylindrical forms, whereas *medusae* are flattened, mouth-down polyps that move by passive drifting and weak contractions of the bell. Both these body forms are present in some cnidarians.

Cnidarians use their ring of tentacles, armed with *cnidocytes* or stinging cells, to capture prey. Cells of the epidermis and gastrodermis have bundles of micro-filaments arranged into contractile fibers. Cells can contract against the hydraulic skeleton of the gastrovascular cavity and coordinate their movement by a nerve net.

The class *Hydrozoa* includes animals that alternate between polyp and medusa forms, as seen in the colonial *Obelia*. Sexual reproduction occurs in the medusa stage, whereas the asexual polyp stage usually is more conspicuous. The common *Hydra* exists only in polyp form, reproducing asexually by *budding*, or sexually when environmental conditions deteriorate.

In the class *Scyphozoa*, the medusa stage is more prevalent. Jellyfishes are scyphozoans. The class *Anthozoa* includes sea anemones and most corals. They occur only as polyps.

The *phylum Ctenophora* includes the comb jellies, the largest animals that use cilia for locomotion. Nerves running from a sensory organ to combs of cilia coordinate movement.

Acoelomates

The flatworms of the *phylum Platyhelminthes* have the simplest organization of all the Bilateria. They develop from a three-layered embryo, have distinct organs, and have a gastrovascular cavity with only one opening.

The class *Turbellaria* includes freshwater *planarians* and other, mostly marine, free-living flatworms. Their branching gastrovascular cavity functions for both digestion and circulation. Gas exchange and diffusion of nitrogenous wastes occurs through the body wall. Ciliated flame cells function in osmoregulation. A planarian has eyespots on its head that detect light, and auricles that function for smell or chemical detection. A pair of ganglia lead into a ladderlike nervous system. Planarians reproduce asexually by regeneration or sexually by copulation between hermaphroditic worms.

Flukes are placed in the class *Trematoda*. Adapted to live as parasites in other animals, flukes have a tough outer covering, suckers, and extensive reproductive glands. The life cycles of flukes are complex, usually including intermediate hosts in which larvae develop.

The tapeworms of the class *Cestoda* are parasites, mostly of vertebrates. Tapeworms consist of a scolex with suckers and hooks for attaching to the host's intestinal lining, and a ribbon of proglottids packed with reproductive organs. Absorption of predigested food from the host eliminates the need for a digestive system. The life cycles of tapeworms may include intermediate hosts.

Most proboscis worms in the *phylum Nemertina* are marine. Their long, retractable *proboscis* is used to explore the environment and capture prey. These worms probably evolved from flatworms, but show two new anatomical features: a digestive tube with separate mouth and anus, and a circulatory system of vessels through which blood flows.

Pseudocoelomates

The *phylum Rotifera* includes the tiny rotifers, which are smaller than many protoctists but have a pseudocoelom, a complete digestive tract, and other organ systems. A crown of cilia draws water into the mouth, and microscopic food organisms are ground in a jawlike organ posterior to the mouth. Many populations of rotifers reproduce by *parthenogenesis*; females produce female offspring from unfertilized eggs. In some species, another type of egg, also produced by parthenogenesis, develops into degenerate males that fertilize eggs, which then form resistant zygotes.

The roundworms in the *phylum Nematoda* are among the most abundant of all animals, found inhabiting water, soil, and the bodies of plants and animals. These cylindrical worms are covered with tough cuticles, have complete digestive tracts, and move by contraction of their longitudinal muscles. Reproduction is exclusively sexual; fertilization is internal, and most zygotes form resistant cells. Many nematodes are serious agricultural pests and animal parasites.

Coelomate Protostomes

There are several phyla of worms with coeloms that function as hydraulic skeletons for burrowing. The *phylum Annelida* includes segmented worms that are found in most aquatic and soil habitats. The earthworm, a well-known example, has a complete digestive system with specialized regions, a complex circulatory system, and both male and female reproductive organs. These hermaphroditic worms exchange sperm during copulation. Respiration occurs across the moist skin. Septa partition the coelom into segments, in which are found a pair of excretory nephridia and ganglia along the ventral nerve cords. The brain is a pair of cerebral ganglia.

The noncompressible coelomic fluid, enclosed by septa within the body segments, serves as a hydraulic skeleton; circular and longitudinal muscles contract alternately to extend the body and to pull it forward while the setae grip the substrate. Segmentation may have first evolved as an adaptation for this type of burrowing or creeping movement.

The class *Oligochaeta* includes the earthworms and other aquatic species. The mostly marine segmented worms in the class *Polychaeta* have parapodia on each segment that function in gas exchange and locomotion. Many marine polychaetes are tube-dwellers. Most of the leeches in the class *Hirudinea* inhabit fresh water. Many are parasites that temporarily attach to animals, slit or digest a hole through the skin, and suck the blood of their host.

The mollusks, in the *phylum Mollusca*, are mostly marine, soft-bodied animals, many of which are protected by a shell. The molluscan body plan has three main parts: a muscular *foot* used for movement, a *visceral mass* containing the internal organs, and a *mantle* that covers the visceral mass and may secrete a shell. A muscular tongue with a rasping *radula* is used for feeding by all mollusks except the bivalves. Most mollusks have separate sexes.

Many mollusks have a life cycle that includes a ciliated larva called the trochophore, also found in many annelids. Mollusks do not share the trait of segmentation with annelids, and many zoologists believe that mollusks and annelids had separate origins from flatworms.

Of the seven molluscan classes, four are considered to be major. Chitons, in the class *Polyplacophora*, are oval marine animals with shells that are divided into eight dorsal plates.

Gastropoda is the largest class. Most members are marine, although there are many freshwater species and some snails and slugs are terrestrial. Land snails lack gills; their vascularized mantle cavity functions as a lung. Many gastropods have distinct heads with eyes at the tips of tentacles. Most gastropods have single spiraled shells. A distinctive feature of this class is *torsion*, the embryonic asymmetric growth that results in a U-turn of the digestive tract, with the anus ending up near the mouth.

The clams, oysters, mussels and scallops of the class *Bivalvia* have the two halves of their shells hinged at the mid-dorsal line and closed by powerful adductor muscles. Bivalves have no heads; most are filter-feeders in which food particles are trapped in the mucus that coats the gills, and swept to the mouth by cilia. Water flows into and out of the mantle cavity through siphons.

The carnivorous squid and octopuses of the class *Cephalopoda* are much faster than are the sluggish gastropods or sedentary or sessile bivalves. The mouth at the center of the foot has beaklike jaws to crush prey. The foot is drawn out to form several long tentacles. The shell is either reduced to an internal cuttlebone (squid) or absent. The giant squid is the largest of all invertebrates.

Cephalopods are the only mollusks with closed circulatory systems. They have well-developed nervous systems and sense organs, and complex brains. The ancestors of octopuses and squid were probably shelled predacious mollusks that eventually reduced their shells. Shelled ammonites were the dominant invertebrate predators until their extinction at the end of the Cretaceous period. The chambered nautilus has survived.

Characteristics of the *phylum Arthropoda* include jointed appendages and a *cuticle* that completely covers the body as an *exoskeleton* composed of chitin and protein. To grow, an arthropod must *molt*, shedding its old exoskeleton and secreting a larger one. Arthropods have extensive cephalization and well-developed sensory organs.

A heart pumps blood through an open circulatory system consisting of short arteries and a network of blood-filled sinuses known as the hemocoel. Respiration may occur through gills, in most aquatic species, or via tracheal systems, in terrestrial arthropods.

The segmented arthropods probably evolved from annelids or from a segmented protostome that was a common ancestor of both segmented phyla. Early arthropods may have resembled onychophorans, a phylum of worms with unjointed walking appendages. There are Cambrian fossils of jointed-legged animals that resemble segmented worms.

The *trilobites*, with pronounced segmentation and uniform appendages, were early arthropods that were common throughout the Paleozoic era. The large eurypterids, or sea scorpions, were marine predators that belonged to an arthropod group called *chelicerates*. The chelicerate body is divided into a cephalothorax and abdomen, with the most anterior appendages modified as pincers or fangs, called *chelicerae*. The horseshoe crab is the one surviving marine chelicerate. Most modern species are in the terrestrial class Arachnida.

Another evolutionary line produced the *mandibulates*, whose most anterior appendages are modified as jawlike *mandibles*. Mandibulates also have a pair of anterior appendages modified as sensory *antennae*, and most have *compound eyes*.

The exoskeleton of early marine arthropods, which probably functioned in protection and anchorage for muscles, helped preadapt certain arthropods for terrestrial life by providing the needed support and protection from water loss. During the early Devonian period, both chelicerates and mandibulates spread onto land. The oldest record of terrestrial animals are millipede fossils from about 400 million years ago.

Scorpions, spiders, ticks, and mites are grouped in the class *Arachnida*. The cephalothorax has six pairs of appendages: two pairs of chelicerae, pedipalps that usually function in feeding, and four pairs of walking legs. In most spiders, *book lungs* are used in gas exchange. Many spiders spin silk webs for capturing prey.

The class *Crustacea* has remained mostly in marine and freshwater habitats. Crustaceans have two pairs of antennae, three or more pairs of mouthpart appendages, walking legs on the thorax, and appendages on the abdomen. The most anterior walking legs are modified as pincers in lobsters and crayfish. Larger crustaceans have gills. In their open type of circulatory system, a heart pumps bloods through arteries into sinuses that bathe the organs. Nitrogenous wastes pass by diffusion through thin areas of the cuticle, and a pair of glands regulates salt balance in the blood.

Sexes usually are separate. Most aquatic crustaceans

go through a swimming larval stage.

Lobsters, crayfish, crabs, and shrimp are relatively large crustaceans called decapods. The dorsal side of the cephalothorax forms a shield called the carapace. Isopods are mostly small marine crustaceans, but also include the terrestrial sow bugs. The small, very numerous copepods are important members of the plankton communities that form the foundation of marine and freshwater food chains. Barnacles are sessile crustaceans that feed by using their feet to sweep food toward their mouths.

The millipedes of the class *Diplopoda* are wormlike, segmented mandibulates with two pairs of walking legs per segment. They eat plant matter. The centipedes of the class *Chilopoda* are terrestrial carnivores with poison claws for paralyzing prey. The head has a pair of antennae and three pairs of appendages modified as mouthparts; each segment of the trunk has one pair of legs.

The class *Insecta* is divided into about 26 orders, and has more species than all other forms of life combined. The study of insects, called *entomology*, is a vast field. The oldest insect fossils date back to the Devonian period, but a major diversification occurred in the Carboniferous and Permian periods with the evolution of flight. Another important radiation occurred as insects developed coevolutionary relationships with flowering plants during the Cretaceous period.

Most insects have one or two pairs of wings emerging from the dorsal thorax, which may have evolved as extensions of the cuticle that helped the insect body absorb heat. Insects flap their wings by the action of muscles that change the shape of the cuticle covering the thorax. Dragonflies were among the first winged insects.

A generalized insect body has three regions: a head, a thorax, and an abdomen. The head consists of several segments that have become fused together. The head has one pair of antennae, a pair of compound eyes and several pairs of mouth parts. The thorax bears wings and three pairs of walking legs.

The digestive tract has several specialized regions. The circulatory system is open; the heart pumps blood through arteries into the hemocoel. *Malpighian tubules* are outpocketings of the gut that function as excretory organs, removing nitrogenous wastes from the blood. Gas exchange is accomplished by a tracheal system of branched, chitin-lined tubes that open to the outside through spiracles and ramify throughout the body.

The nervous system consists of a pair of ventral nerve cords with a ganglion in each segment, and a dorsal brain in the head formed from several fused ganglia. Insects are capable of complex behavior.

Many insects have larval stages in their life cycles. In *incomplete metamorphosis*, the larvae are smaller versions of the adult insect. In *complete metamorphosis*, the larvae, known as grubs or caterpillars, look entirely different from the adult. Larvae are designed to eat and grow; adults to mate and reproduce. Females usually lay their eggs on a food source so that the larvae can begin eating as soon as they hatch.

Reproduction almost always is sexual; fertilization usually is internal. Sperm may be stored inside the female in a spermatheca where they can be used to fertilize more than one batch of eggs. Males may have specialized phallic lobes or other structures that help transfer sperm from the male to the female gonopore during copulation.

Insects have a huge influence on humans as pollinators of food crops, vectors of diseases, and competitors for food.

The Lophophorate Animals

The three phyla of *lophophorate animals, Phoronida, Bryozoa,* and *Brachiopoda,* are difficult to assign as protostomes or deuterostomes on the basis of their embryonic development. One hypothesis holds that the lophophorate animals branched from a common coelomate ancestor before all the developmental features characteristic of deuterostomes had evolved. Their name refers to a distinctive feeding organ, called a *lophophore*—a horseshoe-shaped or circular fold bearing ciliated tentacles—which surrounds the mouth and functions in filter-feeding. The lophophore suggests that all three phyla are related. The U-shaped digestive tract and lack of a distinct head may have evolved separately in each group as adaptations to a sessile life.

Phoronids are marine worms that live buried in sand in chitin tubes with their lophophore extended from the tube for feeding. Bryozoans are tiny animals that live in colonies that resemble mosses and may be encased in a hard exoskeleton. Bryozoans are among the most widespread and numerous sessile animals in the sea. The marine brachiopods, or lamp shells, resemble bivalves. They attach to the substratum by a stalk and open their shell slightly to allow water to flow through the lophophore. *Lingula* is a living brachiopod genus that has changed little in 400 million years.

Deuterostomes

The *phylum Echinodermata* includes sessile or sedentary marine animals with radial symmetry. *Echinoderms* have a thin skin covering an endoskeleton of hard calcareous plates. A *water vascular system*, with a network of hydraulic canals and extensions called tube feet, functions in locomotion, feeding, and gas exchange.

Reproduction is sexual; fertilization is external. Bilateral larvae metamorphosize into radial adults.

The sea stars of the class *Asteroidea* have five arms radiating from a central disc. Sea stars use tube feet to creep slowly and to open bivalves for food. They evert their stomach through their mouth and slip it into a slightly opened shell, where digestive juices begin to digest the bivalve. Sea stars are capable of regeneration from fragments. Brittle stars, of the class *Ophiuroidea*, have smaller central discs and longer arms than do sea stars. They lack tube feet and move by lashing their arms.

The sea urchins and sand dollars of the class *Echinoidea* have no arms, but do have five rows of tube feet by which they move slowly. The mouth is ringed by complex jawlike structures used to eat seaweeds and other foods. Sea lillies, in the class *Crinoidea*, live attached by stalks to the substratum, and extend their arms upward from around the mouth. Their form has changed little in 500 million years of evolution.

The class *Holothuroidea* includes the sea cucumbers. These elongated animals with five rows of tube feet bear little resemblance to other echinoderms. The *phylum Chordata* contains two subphyla of invertebrates plus the subphylum Vertebrata. Echinoderms and chordates share the developmental characteristics of deuterostomes, but they have existed as separate phyla for at least 500 million years.

The Origins of Animal Diversity

Several hypothetical ancestors have been suggested for the origin of the eumetazoa. One theory suggests a colonial flagellate ancestory, perhaps involving a ball of flagellated cells or a planuloid, a swimming, cigar-shaped, solid mass of cells. Another suggested hypothetical ancestor is a plakula, modeled on *Trichoplax* (phylum Placozoa), a contemporary tiny, flat marine creature with a ciliated epidermis. Lacking a gut, this animal hunches up to form a temporary digestive cavity when it crawls over a food particle. The first archenteron may have originated in this fashion.

Another hypothesis is that the eumetazoa developed from a coenocytic ancestor, perhaps a ciliate, that became divided into many cells. Early animals may have resembled acoel worms, an order of flatworms that lack digestive cavities and take food in through a pharynx that leads into a solid coenocytic mass.

There is no fossil record of transitional forms between protoctists and the first animals. The early diversification of animals also is a phylogenetic puzzle. The oldest known fossils come from the end of the Precambrian era, in a span from about 680 to 570 million years ago known as the Ediacaran period. These soft-bodied creatures resembled cnidarian medusae, some colonial cnidarians, and annelids.

A second radiation occurred in the Cambrian period at the beginning of the Paleozoic era about 570 million years ago. The evolution of shells and hard skeletons may have opened new adaptive zones. Nearly all modern phyla, along with many extinct phyla, are present in Cambrian fossils. The Cambrian appearance of so many complex animals is one of the great mysteries of biological history.

Some paleobiologists see the Cambrian radiation as a continuation of the diversification of body forms seen in the Precambrian worms and jellyfish. Seilacher interprets the Ediacaran fossils as discontinuous with the Cambrian fauna. In his view, the two groups represent different solutions to problems associated with increased size. The Ediacaran animals were flat and foliate, exposing most cells to the environment. The Cambrian fauna and their descendents were thicker, with extensive internal surfaces for exchange with the environment. The Cambrian radiation may represent new designs that evolved, in the aftermath of a mass extinction, from burrowing cylindrical animals.

STRUCTURE YOUR KNOWLEDGE

1. The diverse phyla of the kingdom Animalia are structured into broad groups based on a series of general characteristics (symmetry, coelom, and so on). Create a concept map or diagram that shows these major divisions, the criteria used to determine them, and the major phyla included in each group.

2. Why are the origin and phylogeny of the kingdom Animalia a mystery? Describe some of the hypotheses concerning animal evolution.

TEST YOUR KNOWLEDGE

MATCHING: *Match the following organisms with their classes and phyla. Answers may be used more than once or not at all.*

Class	Phy-lum	Organism	Classes	Phyla
1. ___	___	jellyfish	A. Arachnida	a. Annelida
2. ___	___	crayfish	B. Bivalvia	b. Arthropoda
3. ___	___	snail	C. Cestoda	c. Cnidaria
4. ___	___	leech	D. Crustacea	d. Echinodermata
5. ___	___	planarian	E. Echinoidea	e. Mollusca
6. ___	___	cricket	F. Gastropoda	f. Nematoda
7. ___	___	scallop	G. Hirudinea	g. Platyhelminthes

TEST YOUR KNOWLEDGE

MATCHING: *Match the following organisms with their classes and phyla. Answers may be used more than once or not at all.*

Class	Phylum	Organism	Classes	Phyla
1. ___	___	jellyfish	A. Arachnida	a. Annelida
2. ___	___	crayfish	B. Bivalvia	b. Arthropoda
3. ___	___	snail	C. Cestoda	c. Cnidaria
4. ___	___	leech	D. Crustacea	d. Echinodermata
5. ___	___	planarian	E. Echinoidea	e. Mollusca
6. ___	___	cricket	F. Gastropoda	f. Nematoda
7. ___	___	scallop	G. Hirudinea	g. Platyhelminthes
8. ___	___	tick	H. Hydrozoa	h. Porifera
9. ___	___	sea urchin	I. Insecta	i. Rotifera
10. ___	___	hydra	J. Oligochaeta	
			K. Ophiuroidea	
			L. Scyphozoa	
			M. Turbellaria	

MULTIPLE CHOICE: *Choose the one best answer.*

1. Invertebrates include
   a. all animals except for the phylum Vertebrata.
   b. all animals without backbones.
   c. only animals that use hydraulic skeletons.
   d. all of the above.

2. The greatest number of animal phyla are found
   a. in the ocean, where environmental conditions are most stable.
   b. in freshwater habitats, where organisms do not have to deal with problems of salt balance.
   c. on land, where gas exchange and locomotion are easiest.
   d. in the tropical rain forests, where temperatures are warm.

3. A larva
   a. is a sexually immature organism that may or may not resemble the adult.
   b. is transformed into an adult by metamorphosis.
   c. is specialized for eating and growth.
   d. is all of the above.

4. The Parazoa
   a. includes the radially symmetrical sponges, jellyfishes, and their relatives.
   b. includes the only animals that do not feed by ingestion.
   c. is the phylum name for the sponges.
   d. is a subkingdom that appears to have had a separate origin from the protoctists.

5. Cephalization
   a. is the development of bilateral symmetry.
   b. is associated with motile animals that concentrate sensory organs in a head region.
   c. is a diagnostic characteristic of deuterostomes.
   d. is common in radially symmetrical animals.

6. A true coelom
   a. is found in deuterostomes.
   b. is a fluid-filled cavity completely lined by mesoderm.
   c. may be used as a hydraulic skeleton by soft-bodied coelomates.
   d. all of the above are correct.

7. Which of the following is descriptive of protostomes?
   a. radial and determinate cleavage, blastopore becomes mouth
   b. spiral and indeterminate cleavage, coelom forms as split in solid mass of mesoderm
   c. spiral and determinate cleavage, blastopore becomes mouth, schizocoelous development
   d. radial and determinate cleavage, enterocoelous development, blastopore becomes anus

8. Hermaphrodites
   a. contain male and female sex organs but usually cross-fertilize.
   b. include sponges, earthworms, and most insects.
   c. are characteristically found in rotifers.
   d. are all of the above.

9. A gastrovascular cavity
   a. functions in both digestion and circulation.
   b. has only one opening that serves as mouth and anus.
   c. is found in the phyla Cnidaria and Platyhelminthes.
   d. all of the above are correct.

10. Which of the following is not true of cnidarians?
    a. An alternation of medusa and polyp stage is common in the class Hydrozoa.
    b. They use a ring of tentacles armed with stinging cells to capture prey.
    c. They include hydras, jellyfishes, sea anemones, and barnacles.
    d. They have a nerve net that coordinates contraction of microfilaments for movement.

11. Which of the following combinations of phylum and characteristic is incorrect?
    a. Nemertina — proboscis worm, first complete digestive tract
    b. Rotifera — parthenogenesis, crown of cilia, microscopic animals
    c. Nematoda — gastrovascular cavity, tough cuticle, ubiquitous
    d. Annelida — segmentation, closed circulation, hydraulic skeleton

12. Which of the following is either not an excretory structure or is incorrectly matched with its class?
    a. nephridia — Oligochaeta
    b. Malpighian tubules — Insecta
    c. flame cells — Turbellaria
    d. thin region of cuticle — Gastropoda

13. The phylum Mollusca is characterized by
    a. radial symmetry, visceral mass, and hydraulic skeleton.
    b. a foot, mantle, and visceral mass.
    c. shell, mantle, and radula.
    d. tube feet, visceral mass, and mantle.

14. Torsion
    a. is embryonic asymmetric growth that results in a U-shaped digestive tract in gastropods.
    b. is characteristic of molluscs.
    c. is responsible for the spiral growth of snail shells.
    d. is all of the above.

15. Bivalves differ from other molluscs in that
    a. they are predaceous.
    b. they have no heads and are filter feeders.
    c. they have shells.
    d. they have open circulatory systems.

16. The oldest fossils of terrestrial animals are
    a. trilobites.
    b. land snails.
    c. millipedes.
    d. eurypterids.

17. The exoskeleton of arthropods
    a. functions in protection and anchorage for muscles.
    b. is composed of chitin and cellulose.
    c. is absent in millipedes and centipedes.
    d. expands at the joints when the arthropod grows.

18. Which of the following is not true of the chelicerates?
    a. Their body is divided into a carapace and an abdomen.
    b. The horsehoe crab is the one surviving marine chelicerate.
    c. They include ticks, scorpions, and spiders.
    d. The most anterior appendages are modified as pincers or fangs.

19. The lophophorate animals
    a. are named for their locomotive organ called a lophophore.
    b. do not have an embryology characteristic of protostomes or deuterostomes.
    c. are three phyla of sessile animals that resemble moss.
    d. are all of the above.

20. The oldest known animals
    a. were soft-bodied creatures from the Ediacaran period, over 570 million year ago.
    b. were from the Cambrian period at the beginning of the Paleozoic era.
    c. were planuloids, swimming, cigar-shaped, solid masses of cells.
    d. were coenocytic ciliates resembling acoel worms.

CHAPTER 30

VERTEBRATE GENEALOGY

FRAMEWORK

This chapter focuses on the origin and characteristics of the subphylum Vertebrata and its classes: Agnatha, Placodermi, Chondrichthyes, Osteichthyes, Amphibia, Reptilia, Aves, and Mammalia. The major orders or groups within these classes are discussed, and the fossil evidence concerning human evolution is described.

CHAPTER SUMMARY

Phylum Chordata

There are three subphyla of *chordates*, two without backbones—the *cephalochordates* and the *urochordates*—and one with backbones—the *vertebrates*. There are three distinguishing anatomical features present in the embryo stage of chordates: (1) a flexible rod called a *notochord* running the length of the animal, (2) a *dorsal, hollow nerve cord* that develops into the brain and spinal cord of the central nervous system, and (3) *pharyngeal slits* that open from the anterior region of the digestive tube to the outside.

Amphioxus, also called the Lancelet, is a tiny marine animal in the subphylum Cephalochordata. It secretes a mucous net across the pharyngeal slits that traps food particles. Amphioxus swims by coordinated contractions of serial muscle segments that flex the notochord from side to side. Blocks of mesoderm called *somites*, which flank the notochord of chordate embryos, develop into these muscle segments.

Tunicates are sessile, filter-feeding, marine animals in the subphylum Urochordata. Water moves through the saclike, tunic-covered animal from the incurrent siphon, through pharyngeal slits, to the excurrent siphon. The larval traits of notochord and nerve cord are lost in the adult animal.

The Origin of Vertebrates

A fossil resembling amphioxus has been dated at 550 million years old; vertebrates appear 50 million years later. The fossil record does not show the origin of vertebrates from invertebrate ancestors. The vertebrate ancestor may have been a cephalochordate that left its burrowing or sessile existence and became more active, resulting in a need for a distinct head with sensory organs. The vertebrate ancestor also could have been a larval form, perhaps like a tunicate larva, that became sexually mature before undergoing metamorphosis. Mutations in regulatory genes could result in sexual maturity within a larval form, a condition known as *paedogenesis*.

Vertebrate Characteristics

Cephalization, the development of sense organs and a brain, and *vertebrae*, a series of skeletal units enclosing the nerve cord, are the most fundamental differences setting off vertebrate from invertebrate chordates. The skull enclosing the brain and the vertebral column are made of bone and/or cartilage and can grow with the animal.

Vertebrates have a closed circulatory system with a ventral, two-or-more-chambered heart pumping blood through arteries to capillaries and back through veins. The blood, containing hemoglobin in red blood cells, is oxygenated through the skin, gills, or lungs. Waste products are removed as the blood passes through the kidneys. Reproduction almost always is sexual, with separate male and female sexes.

Four classes of the subphylum Vertebrata are called *tetrapods* because they have two pairs of limbs. The other four classes are commonly called fishes.

Class Agnatha

The oldest vertebrate fossils are jawless creatures called *agnathans*, including the fishlike *ostracoderms* that have an armor of bony plates. First traces of agnathans are found in the early Ordovician strata, and a tremendous radiation occurred by the late Silurian, about 400 million years ago. Agnathans were generally small and probably were mud-suckers or filter-feeders. Most disappeared during the Devonian period. The lampreys and hagfishes represent the jawless fishes today.

Class Placodermi

A diversity of armored fishes called *placoderms* replaced the agnathans during the late Silurian and early Devonian periods. These larger fish had paired fins that enhanced swimming, and jaws that evolved from the skeletal supports of the anterior gill slits. Food could be obtained by biting. The remaining gill slits, no longer needed for filter-feeding, maintained their function in gas exchange.

The Devonian period is known as the Age of Fishes. The placoderms radiated in both fresh and salt water, and the sharks and bony fishes made their initial appearances. The placoderms and acanthodians (another group of jawed fishes) disappeared almost completely by the Carboniferous period, 350 million years ago.

Class Chondrichthyes (Cartilaginous Fishes)

Sharks and rays have skeletons made of cartilage, and well-developed jaws and paired fins. Sharks must swim to maintain buoyancy and to move water into their mouths and past their gills. The largest sharks filter-feed on plankton, although most are carnivores. Shark teeth evolved from the jagged scales that cover the skin. A spiral valve in the intestine increases surface area within the relatively short digestive tract.

Sharks have keen senses: sharp eyes, nostrils, and regions in the head that can detect electric fields generated by muscle contractions of nearby fish. The *lateral line system* is a row of microscopic organs, running along the sides of the shark body, that are sensitive to water pressure changes. Sharks also have a pair of auditory organs.

Following internal fertilization, *oviparous* species of sharks lay eggs that hatch outside the mother. *Ovoviviparous* species retain the fertilized eggs until they hatch within the uterus. A few species are *viviparous*; the developing young are nourished by a placenta until born. The cloaca is the common chamber for the openings of the reproductive tract, excretory system, and digestive tract.

Class Osteichthyes (Bony Fishes)

The skeleton of most bony fishes has a hard matrix of calcium phosphate. The skin often is covered by flattened bony scales. The movement of the *operculum*, or flap over the gill chamber, draws water across the gills. In contrast to sharks, fish can breathe while stationary. Bony fishes also have a *swim bladder*, an air sac that controls buoyancy.

The flexible fins of bony fish are better for steering and propulsion than are the stiffer fins of sharks. The fusiform body shape of fish, sharks, and aquatic mammals reduces drag while the animal is swimming. Most species of fish are oviparous, and fertilization is external.

Bony fishes probably originated in fresh water, and the swim bladder was modified from lungs that were used to augment gills. Two distinct subclasses evolved: the *ray-finned fishes*, including most of the familiar fish, many of which spread to the seas during their evolution; and the *fleshy-finned fishes*, which remained in fresh water and continued to use their lungs. These fish had muscular pectoral and pelvic fins that were supported by extensions of the skeleton and may have enabled the fish to "walk" along the sea bottom or, occasionally, on land. Three genera of *lungfishes* living today descended from the fleshy-finned fishes, as well as one species of lobe-fin fish, the coelocanth, that is a lungless species living in the sea.

Class Amphibia

Lobe-finned fishes with lungs, living in freshwater habitats subject to drought, may have been the ancestors of tetrapods. The amphibians were the first vertebrates on land; the oldest fossils date back to the late Devonian, about 350 million years ago. Amphibians have always been predators, eating the insects and other invertebrates that preceded them onto land. The Carboniferous period is known as the Age of Amphibians, during which a diversity of new forms and sizes evolved. Amphibians declined during the late Carboniferous, and by the Mesozoic era the survivors resembled modern amphibians.

Some frogs and their relatives undergo a metamorphosis from a tadpole, an aquatic herbivorous larval form, to the carnivorous terrestrial adult form with legs and lungs. Most members of the class do not have a tadpole stage. Oviparous amphibians generally lay their eggs in aquatic or moist environments. Some species are ovoviviparous or viviparous.

Even with lungs, amphibians carry out much of their gas exchange across their moist skin; they are most abundant in damp habitats.

There are three major orders of amphibians: *urodeles* or salamanders, *anurans* or frogs, and *apodans* or burrowing, limbless amphibians called caecilians. Most salamanders move by a back-and-forth bending of the body. A tadpolelike stage is less common. Paedogenesis occurs in some groups; the mudpuppy retains gills and other larval features.

Anurans are more specialized than urodeles for moving on land, as seen by the powerful hind legs that frogs use for hopping. In mating, the male frog grasps the female, but fertilization is external. Metamorphosis is common among anurans.

Class Reptilia

Reptiles have several adaptations for terrestrial living. The skin is covered with keratinized waterproof scales, and most species lay eggs covered by protective shells on land. The embryo develops within a fluid-filled amniotic sac, and the evolution of the *amniote egg* allowed vertebrates to complete their life cycles on land.

Although reptiles do not use their metabolism to control body temperature, they can regulate their temperature through behavioral adaptations. Reptiles are *ectothermic*; they absorb external heat rather than generating their own, and thus have much lower caloric needs than a mammal has.

The oldest reptilian fossils date from 300 million years ago from the upper Carboniferous period. The *cotylosaurs*, or "stem reptiles," were the small insect-eating, lizardlike ancestors of the various reptilian orders that arose in two major adaptive radiations. During the Permian period, the first radiation gave rise to terrestrial predators called *synapsids*, from which one lineage led to the *therapsids*. Mammals evolved from the therapsid line. Some stem reptiles returned to water, leading to the plesiosaurs and ichthyosaurs. The *thecodonts* also arose during the Permian radiation and, following the mass extinctions of the Permian period, led to the dinosaurs, crocodiles, and birds.

The second great radiation occurred about 200 million years ago during the late Triassic period, when the diverse dinosaurs evolved from thecodonts. The flying pterosaurs and terrestrial dinosaurs were the dominant vertebrates on Earth for millions of years. They included the largest animals ever to live on land. Some scientists postulate that dinosaurs were endothermic and were able to conserve their metabolic heat.

During the Cretaceous period at the end of the Mesozoic era, the climate became cooler and more variable. Within 5 to 10 million years, the dinosaurs declined and became extinct. A few reptilian groups survived and are present today.

The three largest orders of extant reptiles are: *Squamata*, lizards and snakes; *Chelonia*, turtles; and *Crocodilia*, alligators and crocodiles. Lizards are the most numerous and diverse reptiles. Snakes, apparently descended from lizards that adopted a burrowing life style, are known for their absence of limbs. All snakes are carnivorous, with adaptations for locating (heat-detecting organs and olfactory organs), killing (sharp teeth, often with toxins), and swallowing (loosely articulated jaws) their prey.

Turtles evolved from stem reptiles during the Mesozoic era and have changed little since then. Crocodiles and alligators are among the largest living reptiles.

Class Aves (Birds)

Birds evolved from dinosaurs during the reptilian radiation of the Mesozoic era. The amniote egg and scales on the legs are two reptilian traits birds display. The characteristics of birds revolve around their ability to fly. Their bones are honeycombed—strong but light. They are toothless; food is ground in the gizzard. The keratinized beak has a great variety of shapes associated with different diets. Birds are endothermic; feathers and a fat layer help retain metabolic heat. The four-chambered heart segregates oxygenated and oxygen-poor blood, which facilitates a high metabolic rate.

Birds have excellent vision. Their brains, proportionately larger than those of reptiles and amphibians, provide for more complex behavior, and the coordination necessary for flight. Fertilization is internal. Eggs are laid and kept warm during development by the female and/or male birds.

Wings provide for propulsion as well as lift. They are flapped by large pectoral muscles attached to a keel on the sternum. Feathers, made of keratin and evolved from reptilian scales, are extremely light and strong, and shape the wing into an airfoil.

Wings evolved from forelimbs more than once during the Mesozoic radiation. Pterosaurs had wings formed from a membrane of skin stretched between an elongated finger and the body. Wings using the entire forelimb evolved in a line of dinoaurs with feathers. Feathers may have functioned first as insulation in the evolution of endothermy. Fossils of the earliest known bird, *Archaeopteryx*, are found from the Jurassic period, 150 million years ago. Birds evolved from a diverse group of dinosaurs called the *archosaurs*, and many zoologists argue that birds are living archosaurs, which should be regarded as reptiles.

Class Mammalia

Mammals have hair, made of keratin, which helps to insulate these endothermic animals. Their active metabolism is provided for by an efficient respiratory system that uses a *diaphragm* to help ventilate the lungs, and a four-chambered heart that separates oxygenated and oxygen-poor blood. Milk, produced in mammary glands, is used to nourish mammalian young. Fertilization is internal, and the embryo develops in the uterus. In placental mammals, a *placenta*, formed from the uterine lining and embryonic membranes, nourishes the developing embryo. Most mammals bear live young.

Mammals have relatively large brains and are capable of learning. Extended parental care provides time for the young to learn complex skills.

Mammalian teeth come in a variety of shapes and sizes that are specialized for eating a variety of foods.

The oldest fossils of mammals date back to the Triassic period, 190 million years ago. Mammalian ancestors were among the therapsids. Small Mesozoic mammals, which were probably nocturnal and insectivorous, coexisted with dinosaurs. Mammals underwent an extensive radiation in the adaptive zones created by the extinction of the dinosaurs. Endothermy, which may have functioned as an adaptation for a nocturnal life, would have preadapted mammals for the cooler late Cretaceous period and Cenozoic era. There are three major mammal groups today: *monotremes*, egg-laying mammals; *marsupials*, mammals with pouches; and *placental mammals*.

The platypus and spiny anteaters (echidnas) are the only egg-laying mammals today. They have hair and milk. Monotremes seem to have descended from a very early mammal branch.

Marsupials, including opossums, kangaroos, and koalas, complete their embryonic development in a maternal pouch called a marsupium. Born very early, the neonate fixes its mouth to a teat and completes its development while nursing.

Australian marsupials have radiated and filled the niches occupied by placental mammals in other parts of the world. Marsupials apparently originated in what is now North America, spreading southward. After the breakup of Pangaea, South America and Australia became island continents where the marsupials diversified in isolation from placental mammals. Australia has been separated since early in the Cenozoic era, about 65 million years ago. Migration of mammals occurred between North and South America about 12 million years ago and again about 3 million years ago.

A placental mammal completes its development attached to a placenta within the uterus. Placentals and marsupials may have diverged from a common ancestor about 80 to100 million years ago. The major orders of placental mammals radiated during the late Cretaceous and early Tertiary, about 70 to 45 million years ago.

There appear to be four main evolutionary lines of placental mammals. The orders Insectivora and Chiroptera, containing the shrews and bats respectively, resemble early mammals. A second branch from a lineage of medium-sized herbivores gave rise to the lagomorphs (rabbits), perissodactyls (odd-toed ungulates such as horses and rhinos), artiodactyls (even-toed ungulates such as deer and swine), sirenians (seacows), proboscideans (elephants), and cetaceans (porpoises and whales). *Ungulates* are mammals that walk on the tips of the toes.

A third evolutionary line produced the order Carnivora, including cats, dogs, raccoons, and the pinnipeds (seals, sea lions, and walruses). The most extensive radiation produced the primate–rodent complex. The order Rodentia includes rats, squirrels, and beavers. The order Primates includes monkeys, apes, and humans.

The Human Ancestry

The first primates probably were small arboreal animals resembling tree shrews. Dental structure suggests these primates descended from insectivores late in the Cretaceous. By the end of the Mesozoic, 65 million years ago, characteristics of the primates had been shaped by an arboreal existence. These characteristics include limber shoulder joints that permit *brachiation* (swinging from one hold to another), dextrous hands, eye–hand coordination, forward-facing and close-together eyes to provide depth perception, and parental care.

There are two suborders of modern primates: *prosimians* and *anthropoids*. Prosimians, such as lemurs, lorises, and tarsiers, probably resemble early arboreal primates. The anthropoids include monkeys, apes, and humans. The first anthropoids to appear in the fossil record are monkeylike primates that probably evolved from prosimian stock about 40 million years ago in Africa, and possibly earlier in Asia. The New World monkeys of South America have been evolving separately from Old World monkeys for millions of years.

There are four genera of apes: gibbons, orangutans, gorillas, and chimpanzees. Modern apes generally are larger than monkeys, with relatively long legs, short arms, and no tails.

Paleoanthropology is the study of human origins and evolution. Chimpanzees and humans represent two divergent branches from a common, less-specialized ancestor. Human evolution has not occurred as phyletic change within an unbranched hominid line; there have been times when several different species co-existed. Different human features have evolved at different rates, known as *mosaic evolution*. Thus, bipedalism or erect posture evolved before an enlarged brain developed.

The oldest known ape fossils have been assigned to

the genus *Aegyptopithecus*, a cat-sized tree-dweller that lived about 35 million years ago. About 25 million years ago during the Miocene epoch, ape descendants diversified and spread into Eurasia. About 20 million years ago, the Himalayan range arose, the climate became drier, and the forests contracted. Some of the Miocene apes may have begun living on the edge of the forest and foraging for food on the savanna. *Ramapithecus* is an anthropoid known from fossils 8 to 14 million years old. Most anthropologists believe that humans and apes diverged 4 to 5 million years ago.

In 1924, British anthropologist Dart found a skull of an early human that he called *Australopithecus africanus*. It appears that *Australopithecus* was an upright hominid, with humanlike hands and teeth, but a small brain, who foraged on the African savannas for nearly 2 million years, beginning 3 million years ago.

In 1974, a petite *Australopithecus* skeleton was discovered in the Afar region of Ethiopia. Lucy, as this fossil was named, and similar fossils have been designated as a separate species, *Australopithecus afarensis*. There is debate whether Lucy, first estimated at 3.5 million years old, represented the most ancient of the australopithecine group and the common ancestor of two hominid branches (one leading to *A. africanus* and one leading to *Homo*); or whether *A. afarensis* had already diverged from the lineage leading to *Homo*.

Australopithecus walked erect for over 1 million years while there was no substantial enlargement of the brain. Larger-brained fossils date back about 2 million years. Simple stone tools sometimes have been found with these fossils. Some paleoanthropologists place these fossils in the genus *Homo*, calling them *Homo habilis*, or "handy man," whereas other scientists retain them in *Australopithecus*.

Homo habilis and *Australopithecus* co-existed for nearly 1 million years. The former probably hunted animals and gathered fruits and vegetables, whereas the later probably scavenged for food. One theory is that these two were distinct lines of hominids, one ofwhich was an evolutionary dead end, while the other, *Homo habilis*, was on the line to modern humans.

Homo erectus, the first hominid to migrate out of Africa into Europe and Asia, is represented by fossils known as Java Man and Peking Man. They were taller with a larger brain capacity. Fossils range in age from 1.5 million years to 300,000 years. *Homo erectus* survived in the colder climates of the north, living in caves or huts, building fires, wearing animal skins, and using more elaborate stone tools.

The oldest fossils classified as *Homo sapiens* are the Neanderthals, who lived from about 130,000 to 30,000 years ago. They had heavier brow ridges, less pronounced chins, and larger brains than do modern humans. They were skilled tool-makers and left evidence of burials and other rituals.

The oldest fully modern fossils, discovered in the Cro-Magnon caves of France, date back 40,000 years. Neanderthals overlapped modern humans by about 10,000 years. These two forms of *Homo sapiens* may have had separate origins from African ancestors, or modern humans may have evolved from an isolated population of Neanderthals. Neanderthals disappeared about 30,000 years ago, perhaps from competition, war, or interbreeding with modern humans.

The erect stance was the most radical anatomical change in human evolution, requiring remodeling of the foot, pelvis, and vertebral column. Prolonging the period of growth of the brain after birth made possible the enlargement of the human brain. An extended period of parental care helps to make possible the basis of culture—the transmission of knowledge accumulated over the generations. Language, spoken and written, is the major means for this transmission.

Cultural evolution included several stages: the communal efforts of nomads who hunted and gathered food on grasslands, the development of agriculture in Eurasia and the Americas about 10,000 to 15,000 years ago, and the Industrial Revolution, which began in the eighteenth century. The history of life has been biological evolution on a changing planet. Cultural evolution has made *Homo sapiens* a new force in the history of life—a species that changes the environment to meet its needs.

STRUCTURE YOUR KNOWLEDGE

1. Place the major events in the evolution of the Phylum Chordata on the following geological timetable.

Era	Period	Age (millions of years ago)	Major Animal Events	Plant Events
CENOZOIC	Quaternary	0.01		
CENOZOIC		2.5		
CENOZOIC	Tertiary			Dominance of angiosperms
CENOZOIC		65		
MESOZOIC	Cretaceous	135		Flowering plants appear Gymnosperms decline
MESOZOIC	Jurassic	195		
MESOZOIC	Triassic	240		Conifers and cycads dominant
PALEOZOIC	Permian	285		Adaptive radiation of gymnosperms;
PALEOZOIC	Carboniferous	375		Dominance of lycopods, horsetails, ferns; great coal forests
PALEOZOIC	Devonian	420		Seed plants appear, bryophytes; radiation of vascular plants
PALEOZOIC	Silurian	450		Origin of vascular plants
PALEOZOIC	Ordovician	520		Marine algae abundant
PALEOZOIC	Cambrian	570		Cyanobacteria

2. Describe several examples from vertebrate evolution that illustrate the common evolutionary theme that new adaptations usually evolve from preexisting structures.

3. How did various vertebrate groups meet the challenges of a terrestrial habitat?

TEST YOUR KNOWLEDGE

FILL IN THE BLANKS

1. _____ chordate subphylum of sessile, filter-feeding marine animals
2. _____ subphylum that includes amphioxus
3. _____ blocks of mesoderm along notochord that develop into muscles
4. _____ jawless fish with armor of bony plates
5. _____ flap over the gills of bony fishes
6. _____ adaptation that allows reptiles to reproduce on land
7. _____ lizard-like ancestors of the various reptilian orders
8. _____ group from which birds descended
9. _____ egg-laying mammals
10. _____ order in which snakes are placed
11. _____ structure that helps mammals ventilate their lungs

12. _____ mammals that walk on the tips of the toes
13. _____ genus in which the fossil Lucy is placed
14. _____ suborder that includes monkeys, apes and humans
15. _____ major means for the transmission of human culture

MULTIPLE CHOICE: *Choose the one best answer.*

1. Which of the following is not a characteristic of the phylum Chordata?
   a. dorsal, hollow nerve cord
   b. vertebral column
   c. notochord
   d. pharyngeal slits

2. Pharyngeal slits appear to have functioned first as
   a. filter-feeding devices.
   b. gill slits for respiration.
   c. components of the jaw.
   d. portions of the inner ear.

3. Which of the following is not a vertebrate characteristic?
   a. cephalization
   b. vertebrae
   c. paedogenesis
   d. ventral heart and closed circulatory system

4. Modern agnathans are represented by
   a. hagfishes and lampreys.
   b. sharks and rays.
   c. the lobe-fin coelocanth.
   d. lungfishes.

5. Jaws first occurred in the class
   a. Agnatha.
   b. Chondrichthyes.
   c. Osteichthyes.
   d. Placodermi.

6. Sharks need to keep swimming in order to
   a. aerate their lungs.
   b. maintain buoyancy.
   c. use their lateral line system.
   d. do all of the above.

7. Which of the following is incorrectly paired with its gas exchange mechanism?
   a. amphibians — skin and lungs
   b. fleshy-finned fishes — gills and lungs
   c. reptiles — lungs
   d. bony fishes — swim bladder and gills

8. Urodeles include
   a. limbless, burrowing amphibians.
   b. frogs that have a tadpole stage.
   c. frogs and toads that may or may not have a tadpole stage.
   d. salamanders.

9. Reptiles have lower caloric needs than do mammals of comparable size because
   a. they are ectotherms.
   b. they have waterproof scales.
   c. they have an amniote egg.
   d. they move by the bending back and forth of their vertebral column.

10. Which of the following is not an adaptation associated with flight in birds?
    a. hollow or honeycombed bones
    b. a fat layer
    c. feathers
    d. a large keel on the sternum

11. Mosaic evolution refers to the fact that
    a. reptiles had two major periods of adaptive radiation.
    b. preexisting structures are often co-opted for new functions.
    c. different human features evolved at different rates
    d. some fish that evolved in fresh water and moved to the seas have returned to fresh water.

12. In Australia, marsupials fill the niches that placental mammals fill in other parts of the world because
    a. they are better adapted and have out-competed placental mammals.
    b. their offspring complete their development attached to a teat in a marsupium.
    c. after Pangaea broke up, they diversified in isolation from placental mammals.
    d. they evolved from monotremes that migrated to Australia about 12 million years ago.

13. Which of the following characteristics of primates is not associated with an arboreal existence?
    a. shoulder joints that permit brachiation
    b. upright posture
    c. eyes close together on the face to improve depth perception
    d. parental care

14. *Homo erectus*
    a. was the first hominid to migrate out of Africa into Europe and Asia.
    b. was taller and had a larger brain capacity than *Homo habilis*.
    c. is represented by fossils known as Java Man and Peking Man.
    d. All of the above are correct.

15. Which of the following is not true of Neanderthals?
    a. They were the first to be classified as *Homo sapiens*.
    b. They walked erect for over 1 million years with no substantial enlargement of the brain.
    c. They co-existed with modern humans and disappeared about 30,000 years ago.
    d. They were discovered in the Cro-Magnon caves of France.

ANSWER SECTION

CHAPTER 24: EARLY EARTH AND THE ORIGIN OF LIFE

Suggested Answers to Structure Your Knowledge

1. Primitive Earth is believed to have had seas, volcanoes, and large amounts of ultraviolet radiation passing through the thin atmosphere, which probably consisted of H_2O, CO, CO_2, N_2, CH_4, and NH_3. Life was able to evolve in this environment because the reducing (electron-adding) atmosphere and availability of energy from lightning and UV radiation would facilitate the abiotic synthesis of organic molecules, and the hot lava rocks may have aided the polymerization of these monomers. The availability of methane or organic fuels may have provided an energy source for early protobionts.

2. **Step 1:** The abiotic synthesis of organic molecules provided the amino acids, simple sugars, and nucleotides that served as the building blocks of proteins, energy molecules, and precursors for the information molecules of RNA and DNA.

Step 2: In order for life to develop, organic monomers had to be linked into polymers that have emergent properties deriving from their structural organization. Early polypeptides may have functioned as simple enzymes, lipids formed bounding membranes, and RNA molecules provided self-replicating structures capable of ordering amino acids into polypeptides.

Step 3: The aggregation of molecules into discrete packets, separated from the surroundings by selectively permeable membranes, allowed for the concentration of critical molecules and the emerging properties of metabolism, excitability, and primitive reproduction.

Step 4: The development of a precise mechanism for the recording and replication of genetic material allowed for the reproduction and evolution of successful aggregations of organic molecules.

Answers to Test Your Knowledge

Matching:

1. F	4. B
2. D	5. A
3. E	6. C

Multiple Choice:

1. a	6. b
2. b	7. d
3. c	8. c
4. d	9. b
5. c	10. a

CHAPTER 25: PROKARYOTES AND THE ORIGINS OF METABOLIC DIVERSITY

Suggested Answers to Structure Your Knowledge

1.

CHARACTERISTICS OF THE PROKARYOTIC CELL

Characteristic	Description
Cell shape	May be spherical (cocci), rod (bacilli), or spiral (spirilla).
Genome	Circular DNA molecule, has little associated protein, 1/1000 as much DNA as eukaryote, concentrated in nucleoid region. May have plasmids--circular DNA with genes for antibiotic resistance, metabolic enzymes.
Membranes	Few internal membrane systems, some infoldings of plasma membrane used in respiration, thylakoid membranes in cyanobacteria for photosynthesis.
Structures of the cell surface	Cell wall made of peptidoglycan, may have outer membrane (gram-negative bacteria), sticky capsule for adherence, pilli for adherence or conjugation.
Forms of motility	May glide in secreted slime or "swim" with flagella that are either scattered over cell or concentrated at ends of cell. May use alternating runs and tumbles, can result in taxis.
Reproduction and growth	Divide by binary fission, rapidly grow into colony of progeny. Geometric growth slowed by accumulation of toxic wastes or depletion of nutrients. Some genetic exchange during conjugation.

2.

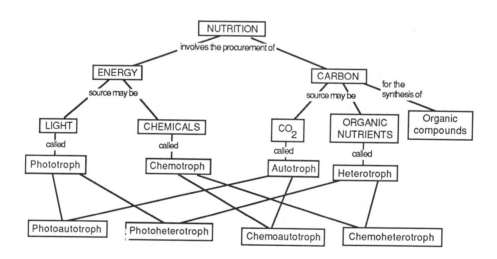

3. • Breakdown of abiotically synthesized organic compounds for energy, led to *glycolysis* and formation of *ATP* by substrate-level phosphorylation.

• *Fermentation* and excretion of acids led to development of *proton pumps* (using ATP) to maintain pH of cell.

• *Electron transport chain* linked oxidation of organic compounds with pumping of H^+ out of cell. Efficient electron transport chain could have formed proton gradient so that H^+ diffused back in, reversing proton pump and generating ATP (*chemiosmosis—anaerobic respiration*).

• Use of light energy to drive electrons from H_2S to $NADP^+$, co-opting electron transport chain, to generate reducing power to fix CO_2 (*photosynthesis*).

• Use of water as source of electrons and hydrogen for fixing CO_2, *evolving oxygen*.

• Use of O_2 to pull electrons from organic molecules down existing electron transport chain (*aerobic respiration*).

Answers to Test Your Knowledge

Multiple Choice:

1. c	6. c	11. a
2. c	7. b	12. c
3. d	8. d	13. d
4. a	9. d	14. d
5. b	10. b	15. c

Fill in the Blanks:

1. bacilli	7. photoautotroph
2. nucleoid region	8. facultative anaerobe
3. gram stain	9. saprophytes
4. pilli	10. archaebacteria
5. taxis	11. exotoxins
6. chemoheterotroph	12. nodules

Matching:

1. D	5. H	9. B
2. E	6. A	10. J
3. L	7. F	11. C
4. I	8. G	12. K

CHAPTER 26: PROTOCTISTS AND THE ORIGIN OF EUKARYOTES

Suggested Answers to Structure Your Knowledge

1. (a) The kingdom Protoctista includes unicellular—and a few colonial and multicellular—eukaryotic organisms. Nearly all protoctists use aerobic respiration and have flagella or cilia at some point in their life. They all reproduce asexually, and some have sexual reproduction. They are found in water or moist habitats (including inside the bodies of hosts), and many form cysts to withstand harsh conditions. Protoctists may be photoautotrophic or heterotrophic or both.

(b) There is controversy over the boundaries of the kingdom Protoctista. Originally Whittaker assigned only unicellular eukaryotes to this kingdom. Some colonial and multicellular eukaryotes, however, lack the distinctive traits of either fungi, plants, or animals, and appear to be more closely related to unicellular protoctists than to these multicellular kingdoms. Margulis and others have advocated an expanded kingdom Protoctista to include these multicellular descendents of unicellular eukaryotes.

2. According to the autogenous hypothesis, eukaryotic cells are thought to have evolved their internal membrane structures from specialization of membranes invaginated from the plasma membrane. The endosymbiotic hypothesis claims that at least the mitochondria and chloroplasts developed from photosynthetic and aerobic heterotrophic prokaryotes that became incorporated into larger cells. The lines of evidence for this model include the following similarities: the chloroplasts of red algae to cyanobacteria; the chloroplasts and pigments of green algae to the bacterium *Prochloron*; the size, membrane enzymes and transport systems, division process, circular DNA molecule, ribosomal RNA sequences, and ribosomes of chloroplasts and mitochondria to those found in eubacteria. The existence of endosymbiotic relationships in modern organisms and the recent discovery of transposons, which may have transferred endosymbiont DNA to the host cell nucleus, provide additional evidence for this model.

Answers to Test Your Knowledge

Matching:

1. L	6. K	11. M
2. N	7. E	12. B
3. J	8. O	13. G
4. I	9. F	14. C
5. D	10. A	15. H

Multiple Choice:

1. c	5. a	9. a
2. d	6. a	10. c
3. b	7. c	11. a
4. c	8. c	12. d

CHAPTER 27: PLANTS AND THE COLONIZATION OF LAND

Suggested Answers to Structure Your Knowledge

1.

Characteristics \ Plant Groups and Their Common Names	Bryophyta — Mosses, Liverworts	Lycophyta Sphenophyta — Club Mosses, Horsetails	Pterophyta — Ferns	Coniferophyta — Conifers, Pines	Anthophyta — Flowering Plants
Cuticle, stomata	X	X	X	X	X
Protected embryo, jacketed reproductive organs	X	X	X	X	X
Vascular tissue for transport and support		X	X	X	X
Protected gametophyte retained on sporophyte plant				X	X
Nonswimming sperm, pollen				X	X
Fibers for support in vascular tissue				X	X
Seed that protects and encloses embryo with food source				X	X
Flowers and fruits, animal pollinators, protected seeds					X

2.

Era	Period	Age (millions of years ago)	Major Plant Events	Animal Events
CENOZOIC	Quaternary		Dominance of angiosperms	Mammals, birds, insects dominant
CENOZOIC	Tertiary	65	Dominance of angiosperms	Mammals, birds, insects dominant
	Climatic Change--Cooler			
MESOZOIC	Cretaceous	135	Flowering plants appear Gymnosperms decline	Dinosaurs become extinct
MESOZOIC	Jurassic	195		Dinosaurs dominant
MESOZOIC	Triassic	240	Conifers and cycads dominant	Dinosaurs dominant
	Climatic Change--Warmer and Drier			
PALEOZOIC	Permian	285	Adaptive radiation of gymnosperms; lycopods and horsetails decline	Amphibians dominant
PALEOZOIC	Carboniferous	375	Dominance of lycopods, sphenopsids (horsetails), ferns; great coal forests	Amphibians dominant
PALEOZOIC	Devonian	400	Appearance of seed plants; first fossils of bryophytes; adaptive radiation of vascular plants	
PALEOZOIC	Silurian	450	Origin of vascular plants	

Answers to Test Your Knowledge

Multiple Choice:

1. c	**5.** c	**9.** b
2. a	**6.** c	**10.** d
3. c	**7.** a	**11.** a
4. b	**8.** d	**12.** d

True or False:

1. F—add *or algae.*
2. F—change *heteromorphic* to *heterosporous,* or change *produce . . .* to *have gametophyte and sporophyte that are morphologically distinct.*
3. T
4. F—change *naked seed* to *seedless,* or change *club mosses and horsetails* to *gymnosperms.*
5. T
6. T
7. F—change *tracheids* to *vessel elements.*
8. T
9. F—change *angiosperm* to *gymnosperm,* or change *haploid cells . . .* to *an embryo sac with eight haploid nuclei.*
10. T

CHAPTER 28: FUNGI

Suggested Answers to Structure Your Knowledge

1.

Division	Common Members	Morphology	Asexual Reproduction	Sexual Reproduction	Miscellaneous Information
Oomycota	water molds, white rusts, downy mildews	coenocytic hyphae, diploid condition, cellulose walls	zoospores or airborne spores	egg and sperm, flagellated zoospores	cause potato blight, downy mildew in grapes
Chitridio-mycota	chytrids	unicellular with rhizoids, or coenocytic hyphae		flagellated gametes and zoospores	parasites and saprophytes, soil and water
Zygomycota	black bread mold, *Pilobolus*	coenocytic hyphae, rhizoids, horizontal aerial hyphae	spores in sporangia at tips of aerial hyphae	resistant zygospores, conjugation	terrestrial
Ascomycota	sac fungi, yeasts, cup fungi	septate hyphae or unicellular	conidia, wind-dispersed spores	heterokaryotic hyphae, ascus, ascospores, ascocarps	yeasts, decomposers, in most lichens, genetic studies
Basidio-mycota	club fungi, mushrooms, puffballs, shelf fungi	heterokaryotic hyphae, mycelia, basidiocarp	no asexual stage	basidium with basidiospores	fairy rings, mycorrhizae
Deutero-mycota	fungi imperfecti, *Penicillum, Dactylaria*		many form conidia	no sexual stage	many produce antibiotics
Lichens	lichens	foliose, crustose, fruticose	soredia or fragments	fungus or alga may sexually reproduce	symbolic association, harsh habitats

2.   The role of fungi as saprophytes and decomposers of organic material is central to the recycling of chemicals between living organisms and their physical environment. Decomposers also prevent the extensive accumulation of litter, feces, and dead organisms. Parasitic fungi have economic and health effects on humans. Plant pathogens cause extensive loss of food crops and trees (late potato blight, chestnut blight, corn smut, wheat rusts). Fungal pathogens of humans can cause annoying to serious problems. An example of a mutualistic symbiont is the mycorrhizae found on the roots of most plants. These associations greatly benefit the plant.

Answers to Test Your Knowledge

Fill in the Blanks:

1.   septum
2.   heterokaryon
3.   chitin
4.   conidia
5.   basidium
6.   asci
7.   mycorrhizae
8.   zygospore
9.   coenocytic
10. haustoria

Multiple Choice:

1.   b
2.   c
3.   c
4.   c
5.   b
6.   d
7.   a
8.   d
9.   c
10. a

CHAPTER 29: INVERTEBRATES AND THE ORIGIN OF ANIMAL DIVERSITY

Suggested Answers to Structure Your Knowledge

1.

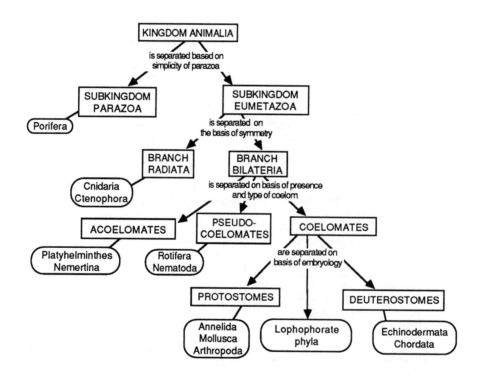

2. Transitional forms between protoctists and the first animals and between the various animal phyla that would shed light on the origin and phylogeny of animals have not been found in the fossil record. Suggested hypothetical ancestors of the eumetozoa include a colonial flagellate, a planuloid, a plakula, or a coenocytic ciliate. The fossil record begins with soft-bodied medusae and annelidlike worms from the Ediacaran period of the Precambrian era. The fossils of the next period include nearly all modern phyla. The Cambrian appearance of so many complex animals remains unexplained.

Answers to Test Your Knowledge

Matching:

1.	L	c	6.	I	b
2.	D	b	7.	B	e
3.	F	e	8.	A	b
4.	G	a	9.	E	d
5.	M	g	10.	H	c

Multiple Choice:

1.	b	6.	d	11.	c	16.	c
2.	a	7.	c	12.	d	17.	a
3.	d	8.	a	13.	b	18.	a
4.	d	9.	d	14.	a	19.	b
5.	b	10.	c	15.	b	20.	a

CHAPTER 30: VERTEBRATE GENEALOGY

Suggested Answers to Structure Your Knowledge

1.

Era	Period	Age (millions of years ago)	Major Animal Events	Plant Events
CENOZOIC	Quaternary	0.01	*Homo sapiens* *Homo erectus* *Homo habilis* *Australopithecus*	Dominance of angiosperms
CENOZOIC	Tertiary	2.5	First anthropoids (monkeys) Mammals, birds, insects dominant Radiation of Mammals	Dominance of angiosperms
	Climatic Change—Cooler	65		
MESOZOIC	Cretaceous	135	Dinosaurs become extinct Primates probably appear	Flowering plants appear Gymnosperms decline
MESOZOIC	Jurassic	195	Dominance of dinosaurs Birds appear	Conifers and cycads dominant
MESOZOIC	Triassic	240	Second reptilian radiation Mammals appear	Conifers and cycads dominant
	Climatic Change—Warmer and Drier			
PALEOZOIC	Permian	285	First reptilian radiation	Adaptive radiation of gymnosperms;
PALEOZOIC	Carboniferous	375	Age of Amphibians Reptiles appear	Dominance of lycopods, horsetails, ferns; great coal forests
PALEOZOIC	Devonian	420	Age of Fishes. Most agnathans disappear Placoderms, Chondrichthyes, Osteichthyes appear	Seed plants appear, bryophytes; radiation of vascular plants
PALEOZOIC	Silurian	450		Origin of vascular plants
PALEOZOIC	Ordovician	520	First vertebrate (Agnatha)	Marine algae abundant
PALEOZOIC	Cambrian	570	First invertebrate (Chordate) Most animal phyla appear	Cyanobacteria

2. The following are examples of new adaptations that evolved from preexisting structures: (1) use of pharyngeal slits, through which water flowed for filter feeding, for gas exchange, development of gills; (2) transformation of lungs, developed in bony fish in stagnant waters for gas exchange, to an air bladder functioning in buoyancy; (3) development of fleshy fins, supported by skeletal elements, into limbs for terrestrial animals; (4) use of feathers, probably functioning for insulation in endothermy, for flight; (5) use of dextrous hands, important for an arboreal life, to manipulate tools.

3. Amphibians remained in damp habitats, burrowing in mud during droughts; some secrete foamy protection for eggs laid on land. Reptiles developed scaly, waterproof skin, an amniote egg that provided aquatic environment for developing embryo, and behavioral adaptations to modulate changing temperatures. All terrestrial groups are tetrapods, using limbs for locomotion on land.

Answers To Test Your Knowledge

Fill in the Blanks:

1. Urochordates
2. Cephalochordates
3. somites
4. ostracoderms
5. operculum
6. amniote egg
7. cotylosaurs, stem reptiles
8. archosaurs, dinosaurs
9. monotremes
10. Squamata
11. diaphragm
12. undulates
13. *Australopithecus*
14. anthropoid
15. language

Multiple Choice:

1. b	6. b	11. c
2. a	7. d	12. c
3. c	8. d	13. b
4. a	9. a	14. d
5. d	10. b	15. b

Plants: Form and Function

CHAPTER 31

ANATOMY OF A PLANT

FRAMEWORK

The body of a plant includes roots, stems, and leaves. These plant organs consist of cells and tissues that are specialized for absorption, anchorage, storage, support, photosynthesis, transport, and protection. There are three basic tissue systems in a plant—the dermal, ground, and vascular tissue systems. The indeterminate growth of plants is produced by the apical meristems, which result in primary growth, and the lateral meristems, which produce secondary growth. A rather large new vocabulary is needed to name the specialized cells and structures in a study of plant anatomy. Focus of this chapter is how the structure of the parts of a plant integrates into the functioning of the entire living organism.

CHAPTER SUMMARY

Botany is the study of plants, one of the oldest and most practical sciences. Angiosperms, characterized by flowers and fruits, are split into two subdivisions: *monocots*, which have one embryonic seed leaf or cotyledon, and *dicots*, which have two. Plants show adaptations to their terrestrial environment in both evolutionary and physiological time scales. The anatomy of plants also illustrates the correlation between structure and function.

The Parts of a Flowering Plant

Plants have an underground *root system* for obtaining water and minerals from the soil and an aerial *shoot system*, consisting of stems and leaves, for absorbing light and carbon dioxide for photosynthesis. Vascular tissue transports material between roots and shoots. *Xylem* carries water and dissolved minerals upward,

whereas *phloem* conducts food made in the leaves to nonphotosynthetic parts of the plant and stored food from the root to regions of active growth.

The structure of roots is adapted for anchorage, absorption, conduction, and storage of food. A *taproot* system is found commonly in dicots, whereas monocots generally have *fibrous root* systems. Most absorption of water and minerals occurs through the tiny *root hairs* that are clustered near the root tips. *Adventitious* roots arise above ground from stems or leaves and may serve as props to help support stems.

The stem consists of *nodes*, at which leaves are attached, and segments between nodes, called *internodes*. An *axillary bud* is found in the angle between a leaf and the stem. A *terminal bud*, consisting of developing leaves, nodes, and internodes, is found at the tip of a shoot. The terminal bud may exhibit *apical dominance* and inhibit the growth of axillary buds. Axillary buds may develop into flowers or *vegetative* branches with their own terminal buds, leaves, and axillary buds.

Modifications of stems include stolons, horizontal stems that grow along the ground; rhizomes, horizontal stems growing underground that may end in enlarged tubers; and bulbs, vertical, underground shoots functioning in food storage.

Leaves, the main photosynthetic organs of most plants, usually consist of a flattened blade and a *petiole*, or stalk. Most monocots lack petioles. Monocot leaves usually have parallel major veins, whereas dicot leaves have networks of branched veins. Leaves emerge from the stem in an alternate, opposite, or whorled arrangement, and they may be simple (single, undivided blade) or compound (divided into seveal leaflets).

Plant Cells and Tissues

Most of the growth of a plant cell occurs when the vacuole and cytoplasm take in water as the cell wall loosens. When a plant cell stops growing, the wall becomes thicker and more rigid.

Parenchyma cells, relatively unspecialized cells that lack secondary walls, function in photosynthesis and food storage. *Collenchyma* cells also may lack secondary walls, but have thickened primary walls and can function in support. Strands or cylinders of collenchyma cells elongate along with the young parts of the plant.

Schlerenchyma cells have thick secondary walls strengthened with lignin. These supporting elements often lose their protoplasts (the living portion of a plant cell) at maturity. *Fibers* are long, tapered schlerenchyma cells that usually occur in bundles. *Sclereids* are shorter and are irregular in shape.

After the cells of xylem form secondary walls and mature, the protoplast disintegrates, and a *vessel* or conduit is left behind through which water flows. *Tracheids* are long, thin, tapered xylem cells with lignin-strengthened walls. *Vessel elements* are wider, shorter, and thinner walled. Gymnosperms have only tracheids, whereas angiosperms have both types of xylem cells.

Phloem consists of chains of cells called *sieve-tube members*, which remain alive but lack nuclei, ribosomes, and vacuoles at functional maturity. Fluid flows through pores in the sieve plates in the end walls between cells. The nucleus and ribosomes of a *companion cell*, which is connected to a sieve-tube member by numerous plasmodesmata, may serve both cells.

Simple tissues are groups of cells of a single cell type, such as parenchyma or schlerenchyma. *Complex tissues*, such as xylem or phloem, are composed of more than one type of cell.

Plants consist of three tissue systems. The *dermal tissue system*, or *epidermis*, is a single layer of cells that covers and protects the young parts of the plant. Root hairs are extensions of epidermal cells that increase the surface area for absorption. The epidermis of leaves and some stems is covered with a *cuticle*, a waxy coating that prevents excess water loss.

The *vascular tissue system* consists of xylem and phloem and functions in transport and support. The *ground tissue system*, making up the bulk of a young plant, is predominately parenchyma. Ground tissue functions in synthesis of organic compounds, support, and storage.

Primary Growth

Plants exhibit *indeterminate growth*: they continue to grow as long as they live. Animals have *determinate growth*: they stop growing after reaching a certain size. Plants may be *annuals*, which complete their life cycle in one growing season or year; *biennials*, which have a life cycle spanning two years; or *perennials*, which live many years.

Plants have tissues called *meristems* that remain embryonic and divide to form new cells at the growing points of the plant. *Apical meristems*, located at the tips of roots and in the buds of shoots, produce *primary growth*, resulting in the elongation of roots and shoots and the formation of the primary tissues.

The meristem of the root tip is protected by a *root cap*, which secretes a polysaccharide slime to lubricate the growth route. The *zone of cell division* includes the apical meristem and the primary meristems of the protoderm, procambium, and ground meristem. These meristems divide and give rise to the three tissue systems of the root. In the *zone of cell elongation*, the cells lengthen to more than ten times their original size. Cells specialize in structure and function in the *zone of cell differentiation*.

The *protoderm* gives rise to the single cell layer of the epidermis. The *procambium* forms a central vascular cylinder, or *stele*. In most dicots, xylem cells radiate from the center of the stele in spokes, with phloem in between. The stele of a monocot may have *pith*, a central core of parenchyma cells, inside the xylem and phloem.

The *ground meristem* produces the ground tissue system, filling the *cortex* or region of the root between stele and epidermis. A one-cell–thick *endodermis* forms the boundary between cortex and stele and regulates the passage of materials into the stele. *Lateral roots* may develop from the *pericycle*, the outermost layer of the stele, which remains capable of cell division.

In the shoot, the dome-shaped mass of the apical meristem in the terminal bud also forms the primary meristems: protoderm, procambium, and ground meristem. Leaf primordia form as tiny bulges on the flanks of the apical dome and axillary buds develop from clumps of meristematic cells left at the bases of the leaf primordia. Although most of the growth in length of dicot shoots occurs at the tip, the shoots of monocots may elongate at each node, where meristematic cells continue to divide.

Vascular tissue runs through the stem in *vascular bundles*. In dicots, the vascular bundles may be arranged in a ring, with pith internal and cortex external to the ring. The pith and cortex are connected by pith rays. Xylem is located internal to the phloem in the vascular bundles. In most monocot stems, the vascular bundles are scattered throughout the ground tissue, and no definite pith region exists.

The leaf is covered by the waxy-coated, tightly interlocking cells of the epidermis. *Stomata*, tiny pores flanked by *guard cells*, permit gas exchange to the inside of the leaf and are the site of *transpiration*, the evaporation of water from the leaf.

The *mesophyll* consists of parenchyma ground tissue cells containing chloroplasts. In most dicots, the upper half of the leaf has columnar cells in a layer called the palisade mesophyll, and the lower half, called the spongy mesophyll, has loosely packed, irregularly shaped cells surrounding many air spaces.

A *leaf trace*, branching from a vascular bundle at a node, continues into the petiole as a vein, which divides repeatedly within the blade of the leaf, providing support and vascular tissue to the photosynthetic mesophyll.

Secondary Growth

Secondary growth is the increase in diameter created by new cells produced by two *lateral meristems*, the vascular cambium and cork cambium. Most monocots do not produce secondary tissues and have only primary growth.

A band of parenchyma cells between the xylem and phloem of each vascular bundle and extending laterally into the ground tissue forms the *vascular cambium*, a continuous cylinder of meristematic cells. Cells produced internally to the vascular cambium differentiate into secondary xylem, whereas externally produced cells become secondary phloem. Wood is the accumulated secondary xylem cells with their thick, lignified walls. Rays of parenchyma cells are produced by the vascular cambium that function in storage and lateral transport through the secondary xylem. Annual growth rings result from the seasonal cycle of xylem production.

The epidermis splits off during secondary growth and is replaced by new protective tissues produced by the *cork cambium*. Cork cells, with waxy walls impregnated with suberin, are produced externally by the cork cambium, and parenchyma tissue called phelloderm is produced internally. The protective coat formed by these three layers (cork, cork cambium, phelloderm) is called the *periderm*. As secondary growth continually splits the periderm, a new cork cambium develops, eventually forming from parenchyma cells in the secondary phloem. Older secondary phloem, outside the cork cambium, also helps protect the stem as part of the bark. The only living tissues in tree trunks and large branches are the youngest secondary phloem, differentiating xylem cells, xylem rays, and the vascular and cork cambiums.

The two lateral meristems also function in the secondary growth of roots. Vascular cambium produces xylem internal and phloem external to itself, and a cork cambium forms from the pericycle and produces the periderm.

STRUCTURE YOUR KNOWLEDGE

1. In the following table, list the requirements or functions of a plant and the specialized cells, tissues, or structures that perform those functions within the roots, stems, and leaves.

| Function | Cells, Tissues, Or Structures That Perform Function | | |
	Root	Stem	Leaf

2. Compare and contrast primary and secondary growth.

TEST YOUR KNOWLEDGE

MATCHING: *Match the plant structure with its description.*

1. _____ schlerenchyma
2. _____ collenchyma
3. _____ tracheid
4. _____ fibers
5. _____ stele
6. _____ pericycle
7. _____ mesophyll
8. _____ periderm
9. _____ endodermis
10. _____ pith

A. chains of cells that transport food materials

B. tapered xylem cells with lignin in cell walls

C. parenchyma cells with chloroplasts in leaves

D. protective coat of woody stems and roots

E. cell layer in root regulating movement into central vascular cylinder

F. supporting cells with thickened primary walls

G. parenchyma cells inside vascular ring in stem

H. layer from which lateral roots originate

I. unspecialized cells functioning in food storage or photosynthesis

J. central vascular cylinder of root

K. bundles of long schlerenchyma cells

L. supporting cells with thick secondary walls

MULTIPLE CHOICE: *Choose the one best answer.*

1. Which of these is not a difference between monocots and dicots?
   a. simple leaves versus compound leaves
   b. parallel veins versus branching venation in leaves
   c. usually only primary growth versus primary and secondary growth
   d. vascular bundles scattered in stem versus vascular bundles in a ring

2. Most absorption of water and dissolved minerals occurs through
   a. adventitious roots.
   b. fibrous roots.
   c. root hairs.
   d. root caps.

3. Axillary buds
   a. may not develop due to apical dominance of terminal bud.
   b. form at nodes in angle where leaves join stem.
   c. have their own apical meristems and leaf primordia.
   d. All of the above are correct.

4. A leaf trace is
   a. a petiole.
   b. the outline of the vascular bundles in a leaf.
   c. a branch from a vascular bundle at a node that extends into a leaf.
   d. a tiny bulge on the flank of the apical dome that grows into a leaf.

5. Ground meristem
   a. produces the root system.
   b. produces the gound tissue system.
   c. produces secondary growth.
   d. is meristematic tissue found at the nodes in monocots.

6. The zone of cell elongation
   a. is responsible for pushing a root through the soil.
   b. comes between the zone of cell division and the zone of cell enlargement.
   c. produces the protoderm, procambium, and ground meristem tissues.
   d. does all of the above.

7. Which of the following cell types is incorrectly paired with its meristematic tissue?
   a. epidermis—protoderm
   b. secondary phloem—cork cambium
   c. stele—procambium
   d. cortex—ground meristem

8. The function of stomata is to
   a. allow transport of water vapor into the leaf.
   b. allow transport of sugars out of the leaf.
   c. allow gas exchange between the leaf and the atmosphere.
   d. open and close the guard cells.

9. Lateral meristems
   a. are responsible for secondary growth.
   b. usually are not present in dicots.
   c. produce protoderm, procambium, and ground meristem tissues.
   d. All of the above are correct.

10. Sieve-tube members
   a. are responsible for lateral transport through secondary xylem.
   b. control the activities of phloem cells that have no nuclei or ribosomes.
   c. are transport cells in which fluid passes through sieve plates in the end walls between cells.
   d. have spiral thickenings that allow the cell to grow and function in support.

FILL IN THE BLANKS: *Starting from the outside, place the letter of the tissues in the order in which they are located in a young woody tree trunk.*

___ ___ ___ ___ ___ ___ ___ ___ ___

A. primary phloem
B. secondary phloem
C. primary xylem
D. secondary xylem
E. pith

F. cork cambium
G. vascular cambium
H. cork cells
I. parenchyma tissue called phelloderm

CHAPTER 32

TRANSPORT IN PLANTS

FRAMEWORK

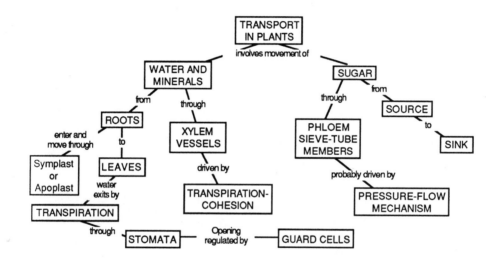

CHAPTER SUMMARY

Transport in plants involves the absorption of water and solutes by individual cells, the movement of substances from cell to cell, and the long-distance transport of sap in xylem and phloem.

Absorption of Water and Minerals by Roots

The soil solution soaks into the hydrophilic walls of epidermal cells and along the matrix of walls into the root cortex, exposing a large surface area of plasma membrane for absorption of water and minerals. Minerals may enter cortex cells by diffusion down their

concentration gradients or by the active transport against a gradient involving specific carrier proteins in the plasma membrane and the expenditure of ATP. The *proton pump* found in plant cell membranes may influence the transport of many different minerals simultaneously. This pump uses ATP to force hydrogen ions (H^+) out of the cell, generating a voltage or membrane potential across the membrane. The resulting negative charge inside the cell helps drive positively charged minerals into the cell through their specific carriers.

Water and minerals may cross the root cortex through the *symplast*, the living connection of cytoplasm extending through plasmodesmata from cell to cell. Or water and minerals may cross through the *apoplast*, the extracellular pathway along the continuous matrix of cell walls. The ring of suberin around each endodermal cell, called the *Casparian strip*, prevents water from the apoplast from entering the stele. The water and minerals must cross the selective plasma membrane of an endodermal cell to move into the stele.

Transfer cells within the stele selectively pump minerals out of their cytoplasm and into their walls. The minerals enter the nonliving xylem tracheids and vessel elements and move again through the apoplast system. Transfer cells are found where there is extensive traffic of solutes between symplast and apoplast.

Ascent of Xylem Sap

Through transpiration, plants lose a tremendous amount of water that must be replaced by water transported up from the roots. *Water potential*, abbreviated by ψ (psi), is a useful measurement for predicting the direction that water will move from one plant part to another. The unit used for water potential is the bar, about the same as one atmosphere of pressure.

The water potential of pure water in an open container is defined as zero. Water will flow from a region of higher water potential to one of lower water potential. The addition of solutes lowers the water potential; a solution has a negative water potential. Pressure increases water potential. Pushing down with a piston on (adding pressure to) a solution in an osmometer will prevent the movement of water into the solution. The positive effect of the pressure on the water potential of the solution balances out the negative effect of the solute, and the solution's water potential is zero ($\psi = 0$). There will be no net movement of water, no osmosis, in the osmometer.

A plant cell bathed in a more concentrated solution (hypertonic) will lose water by osmosis because the solution has a lower ψ. In pure water, the cell has the lower ψ. Water will enter the cell until enough turgor pressure builds up to bring $\psi = 0$ in the cell, and net movement of water will stop. Differences in water potential affect movement of water both between individual plant cells and between different regions of a plant in long-distance transport.

Accumulation of minerals in the stele (separated from the cortex by the selectively permeable membranes of the endodermal cells) lowers the stele's water potential. Water flows in by osmosis, resulting in *root pressure*, which forces fluid up the xylem. The root must expend energy to pump minerals into the xylem; thus, the creation of root pressure is an active process. In small plants, root pressure may cause *guttation*, the exudation of water droplets through specialized structures in the leaves when more water is forced up the xylem than is transpired by the plant. In most plants, root pressure is not a major mechanism in the movement of xylem sap.

The transpiration–cohesion theory is the most widely-accepted explanation of the ascent of xylem sap. Transpiration, caused by differences in water potential, provides the pull on the sap in the xylem. Water moves along a gradient of decreasing water potential from xylem to neighboring cells to air spaces to the drier air outside the leaf.

The transpirational pull on xylem sap is transmitted from the leaves to the root tips by the cohesiveness of water that results from the hydrogen bonds between molecules. The adhesion of water molecules to the hydrophilic walls of the xylem also helps the upward movement.

The upward transpirational pull on the cohesive sap creates tension within the xylem, further decreasing water potential and providing for the passive flow of water from the soil, across the cortex, and into the stele.

Cavitation, or a break in the chain of water molecules by the formation of an air pocket in a xylem vessel, breaks the transpirational pull, and the vessel cannot function in transport.

The Control of Transpiration

A plant's tremendous requirement for water is partly a function of its making food by photosynthesis. To obtain sufficient CO_2 for photosynthesis, leaves must exchange gases through the stomata, exposing a great surface area on the cells surrounding the air spaces of the leaf from which water may evaporate. Although stomata account for only 1% to 2% of the external leaf surface, about 90% of the water a plant loses passes through these small pores. Transpiration is certainly not all bad; it brings minerals to the leaves and provides evaporative cooling.

A common ratio of transpiration to photosynthesis is 600 grams of water transpired for each gram of CO_2 that is incorporated into carbohydrate. Plants that use the C_4 pathway for photosynthesis have ratios of 300 or less.

The regulation of the size of stomatal openings can control the rate of transpiration. Each stoma is flanked by a pair of kidney- or dumbbell-shaped guard cells, suspended by subsidiary cells. When guard cells become turgid and swell, they buckle outward and increase the size of the gap between them. When they become flaccid, they sag and close the space between them. The reversible active absorption and the loss of potassium ions (K^+) by the guard cells result in this increase and decrease of turgor pressure. The movement of K^+ across the guard cell membrane probably is coupled to the generation of membrane potentials by proton pumps.

Stomata usually are open during the day and closed at night. The opening of stomata at dawn is related to at least three factors. First, light stimulates guard cells to accumulate K^+, perhaps triggered by the illumination of a blue-light receptor on the tonoplast of guard-cell vacuoles. Light also may stimulate stomatal opening by driving photosynthesis in the guard-cell chloroplasts, making ATP available for the active transport of ions. Second, the depletion of CO_2 within air spaces of the leaf, occurring when photosynthesis begins in the mesophyll, is another factor in stomatal opening. The third factor is a daily rhythm of opening and closing that is endogenous to guard cells. Cycles that have intervals of approximately 24 hours are called *circadian rhythms*.

Environmental stress can cause stomata to close during the day. A hormone called abscisic acid signals guard cells to close stomata when the plant experiences a water deficiency. High temperatures induce closing, probably by increasing cellular respiration, which raises CO_2 concentrations in the air spaces in the leaf.

Xerophytes are plants adapted to arid climates. Their leaves may be small and thick, limiting water loss by reducing surface area relative to volume. Cuticles are thick, and the stomata, concentrated on the lower leaf surface, may be located in depressions. Some desert plants lose their leaves in the driest months.

The succulent plants of the family Crassulaceae assimilate CO_2 by a pathway known as CAM, for <u>C</u>rassulacean <u>A</u>cid <u>M</u>etabolism. CO_2 is assimilated into organic acids during the night and then is transferred to mesophyll cells for photosynthesis during the day. Thus, the stomata are open at night and can close during the day when transpiration would be greatest.

Transport in Phloem

The transport of food in the plant through phloem is called *translocation*. Phloem is a complex tissue consisting of parenchyma cells, fiber cells, sieve-tube members, and companion cells. Sap flows between sieve-tube members through porous sieve plates. Phloem sap may have a sucrose concentration as high as 30% and may also contain minerals, amino acids, and hormones. Girdling of a tree disrupts the downward flow of sugar through the phloem, which eventually kills the roots.

The sieve tubes of phloem carry food from a sugar *source*, where it is being produced by photosynthesis or the breakdown of starch, to a sugar *sink*, an organ that consumes or stores sugar. When a storage organ, such as a tuber, is storing carbohydrates during the summer, it is a sugar sink; when its starch is broken down to sugar in the spring, it is a sugar source. The direction of transport in any one sieve tube depends on the locations of the source and sink connected by that tube.

Sucrose probably is loaded by active transport into phloem at the source. Sugar in the leaf appears to move through the symplast of the mesophyll cells. Mesophyll cells bordering veins apparently secrete sugar into the apoplast, from which it is actively accumulated into sieve-tube members, probably being moved through companion cells. This energy-requiring process is believed to be driven indirectly by the proton pumps of plant membranes. Sucrose may be pumped by active transport out of sieve tubes at the sink end.

The rapid movement of phloem sap may be due to a pressure–flow mechanism. The high solute concentration at the source lowers the water potential and the resulting movement of water into the sieve tube produces hydrostatic pressure. At the sink end, the osmotic loss of water following the exodus of sucrose results in a lower pressure. The difference in these pressures causes water to flow from source to sink, transporting sugar. This model, although difficult to test, is the best explanation for the flow of sap in phloem.

STRUCTURE YOUR KNOWLEDGE

1. Discuss the models that explain the movement of xylem and phloem sap throughout the plant.

TEST YOUR KNOWLEDGE

MULTIPLE CHOICE: *Choose the one best answer.*

1. Water and minerals that cross the cortex through the symplast move
   a. up a water potential gradient.
   b. extracellularly along the continuous matrix of cell walls.
   c. from cell to cell through the plasmodesmata.
   d. by active transport.

2. Proton pumps in the plasma membranes of plant cells may
   a. generate a membrane potential that helps drive positively charged minerals into the cell through their specific carriers.
   b. be coupled to the movement of K^+ into guard cells.
   c. indirectly drive the accumulation of sucrose in sieve-tube members.
   d. be involved in all of the above.

3. In order to enter the stele, water must
   a. pass through the Casparian strip.
   b. pass through transfer cells.
   c. pass through an endodermal cell.
   d. pass through xylem vessels.

4. The water potential of a solution
   a. is less than the water potential of pure water.
   b. is negative when the solution is under pressure.
   c. is greater than that of a second solution if water moves from the second solution into the first solution.
   d. has a value of –2.3 bars.

5. Root pressure
   a. is responsible for the movement of sap up through xylem.
   b. results from the accumulation of minerals in the stele and the resulting osmosis of water.
   c. occurs because the water potential of the root is less than that of the leaves.
   d. involves all of the above.

6. Which of these is not a major factor in the movement of xylem sap?
   a. transpiration
   b. active transport using a proton pump
   c. adhesion
   d. cohesion

7. Cavitation may result in
   a. the girdling of a tree.
   b. guttation.
   c. a nonfunctional xylem vessel.
   d. stomata occurring in pits or depressions on the bottoms of leaves.

8. An increase in the efficiency of CO_2 fixation could result in
   a. less water lost through transpiration.
   b. a higher transpiration-to-photosynthesis ratio.
   c. more carbohydrate for each gram of CO_2 fixed.
   d. all of the above.

9. Which of these is not a factor in the opening of stomata?
   a. a decrease in the turgor of guard cells
   b. depletion of CO_2 in the air spaces of the leaf
   c. movement of K^+ into the guard cells
   d. circadian rhythm of guard-cell opening

10. The movement of phloem from a sugar source to a sugar sink
    a. is a passive process.
    b. occurs through the companion cells.
    c. may be due to a pressure–flow mechanism.
    d. takes place in the apoplast.

CHAPTER 33

PLANT NUTRITION

FRAMEWORK

The nutritional requirements of plants, including essential macronutrients and micronutrients, are discussed in this chapter. Carbon dioxide enters the plant through the leaves, but the other inorganic raw materials, including water and minerals, must be absorbed through the roots. The fertility of soil is influenced by its texture and composition. Nitrogen assimilation by plants is made possible by the nitrogen fixation of bacteria and microbes. Mycorrhizae are important associations between fungi and the roots of plants that increase mineral and water absorption.

CHAPTER SUMMARY

Plants, as photosynthetic autotrophs, can make all of their own organic compounds, but must obtain carbon dioxide, water, and a variety of minerals or inorganic ions to do so. Plants also need oxygen for cellular respiration. Roots, shoots, and leaves are structurally adapted to obtain essential nutrients from the soil and air.

Nutritional Requirements of Plants

Aristotle thought that soil provided the substance for plant growth. In the seventeenth century, van Helmont concluded that the 169-pound willow tree, grown in a pot in which only 2 ounces of soil had disappeared in five years, had developed mainly from the water that was added. Hales, an English physiologist of the eighteenth century, thought that plants were nourished mostly by air. All three of these ideas are partly correct.

Minerals, although essential, make only a small contribution to the mass of a plant. Water supplies most of the hydrogen and some oxygen incorporated into organic compounds by photosynthesis, but 90% of the water absorbed is lost by transpiration, and most of the water retained is used to make cell elongation possible. By weight, CO_2 is the source of the bulk of organic material of a plant.

Organic substances, most of which are carbohydrates (including cellulose), make up 95% of the dry weight of plants. Carbon, oxygen, and hydrogen, which make up carbohydrates, are the most abundant elements. Nitrogen, sulfur, and phosphorus, also ingredients of organic compounds, are relatively abundant.

Essential nutrients are those required for the plant to complete its life cycle from a seed to an adult that produces more seeds. *Hydroponic culture* has been used to determine which of the mineral elements are essential nutrients. Sixteen elements have been identified as essential in all plants; a few others are necessary for certain plant groups.

Macronutrients are those elements required by plants in relatively large amounts. The six major elements of organic compounds, as well as calcium, potassium, and magnesium, are macronutrients. Calcium combines with pectins in the middle lamella to glue plant cells together, and also has other regulatory functions in the plant. Potassium is the major solute for osmotic regulation and an activator of several enzymes. Magnesium is a basic component of chlorophyll and a co-factor of several enzymes.

Seven elements have been identified as *micronutrients*, which are needed by plants in very small amounts. They function in the plant mainly as co-factors of enzymatic reactions. Additional micronutrients have not been identified, perhaps because the slightest traces in a seed or as a contaminant of the glassware, water, or other chemicals used in hydroponic culture may satisfy the needs of the plant.

The symptoms of a mineral deficiency may depend

on the function of that nutrient. A magnesium deficiency results in less chlorophyll and causes chlorosis, or yellowing of the leaves. A deficiency of calcium, needed for cell wall synthesis, retards the growth of roots and shoots. The mobility of a nutrient within the plant also affects the symptoms of a deficiency. A mobile nutrient will move to young, growing tissues, and a deficiency will show up first in older parts of the plant. A deficiency in a relatively immobile nutrient may show up first in young parts of the plant, if older tissues have an adequate supply of the mineral.

Symptoms of a mineral deficiency may be distinctive enough for the cause to be diagnosed by a plant physiologist or farmer. Soil and plant analysis can confirm a specific deficiency. Nitrogen, potassium, and phosphorus deficiencies are most common.

Soil

The kinds of plants that can grow well in a particular region are determined to a great extent by the texture and chemical composition of the soil.

The formation of soil begins with the weathering of rock and is accelerated when lichens, fungi, bacteria, and roots secrete acids and when roots expand in fissures, thus breaking rocks into smaller pieces. *Topsoil* is a mixture of decomposed rock, living organisms, and *humus* (decomposing organic matter).

Topsoil varies in the size of its particles from coarse sand to fine clay. *Loams*, made up of a mixture of sand, silt, and clay, are the most fertile soil, having enough fine particles to provide a large surface area for retaining water and minerals, but enough coarse particles to provide air spaces that hold oxygen for respiring roots.

The activities of the numerous soil inhabitants, such as bacteria, fungi, algae, other protoctists, insects, worms, nematodes, and plant roots, affect the physical and chemical properties of soil. Humus builds a crumbly soil that retains water, provides good aeration of roots, and supplies mineral nutrients as microorganisms decompose the organic matter.

Water, containing dissolved minerals, is available to roots from small spaces in soil where it is bound to hydrophilic soil particles. Many positively charged minerals, such as K^+, Ca^{2+}, Mg^{2+}, adhere to the negatively charged surfaces of finely divided clay particles. Negatively charged minerals, such as nitrate (NO_3^-), phosphate (PO_4^{3-}), and sulfate (SO_4^{2-}), tend to leach away more quickly. Roots release acids that add hydrogen ions to the soil solution and facilitate *cation exchange*, in which H^+ ions make positively charged mineral ions available for absorption by displacing them from the clay particles.

Without good *soil management*, agriculture can greatly reduce the fertility of soil that has built up over centuries. Agriculture diverts essential elements from the chemical cycles when crops are harvested.

Historically, farmers used manure to fertilize their crops. Today in developed nations, commercially produced fertilizers, usually enriched in nitrogen, phosphorous, and potassium, are used. Manure, fishmeal, and compost are called organic fertilizers because they contain organic material in the process of decomposing. These materials must be decomposed to inorganic nutrients, in the same form supplied by commercial fertilizers, in order for the roots to absorb them. A disadvantage of commercial fertilizers is that they are rapidly leached from the soil and may pollute streams and lakes.

The acidity of the soil affects cation exchange and influences the chemical form of minerals. Managing the pH level of soil is an important element of maintaining fertility.

Irrigation can make farming possible in arid regions, but it places a huge drain on water resources and raises the salinity of the soil. New methods of irrigation and new varieties of plants that can tolerate less water or more salinity may reduce some of these problems.

Thousands of acres of farmland are lost to water and wind erosion each year. Agricultural practices that minimize erosion need to be used.

Nitrogen Assimilation by Plants

Nitrogen is the mineral that usually limits plant growth and yields. To be absorbed by plants, nitrogen must be converted, by the action of certain soil bacteria or microbes decomposing humus, to nitrate or ammonium. Bacteria capable of *nitrogen fixation* contain *nitrogenase*, an enzyme that reduces N_2 by adding H^+ and electrons to form ammonia (NH_3). In the soil solution, ammonia forms ammonium (NH_4^+), which plants can absorb. Bacteria oxidize NH_4^+ to nitrate (NO_3^-), the form of nitrogen most readily absorbed by roots. A plant enzyme reduces nitrate back to ammonium, which is then incorporated into amino acids that are transported through xylem to the rest of the plant.

Plants of the legume family have root swellings, called *nodules*, composed of plant cells containing nitrogen-fixing bacteria of the genus *Rhizobium*. The symbiotic relationship between *Rhizobium* and legumes is illustrated by their cooperative synthesis of leghemoglobin. This oxygen-binding protein releases oxygen for the intense respiration needed to supply energy for nitrogen fixation, and keeps the free-oxygen concentration low in the nodules, which is important because the enzyme nitrogenase is inhibited by oxygen.

Root nodules may secrete excess ammonium into the soil and increase the fertility of the soil for nonlegumes, which are grown in rotation with the legumes. In their rice paddies, rice farmers culture a water fern that has symbiotic cyanobacteria that fix nitrogen.

Protein deficiency is the most common form of human malnutrition. Improving the quality and quantity of proteins in crops is a major goal of agricultural research. Varieties of corn, wheat, and rice have been developed that are enriched in protein, but they require the addition of large quantities of expensive nitrogen fertilizer. Incorporating mutant strains of *Rhizobium*, which do not shut off production of nitrogenase when fixed nitrogen accumulates in the nodules, and developing legumes and bacteria that use less energy to fix nitrogen, could improve the productivity of symbiotic nitrogen fixation and increase protein yields in crops.

Genetic engineering is being used to create varieties of *Rhizobium* that can infect nonlegumes and to transplant the genes for nitrogen fixation directly into plant genomes.

Some Nutritional Adaptations of Plants

Carnivorous plants, living in acid bogs or other poor soils, obtain nitrogen and minerals by killing and digesting insects. The Venus flytrap, pitcher plants, and sundews have evolved insect traps made from modified leaves.

Parasitic plants, such as the mistletoe, may produce haustoria that invade a host plant and siphon sap from its vascular tissue. *Epiphytes* are plants that grow on the surface of another plant but do not rely on it for nourishment.

Many plants have modified roots called *mycorrhizae*, which are symbiotic associations between the roots and fungi. The hyphal extensions of the fungi provide a large surface area for the absorption of minerals and water, which are shared with the plant. Photosynthetic products from the plant nourish the fungus. Mycorrhizae are common on plants growing in poor soils. The manipulation of mycorrhizae may have important agricultural applications. The fossil record shows that the earliest land plants had mycorrhizae.

STRUCTURE YOUR KNOWLEDGE

1. Develop a concept map that shows the basic nutritional requirements of plants.

2. List the key properties of a fertile soil.

3. Explain how nitrogen assimilation by plants depends on bacteria and microbes. Why is the nitrogen assimilation of plants so important in agriculture?

TEST YOUR KNOWLEDGE

MULTIPLE CHOICE: *Choose the one best answer.*

1. Most of the mass of organic material of a plant comes from
   a. water.
   b. carbon dioxide.
   c. soil minerals.
   d. oxygen.

2. Hydroponic culture
   a. involves growing plants in mineral solutions.
   b. can supply the exact mineral requirements of plants, but is an expensive commercial technique for growing food.
   c. can be used to identify essential plant nutrients.
   d. is all of the above.

3. Micronutrients are needed in very small amounts
   a. because most of them are mobile.
   b. because most function as co-factors in enzymes.
   c. because most are supplied in large enough quantities in seeds.
   d. because not all of them are essential.

4. Humus
   a. is an organic fertilizer.
   b. is decomposing organic material.
   c. is made up of a mixture of sand, silt, and clay.
   d. facilitates cation exchange.

5. The most fertile type of soil is
   a. sand, because its large particles allow room for air spaces.
   b. loam, which has a mixture of fine and coarse particles.
   c. clay, because the fine particles provide much surface area to which positively charged minerals and water adhere.
   d. topsoil, which is a mixture of decomposed rock, living organisms, and humus.

6. Chlorosis is
   a. a symptom of a mineral deficiency.
   b. the uptake of the micronutrient chlorine by a plant.
   c. the formation of chlorophyll by a plant.
   d. a contamination of glassware in hydroponic culture.

7. Negatively charged minerals
   a. are released from clay particles by cation exchange.
   b. are reduced by cation exchange before they can be absorbed.

c. are leached away by the action of rainwater more easily than are positively charged minerals

d. are bound when roots release acids into the soil.

8. Nitrogenase
   a. is an enzyme that reduces atmospheric nitrogen to ammonia.
   b. is found in *Rhizobium* and other nitrogen-fixing bacteria.
   c. is inhibited by high concentrations of oxygen.
   d. is all of the above.

9. Mycorrhizae
   a. are nitrogen-fixing nodules found on legumes.
   b. are parasitic plants.
   c. are symbiotic associations between roots and fungi.
   d. were the first land plants.

10. Nitrogen is transported throughout the plant in the form of
    a. amino acids.
    b. ammonium.
    c. nitrates.
    d. ammonia.

CHAPTER 34

PLANT REPRODUCTION

FRAMEWORK

This chapter describes the sexual and asexual reproduction of flowering plants. The flower, with leaves modified for sexual reproduction, produces the haploid gametophyte stages of the life cycle: microspores in the anther develop into pollen grains and a megaspore in the ovule produces an embryo sac. Pollination and double fertilization of the egg and polar nuclei are followed by the development of a seed with a quiescent embryo and endosperm, protected in a seed coat and housed within a fruit. Seed dormancy is broken following proper environmental cues and the imbibition of water.

Vegetative propagation allows successful plants to clone themselves. Agriculture makes extensive use of this type of plant reproduction, including cuttings, grafts, and test-tube cloning.

CHAPTER SUMMARY

Adaptations in reproduction were a key to the spread of plants into a variety of terrestrial habitats. In the conifers and angiosperms, pollen transferred by wind or animals has replaced flagellated sperm, and zygotes develop into embryos protected by seeds.

Sexual Reproduction of Flowering Plants

The flower, specialized for sexual reproduction, is a compressed shoot with four whorls of modified leaves. *Sepals*—the outermost, usually green leaves—enclose and protect a floral bud before it opens. *Petals* are generally brightly colored and advertise the flower to pollinators. Sepals and petals are sterile floral parts.

A *stamen* consists of a stalk, the *filament*, and a pollen sac called the *anther*. A pollen grain contains a male gamete in the form of a sperm nucleus. A *carpel* consists of a sticky *stigma* at the top of a slender neck or *style*, which leads to an *ovary* in which one or more *ovules* develop. An egg, the female gamete, develops in the ovule. Two or more carpels may be fused to form an ovary with more than one chamber. *Pistil* is a term sometimes used for single or fused carpels.

A *complete flower* has sepals, petals, stamens, and carpels. *Incomplete flowers* are missing one or more parts. A *perfect* flower has both stamens and carpels; an *imperfect* flower is missing one of these two. Imperfect flowers may be either staminate or carpellate. A *monoecious* plant species has both staminate and carpellate flowers on the same plant; a *dioecious* species has these flowers on separate plants.

Plants exhibit an alternation between haploid and diploid generations. The diploid plant, the *sporophyte*, produces haploid spores by meiosis. The haploid multicellular male or female *gametophyte* produces gametes by mitosis. Zygotes grow into new sporophyte plants.

In angiosperms, the gametophytes are reduced in size and remain dependent on the sporophyte plant. The flower is part of the sporophyte generation. Although stamens and carpels are referred to as male and female sex organs, they actually produce spores, not gametes. The spores develop into tiny gametophytes while retained within the flower. A pollen grain is the male gametophyte. The embryo sac that develops within an ovule is the female gametophyte. These gametophytes produce sperm nuclei and eggs.

Diploid cells called microsporocytes undergo meiosis to form four haploid microspores. The nucleus of a microspore divides once by mitosis to produce a generative nucleus and a tube nucleus. The microspore wall thickens into the durable coat of the pollen grain, with an elaborate pattern characteristic of the plant species. The generative nucleus divides to form two *sperm nuclei*.

Ovules, which contain the female gametophytes, form within the ovary. A megasporocyte in each ovule grows and undergoes meiosis to form four haploid megaspores. Usually only one survives. This megaspore divides by mitosis three times, forming one large cell with eight haploid nuclei. Membranes form between nuclei, and this multicellular structure, called the *embryo sac*, is the female gametophyte. An egg cell is at one end between two synergid cells. Three antipodal cells are at the other end: two nuclei, called polar nuclei, share the cytoplasm of the large central cell. Protective layers called *integuments* form from sporophyte tissue around the embryo sac. The *micropyle* is an opening through the integuments.

Pollination is the placement of pollen onto the stigma. The majority of angiosperms have mechanisms that prevent self-pollination, or at least prevent self-fertilization. The pollen grain germinates and grows a tube down the style toward the ovary. The pollen tube, directed by a chemical attractant, grows through the micropyle and releases its two sperm nuclei within the embryo sac. By *double fertilization*, one sperm fertilizes the egg and the other combines with the polar nuclei to form a triploid nucleus.

The first mitotic division divides the zygote into a basal cell and a terminal cell. The basal cell divides to produce a thread of cells, called the suspensor, that anchors the embryo and transfers nutrients from the parent plant. The terminal cell divides to form a spherical proembryo. The *cotyledons* begin to form as bumps on the proembryo. The embryo elongates to form the torpedo stage with apical meristems at the apexes of the embryonic shoot and the root, where the suspensor attaches.

The root–shoot polarity of the embryo is determined by the first cytoplasmic division of the zygote. Egg organelles and chemicals are unevenly distributed in the cytoplasm, so the two cells differ in cytoplasmic composition. The cytoplasmic environment and position of an embryonic cell influence the selective expression of genes during cellular differentiation.

The triploid nucleus divides to form the *endosperm*, a multicellular mass rich in nutrients, which it provides to the developing embryo and stores for use when the seed germinates.

During maturation, the seed dehydrates to a water content about 5% to 15% of its weight. Growth stops, and the embryo remains dormant until the seed germinates. The embryo and its food supply, the endosperm or enlarged cotyledons, are enclosed in a *seed coat* formed from the ovule integuments.

In a dicot seed such as a bean, the embryo is an elongate embryonic axis, attached to fleshy cotyledons. The axis below the cotyledonary attachment is called the *hypocoty*; it terminates in the *radicle*, or embryonic root.

The upper axis is the *epicotyl*; it terminates as a *plumule*, a shoot tip with a pair of miniature leaves. In some dicots, the cotyledons remain thin and absorb nutrients from the endosperm when the seed germinates.

A monocot seed, such as a corn kernel, has a single thin cotyledon, called a scutellum, which absorbs nutrients from the endosperm during germination. A sheath covers the root, and a *coleoptile* encloses the embryonic shoot.

The ovary of the flower ripens into a *fruit*. Other floral parts may contribute to what we call the fruit. Hormonal changes following pollination cause the ovary to grow tremendously. Its thickened wall becomes the *pericarp* of the fruit. If a flower has not been pollinated, fruit usually does not set. Some *parthenocarpic* plants, such as bananas, do produce fruit without fertilization.

A *simple fruit* develops from a single ovary. It may be fleshy or dry. An *aggregate fruit* results from a single flower that has several separate carpels. A *multiple fruit* develops from an *inflorescence*, a group of tightly clustered flowers. Fruits usually ripen as the seeds are completing their development. Ripening of fleshy fruits may include softening of the pulp, a change in color, and an increase in sugar content.

Fruits are adapted to disperse seeds, by being either blown by wind or eaten by animals.

Humans have selectively bred edible fruits. The cereal grains—the wind-dispersed fruits of grasses, which have dry pericarps—are the staple foods for humans.

At germination, the plant resumes the growth and development that were suspended when the seed matured. Some seeds germinate as soon as they are in a suitable environment. Others need specific environmental cues to break dormancy.

Dormancy is an adaptation that increases the chances that the seed will germinate when and where the embryo has a good chance of surviving. The conditions for breaking dormancy vary with the environment, including heavy rain in deserts, intense heat in areas of frequent fires, and cold in areas with harsh winters. Some seeds must have their seed coats weakened by abrasion or chemical attack before they will germinate. The viability of a dormant seed may vary from a few days to decades or longer.

Imbibition, the absorption of water by the dry seed, causes the seed to expand and rupture its coat, and to begin a series of metabolic changes that cause the embryo to resume growth. Storage materials are digested by enzymes, and nutrients are sent to growing regions. Soon after hydration in barley or cereal seeds, the thin outer layer of the endosperm, called the aleurone, begins making α-amylase and other enzymes that digest starch stored in the endosperm. The embryo sends gibberellic acid to the aleurone to initiate enzyme production.

The radicle emerges from the seed first, followed by the shoot tip. In many dicots, a hook that forms in the hypocotyl is pushed up through the ground, pulling the delicate shoot apex and cotyledons behind it. Light stimulates the straightening of the hook, and the elevated epicotyl spreads its plumule and photosynthesis begins.

Light seems to be the main cue that the seedling has broken ground. An etiolated seedling, grown in the dark, continues to extend its hypocotyl until it exhausts its food reserves. In peas, a hook forms in the epicotyl, lifting the shoot tip out of the soil while the pea cotyledons remain in the ground. In monocots, the coleoptile pushes up through the soil, and the shoot tip is protected as it grows up through the tubular sheath.

Germination of a plant seed is a critical and fragile stage in the life cycle. Only a small fraction of seedlings endure to produce seeds themselves.

Asexual Reproduction

Many plant species can clone themselves through asexual or *vegetative reproduction*. This type of reproduction is an extension of the indeterminate growth of plants, in which meristematic tissues can grow indefinitely, and parenchyma cells can divide and differentiate into specialized cells. A common type of vegetative reproduction is *fragmentation*, the separation of a plant into parts that then form whole plants. Some plants can produce seeds asexually, a process called *apomixis*. Diploid cells in the ovules of dandelions develop into embryos without fertilization.

Humans have developed many artifical methods of vegetative propagation of trees, crops, and ornamental plants. New plants develop from shoot or stem cuttings when a *callus*, or mass of dividing cells, forms at the cut end of the shoot, and adventitious roots develop from the callus. Cuttings sometimes can be taken from leaves or storage stems.

Twigs or buds of one plant can be grafted onto a related plant. The plant that provides the root system is called the stock, and the twig graft is called the scion. Grafting combines the best qualities of different plants.

Whole plants can be grown from pieces of tissue on artificial medium in test-tube cloning. A single plant can be cloned into thousands of plants by subdividing the undifferentiated calluses as they grow in tissue culture.

Protoplast fusion is being coupled with tissue culture to create new plant varieties. Protoplasts can be screened for agriculturally beneficial mutations and then cultured. In some cases, two protoplasts from different plant species can be fused and cultured. Direct genetic manipulation of cultured cells and protoplasts using recombinant DNA techniques is now possible.

Genetic variability in crops and orchards has been consciously eliminated so that plants grow at the same rate, fruits ripen in unison, and yields at harvest time are predictable. Vegetative propagation is used to clone exceptional plants, and self-pollinating varieties are used when possible. *Monoculture*, the cultivation of a single plant variety on large areas of land, produces a fragile ecosystem, in which there is little genetic variability and little adaptability. Clones are genetically equivalent to one individual; a condition that becomes harmful will affect the entire monoculture. "Gene banks" have been created where plant breeders maintain seeds of many different plant varieties, so that new varieties can be developed if current ones fail.

Sexual reproduction in plants generates variation in a population, an advantage when the environment changes. Sexual reproduction also produces seeds, a means of dispersal to new locations and dormancy during harsh conditions. Plants well-suited to a stable environment can clone exact copies, and the progeny of vegetative propagation usually are not as frail as seedlings. Both modes of reproduction are useful in the adaptation of plant populations to their environments.

STRUCTURE YOUR KNOWLEDGE

1. Create a diagram or concept map that sketches out the major events in the life cycle of angiosperms.

2. List the advantages and disadvantages of sexual and asexual reproduction in plants. Which of these types of reproduction has contributed to the adaptation of plants to a variety of terrestrial habitats?

TEST YOUR KNOWLEDGE

FILL IN THE BLANKS

1. _____ flower part modified as male reproductive structure

2. _____ sticky top of carpel

3. _____ species with male and female flowers on same plant

4. _____ female gametophyte of angiosperms

5. _____ outer modified leaves of a flower

6. _____ embryonic axis above attachment of cotyledon

7. _____ sheath covering embryonic shoot in monocot

8. _____ plants that produce fruit without fertilization

9. _____ fruit formed from a flower with several separate carpels

10. _____ mass of dividing cells at cut end of a shoot

MULTIPLE CHOICE: *Choose the one best answer.*

1. Which of these describes an imperfect flower?
   a. a flower that has no sepals
   b. a flower that has fused carpels
   c. a flower on a dioecious plant
   d. a flower with no endosperm

2. Which of these is incorrectly paired with its life-cycle generation?
   a. anther — sporophyte
   b. pollen — gametophyte
   c. ovary — gametophyte
   d. pistil — sporophyte

3. The basal cell in a zygote
   a. develops into the root of the embryo.
   b. forms the suspensor that anchors the embryo and transfers nutrients.
   c. results from the fertilization of the polar nuclei by a sperm nucleus.
   d. divides to form the cotyledonary bumps on the proembryo.

4. A unique feature of fertilization in angiosperms is that
   a. it is double; one sperm fertilizes the egg, one combines with the polar nuclei.
   b. the sperm may be carried by the wind to the female organ.
   c. a pollen tube carries two sperm nuclei into the female gametophyte.
   d. a chemical attractant guides the sperm toward the egg.

5. The endosperm
   a. may be absorbed by the cotyledons in the seeds of dicots.
   b. is a triploid tissue.
   c. is digested by enzymes in monocot seeds following hydration.
   d. is all of the above.

6. Which of these is needed by all seeds to break dormancy?
   a. light
   b. period of cold or intense heat
   c. abrasion of seed coat
   d. imbibition

7. Which of these is not true of the hypocotyl hook?
   a. It is the first structure to emerge from a dicot seed.
   b. It pulls the cotyledons and shoot apex up through the soil.
   c. It straightens when exposed to light.
   d. It becomes very long in an etiolated seedling.

8. A common form of vegetative reproduction in plants is
   a. apomixis.
   b. grafting.
   c. fragmentation.
   d. budding.

9. A disadvantage of monoculture is that
   a. the whole crop ripens at one time.
   b. because there is no genetic variability, the whole crop could be annihilated by a change in conditions, a new pest, or disease.
   c. it predominately uses vegetative propagation.
   d. all of the above are disadvantages.

10. Protoplast fusion
    a. is used to develop gene banks to maintain genetic variability.
    b. is the method of test-tube cloning.
    c. can be used to form new plant species.
    d. occurs within a callus.

CHAPTER 35

CONTROL SYSTEMS IN PLANTS

FRAMEWORK

This chapter describes the current understanding of the complex systems in plants that control growth, development, movement, flowering, and senescence, as plants respond to and adapt to their environments. The functions and interactions of the plant hormones—auxin, cytokinins, gibberellins, abscisic acid, and ethylene—are detailed. Plant movements in response to enrivonmental stimuli include phototropism, gravitropism, thigmotropism, and turgor movements. The biological clock of plants controls circadian rhythms, such as stomatal opening and sleep movements, and times nightlength in the photoperiodic control of flowering.

CHAPTER SUMMARY

Plants are able to sense and respond adaptively to their environments, generally by altering their patterns of growth and development. The plasticity in the program for development allows plants of the same species to vary much more in body form than do animals of the same species. The control systems that govern a plant's responses to environmental factors include chemical messengers called hormones.

The Search for a Plant Hormone

Hormones, chemical signals that coordinate various components of an organism, are produced by one part of the body and translocated to other parts, where they trigger responses in target cells and tissues.

The growth of a shoot toward light is called positive phototropism. A coleoptile, enclosing the shoot of a grass seedling, grows straight when in the dark, but bends toward the light when illuminated from one side. The cells on the darker side elongate faster and produce the bending.

Darwin and his son discovered that a grass seedling would not bend toward light if its tip were removed or covered by an opaque cap. They speculated that some signal must be transmitted from the tip down to the elongating region of the coleoptile. Boysen-Jensen demonstrated that the signal was a mobile substance, capable of transmission through a block of gelatin separating a tip from the rest of the coleoptile. A tip segregated by an impermeable barrier of mica did not induce the phototropic response.

In 1926, Went placed coleoptile tips on blocks of agar to extract the chemical messenger. When he placed the agar blocks off-center on decapitated coleoptiles kept in the dark, the coleoptiles elongated and bent away from the side with the block. Went concluded that the chemical produced in the tip, which he called auxin, promoted growth as it passed down the coleoptile and was in a higher concentration on the side away from the light. Thimann and his colleagues purified auxin and determined its structure.

Functions of Plant Hormones

Five classes of plant hormones have been identified: *auxin, cytokinins, gibberellins, abscisic acid*, and *ethylene*. These hormones affect cell division, elongation, and differentiation. They have a variety of effects, depending on the site of action, the developmental stage of the plant, and relative hormone concentrations. Hormones are effective in very small concentrations; their signal must be amplified in the cell in some way. They may act by affecting the expression of genes, the activity of enzymes, or the properties of membranes.

Auxin refers to any compound, including a synthetic one, that causes a phototropic response in the Went oat-seedling test. The natural auxin extracted from plants is indoleacetic acid (IAA).

A major site of auxin synthesis is the apical meristem of a shoot. Within a range of concentrations, auxin stimulates growth of cells in the region of cell elongation in the shoot. Above a certain concentration, auxin inhibits cell elongation, probably because it induces the synthesis of ethylene, which inhibits plant growth. Auxin from the shoot reaches root cells, which are more sensitive to auxin and elongate at concentrations too low to stimulate shoot growth. Auxin illustrates that the same hormone has varying effects on different target cells and, at different concentrations, can have different effects on the same target cells.

Auxin appears to be transported through parenchymal tissue in one direction, from the shoot tip to base. Such polar transport requires energy and probably involves auxin carriers at the basal ends of cells.

According to the acid-growth hypothesis, auxin loosens cell walls by stimulating proton pumps. The resulting lower pH value in the cell wall activates enzymes that break cross-links between cellulose microfibrils. The turgor pressure of the cell then exceeds the restraining wall pressure, and the cell elongates.

Auxin also induces cell division in the vascular cambium, influences the differentiation of secondary xylem, and initiates the formation of adventitious roots at the cut bases of stems. The development of leaf traces, which connect leaves to vascular bundles of the stem, is initiated by auxin produced by leaf primordia. Auxin produced by developing seeds promotes growth of fruit. Synthetic auxins sprayed on tomato vines induce parthenocarpic fruit development. The herbicide 2,4-D, a synthetic auxin, is used to disrupt the normal balance of plant growth in "broad-leaf" dicot weeds.

The active ingredients of the coconut milk and degraded DNA that were found to induce plant cell growth in tissue culture were modified forms of adenine. These growth regulators were called cytokinins, for their ability to stimulate cytokinesis. Cytokinins are produced in actively growing roots, embryos, and fruits. Acting along with auxin, they stimulate cell division and affect differentiation.

When an explant, or piece of tissue, is grown in tissue culture in the absence of cytokinins, the cells grow large but fail to divide. If cytokinins are added along with auxin, the cells divide. The ratio of the two hormones controls differentiation of the cells: equal concentrations produce undifferentiated cells, more cytokinin results in shoot buds, whereas more auxin leads to root formation.

Cytokinins stimulate DNA replication and protein synthesis; but the mechanism of these actions is not known.

The control of apical dominance involves an interaction between auxin, transported down from the bud, which restrains axillary bud development, and cytokinins, transported up from the roots, which stimulate bud growth. This interaction may coordinate the growth of shoot and root systems, signaling the formation of branches when roots become extensive enough to supply and support them. The interaction works the opposite way in the development of lateral roots. Both auxin and cytokinins probably regulate axillary bud growth indirectly by changing the concentration of ethylene.

Cytokinins, which may prolong protein synthesis and mobilize nutrients, can retard aging of some plant organs. Cytokinin sprays are used to keep cut flowers fresh, and to prolong the shelf-life of fruits and vegetables.

In the 1930s, Japanese scientists determined that the fungus causing "foolish seedling disease" was secreting a chemical, given the name gibberellin, that was producing the hyperelongation of rice stems. Over a hundred different gibberellins have been identified; many occur naturally in plants.

Roots and young leaves near the shoot apex are sites of gibberellin production. Gibberellins stimulate growth in both leaves and stem. The application of different concentrations of gibberellin to dwarf plants has demonstrated a positive correlation between growth and the concentration of hormone added. *Bolting,* the growth of an elongated floral stalk, is caused by a surge of gibberellins. In some plants, both auxin and gibberellins must be present for fruit to set. The commercial spraying of these two hormones will induce parthenocarpic fruit development, important in the production of Thompson's seedless grapes.

The release of gibberellins from the embryo signals the seed to break dormancy. The germination of grain is triggered when gibberellins stimulate the synthesis of digestive enzymes that break down stored nutrients. Gibberellin also stimulates the synthesis of messenger RNA. Gibberellins function in the breaking of dormancy of apical buds in spring.

The hormone abscisic acid (ABA), which is produced in the bud, slows growth, inhibits cell division in the vascular cambium, and induces leaf primordia to develop into scales that protect the dormant bud during winter. No clearcut role has been shown for abscisic acid in the *abscission* of leaves from deciduous trees.

Abscisic acid acts in the growth inhibition of the embryo when the seed becomes dormant. ABA must be removed or inactivated, or the ratio of gibberellins to ABA must increase, for dormancy to be broken. Abscisic acid also acts to help the plant cope with adverse conditions. In a water-stressed plant, ABA accumulates in leaves and causes stomata to close.

Plants produce the gas ethylene, which acts as a

hormone to promote fruit ripening, to inhibit growth in roots and axillary buds, and to contribute to the aging, or *senescence*, of parts of the plant. Senescence is a normal part of plant development and may occur on the level of individual cells, of organs, or of the whole plant.

Ethylene initiates or hastens the degradation of cell walls and the decrease in chlorophyll content that is associated with fruit ripening. Many commercial fruits are ripened in huge containers into which ethylene gas is piped.

Before leaves are abscised in the autumn, many essential elements are transported to storage tissues in the stem. The leaf stops making chlorophyll; fall colors are a combination of pigments that had been concealed by chlorophyll and new pigments made during autumn.

Shortening days and cooler temperatures are the stimuli for leaf abscission. The small parenchyma cells of the abscission zone located near the base of the petiole have very thin walls. Less auxin is produced by an aging leaf; this drop initiates changes in the abscission zone. Cells in the abscission zone produce ethylene, which induces the synthesis of enzymes that hydrolyze polysaccharides in the cell walls. The weight of the leaf finally causes a separation within the abscission zone. A layer of cork forms a protective covering over the twig even before the leaf falls.

There are so many unanswered question about the internal chemical signals of plants that some plant physiologists think it is premature to call these growth regulators hormones. Some scientists suggest that responses to growth regulators may be more a product of increased sensitivity of cells to present regulators than the cells' reaction to messages from other parts of the plant.

Plant Movements

Tropisms are growth responses in which a plant organ curves toward or away from a stimulus, as a result of differential growth rates on opposite sides of the organ. These movements may occur in response to light (phototropism), gravity (gravitropism), or touch (thigmotropism).

In *phototropism*, as previously discussed, cells on the darker side of a stem elongate faster because of a larger concentration of auxin moving down from the shoot tip. In a mechanism not yet understood, light causes auxin to migrate laterally across the tip toward the dark side. The photoreceptor is believed to be a yellow pigment that is also involved in stomatal opening.

Roots exhibit positive *gravitropism*, whereas shoots show negative gravitropism. The settling of *statoliths*, dense cellular components such as starch grains, somehow causes hormones to accumulate on the low

side of the plant organ. In stems, a higher concentration of auxin and gibberellins cause the elongation of cells on the lower side. In roots, abscisic acid accumulates on the lower side, inhibiting growth, so that the root bends downward.

Climbing plants have tendrils that coil around supports. The directional growth in response to touch is *thigmotropism*. A more general response to mechanical stimulation is the stunting of growth in height and increase in girth of plants that are exposed to wind or rubbing.

Turgor movements are reversible movements caused by changes in turgor pressure of specialized cells in response to stimuli. The fluctuations in guard cells and the sweeping daily movements of leaves are examples.

More rapid movements occur in the sensitive plant, *Mimosa*, the leaves of which collapse and fold after being touched. The folding motion is due to the rapid loss of turgor by cells in specialized motor organs called pulvini, located at the joints of the leaf. Evidence shows that the motor cells lose potassium when stimulated, resulting in osmotic changes. The message travels through the plant from the point of stimulation, perhaps as the result of chemical messengers and electrical impulses. *Action potentials* resemble nervous messages in animals, but they are thousands of times slower. These electrical messages may be used as a form of internal communication.

Many members of the legume family exhibit sleep movements, the daily opening and folding of leaves in the morning and evening, caused by daily changes in the turgor pressure of motor cells in pulvini. Massive migration of potassium ions from one side of the pulvinus to the other may lead to the reversible osmosis in the motor cells.

Circadian Rhythms and the Biological Clock

Many physiological processes in animals and plants fluctuate with the time of day and are controlled by biological clocks—internal oscillators that keep accurate time. A *circadian rhythm* is a physiological cycle with about a 24-hour frequency. These rhythms persist, even when the organism is sheltered from environmental cues. Research indicates that the oscillator for circadian rhythms is endogenous, although the clock is set (entrained) to a precise 24-hour period by daily environmental signals.

Free-running periods, determined in the absence of environmental cues, vary from 21 to 27 hours, depending on the particular rhythmic response. The biological clocks in free-running periods still keep perfect time, but they are not synchronized with the outside world. Internal clocks are reset when an organism is returned to normal environment cues or crosses time zones. The mechanism for circadian rhythms is unaffected by temperature.

The nature of the oscillator is a mystery. Most scientists who study circadian rhythms place the biological clock at the cellular level.

Photoperiodism

Seasonal events in the life cycles of plants usually are cued by photoperiod, the relative lengths of night and day. A physiological response to daylength is called *photoperiodism*.

Garner and Allard, in 1920, discovered that an exceptionally tall variety of tobacco plant flowered only when the day length was 14 hours or shorter. They termed the Maryland Mammoth a *short-day plant*, because it needed a photoperiod of short days and long nights to flower. *Long-day plants* flower when days are 14 hours or longer; *day-neutral plants* are unaffected by photoperiod.

In the 1940s, researchers found that nightlength, not daylength, controls flowering and other photoperiod responses. If the nighttime portion of the photoperiod is interrupted by even a few minutes of light, a short-day plant such as the cocklebur will not flower. Photoperiodic responses thus depend on a critical nightlength, the maximum (long-day plants) or minimum (short-day plants) number of hours of uninterrupted darkness required for flowering.

Some plants bloom after a single exposure to the required photoperiod. Some will respond to photoperiod only after exposure to some other environmental stimulus, such as cold. The need for pretreatment with cold before flowering is called *vernalization*.

Leaves detect the photoperiod. In some species, exposure of a single leaf to the proper photoperiod will induce flowering in the whole plant. A hormone, named florigen, is believed to be the flowering signal that has been shown experimentally to travel from leaves to buds. The signal has yet to be identified; it may be a mixture of several hormones, or even the absence of inhibitors.

Red light is the most effective in interrupting nightlength. A brief exposure to red light breaks a dark period of sufficient length and prevents short-day plants from flowering, whereas a flash of red light during a dark period over the critical length will induce flowering in a long-day plant. A subsequent flash of light from the far-red part of the spectrum negates the effect of the red light.

The photoreceptor responsible for the reversible effects of red and far-red light is the pigment *phytochrome*, which alternates between two forms, one of which absorbs red light, whereas the other absorbs far-red light. These two variations of phytochrome are said to be photoreversible. The P_r to P_{fr} interconversion controls various events in the life of the plant.

Plants synthesize P_r, which is converted to P_{fr} when the phytochrome is illuminated in sunlight. P_{fr} triggers many plant responses to light, such as the breaking of seed dormancy. In darkness, the P_{fr} molecules gradually revert to P_r. At sunrise, the P_{fr} level suddenly increases. Nightlength is measured by the biological clock. The role of phytochrome may be to synchronize the clock by signalling when the sun sets and rises.

Internal and external stimuli influence the complex interactions of the many components that control plant growth and development. Seasonal events and flowering are influenced by genes, hormones, an endogenous clock, photoreceptors, and the length of day.

STRUCTURE YOUR KNOWLEDGE

1. Individual plant hormones and combinations of hormones have various effects on the growth and development of plants, depending on their relative concentrations and the target cells. The table below lists the five identified classes of plant hormones and some of the effects they may have. Place checks in the appropriate boxes to indicate which hormones play a role in each process.

Function ↓ Hormone→	Auxin	Cytokinins	Gibberellins	Abscisic Acid	Ethylene
Promote stem elongation					
Promote root growth					
Promote fruit development					
Promote fruit ripening					
Promote seed and bud germination					
Promote cell division and growth					
Delay senescence					
Contribute to senescence					
Inhibit growth					
Promote bud and seed dormancy					

2. Plants exhibit tropisms and turgor movements. Create a concept map that organizes your understanding of these plant movements.

TEST YOUR KNOWLEDGE

TRUE or FALSE: *Indicate T or F and then correct false statements.*

1. _____ Target cells are the site of hormone action.

2. _____ Went's oat-seedling test is used to identify gibberellins.

3. _____ Shoots exhibit positive phototropism and negative gravitropism.

4. _____ Cytokinins, synthesized in the root, counteract apical dominance.

5. _____ Bolting occurs when ethylene stimulates the degradation of cell walls and decreases chlorophyll content.

6. _____ ABA induces the synthesis of enzymes that hydrolyze polysaccharides in cell walls of the abscission zone.

7. _____ Action potentials are electrical impulses that may be used as internal communication in plants.

8. _____ Thigmotropism is involved in the rapid movements of *Mimosa* leaves.

9. _____ A physical response to daylength is called a circadian rhythm.

10. _____ Vernalization is the need for pretreatment with cold before flowering.

MULTIPLE CHOICE: *Choose the one best answer.*

1. The body form of plants may vary more within a species than that of animals within the same species does because
   a. growth in animals is indeterminate.
   b. plants respond adaptively to their environments by altering their patterns of growth and development.
   c. plant growth is governed by hormones.
   d. of all of the above.

2. A concentration of auxin that is too low to stimulate growth in a shoot
   a. may induce the synthesis of ethylene,

which inhibits growth.
   b. may still be high enough to break dormancy in a seed.
   c. may still be high enough to stimulate cell elongation in the root.
   d. may still cause bolting.

3. Which of these is not part of the acid-growth hypothesis?
   a. the stimulation of pigments that control potassium pumps that increase osmotic pressure in the cell
   b. the stimulation of proton pumps in cell membranes
   c. a lowered pH value, which activates enzymes in the wall that break cross-links between cellulose microfibrils
   d. the turgor pressure of the cell exceeding wall pressure

4. Which of the following is not true of cytokinins?
   a. Their active ingredient is a modified form of adenine.
   b. Their effects were first observed when coconut milk was added to culture medium.
   c. The ratio of cytokinins to auxin determines cell differentiation in tissue culture.
   d. They seem to have an anti-aging effect, perhaps because they maintain cell wall integrity and high chlorophyll content.

5. Spraying plants with a combination of auxin and gibberellins
   a. prevents senescence.
   b. promotes formation of parthenocarpic fruit.
   c. will induce flowering.
   d. is used to treat dwarfism in plants.

6. The growth inhibitor in seeds usually is
   a. abscisic acid.
   b. ethylene.
   c. gibberellin
   d. a small amount of ABA combined with a larger ratio of gibberellins.

7. Statoliths are involved in
   a. thigmotropism.
   b. stress reactions of plants.
   c. gravitropism.
   d. turgor movements of plants.

8. A circadian rhythm
   a. is controlled by an internal oscillator.
   b. is a physiological cycle that has approximately a 24-hour frequency.
   c. involves a biological clock that is set by daily environmental signals.
   d. all of the above are correct.

9. A flash of far-red light during a critical-length dark period
   a. will induce flowering in a long-day plant.
   b. will induce flowering in a short-day plant.
   c. will not influence flowering.
   d. will increase the P_{fr} level suddenly.

10. Evidence indicates that
   a. a hormone that travels from leaves to buds induces flowering.
   b. flowering is induced by phytochrome levels in the leaves.
   c. short-day plants need vernalization before photoperiod affects them.
   d. a biological clock in the bud determines photoperiod.

11. Turgor movements of plants
   a. may involve movement of potassium ions and osmotic changes.
   b. are involved in the sleep movements shown by many legume plants.
   c. may involve pulvini, specialized motor organs in which turgor pressure changes.
   d. may involve all of the above.

12. Which of the following is not true of free-running periods?
   a. They are entrained by environmental signals.
   b. They are of a set period for each rhythmic response.
   c. They may vary in length from 20 to 27 hours for various responses.
   d. They provide evidence that the biological clock is endogenous.

ANSWER SECTION

CHAPTER 31: ANATOMY OF A PLANT

Suggested Answers to Structure Your Knowledge

1.

Function	Cells, Tissues, Or Structures That Perform Function		
	Root	Stem	Leaf
Anchorage	Tap or fibrous root system		
Absorption	Root hairs on epidermis		Some carnivorous plants
Storage	Parenchyma cells of pith and cortex	Parenchyma cells	
Support	Adventitious roots	Collenchyma, schlerenchyma xylem (wood)	Veins, xylem
Protection	Epidermis, root cap, periderm	Epidermis with cuticle, periderm, secondary phloem	Epidermis with cuticle
Transport	Stele with xylem, phloem	Vascular bundles or ring of xylem and phloem	Veins (parallel or branching)
Photosynthesis		Some parenchyma cells with chloroplasts	Mesophyll with chloroplasts, (pallisade and spongy)
Growth	Apical meristem (primary), vascular and cork cambiums (secondary growth)	Apical meristem (primary), vascular and cork cambiums (secondary growth)	Develop from leaf primordia
Gas exchange	Through epidermis, air tubes in swamp trees	Lenticles	Stomates with guard cells

2. Primary growth results in growth at the tips of roots and stems, whereas secondary growth produces greater girth. The apical meristems produce new cells that form the primary meristems: protoderm, procambium, and ground meristem. These cells elongate and differentiate into the dermal, vascular, and dermal tissue systems. Secondary growth involves a vascular cambium, which produces new xylem and phloem cells, and a cork cambium, which produces cells that form the periderm.

Answers to Test Your Knowledge

Matching:

1.	L	6.	H
2.	F	7.	C
3.	B	8.	D
4.	K	9.	E
5.	J	10.	G

Multiple Choice:

1.	a	6.	a
2.	c	7.	b
3.	d	8.	c
4.	c	9.	a
5.	b	10.	c

Fill in the Blanks:

H F I A B G D C E

CHAPTER 32: TRANSPORT IN PLANTS

Suggested Answers to Structure Your Knowledge

1. The transpiration–cohesion theory explains the ascent of xylem sap. The lower water potential of the air surrounding the leaves compared to that of the cortex cells of the root and soil solution results in the passive movement of water down its water potential gradient in a continuous column in xylem vessels. The cohesion of water molecules transmits the pull resulting from transpiration throughout the column, and the adhesion of water to the hydrophillic walls of the xylem vessels aids the flow. The tension created within the xylem by the upward pull also helps the passive flow of water.

A pressure–flow mechanism explains the movement of phloem sap. The active accumulation of sugars in sieve-tube members greatly decreases the water potential at the source end and results in an inflow of water. The removal of sugar from the sink end of a phloem tube results in the osmotic loss of water from the cell. The difference in hydrostatic pressure between the source and sink ends of the phloem tube causes the flow of water with the accompanying transport of sugar.

Answers to Test Your Knowledge

1.	c	6.	b
2.	d	7.	c
3.	c	8.	a
4.	a	9.	a
5.	b	10.	c

CHAPTER 33: PLANT NUTRITION

Suggested Answers to Structure Your Knowledge

1.

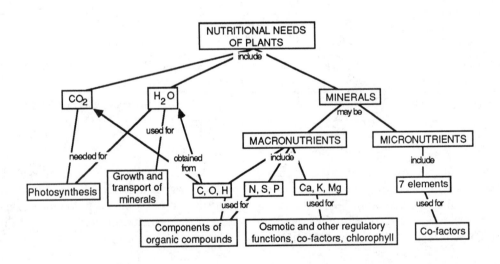

2. The key properties of a fertile soil include:

- *Texture*: mixture of sand, silt, and clay so that air spaces containing oxygen are provided as well as sufficient surface area for the binding of water and minerals

- *Humus*: prevents clay from packing so soil retains water and has air spaces, provides reservoir of mineral nutrients

- *Minerals*: adequate supply of nitrogen, phosphorus, and potassium as well as other minerals

- *pH*: proper pH level so that minerals are in a form that can be absorbed, and cation exchange can take place

3. Nitrogen must be in the form of nitrate or ammonium in order to be absorbed by plants. Nitrogen-fixing bacteria (using the enzyme nitrogenase) and microbes decomposing humus convert nitrogen into these forms. Bacteria in the soil oxidize the ammonium formed by nitrogen-fixing bacteria into nitrate, the form in which plants acquire most of their nitrogen. Nitrogen is essential to plant growth because it is needed for protein formation. Productive plant growth and high protein content of crops are important in agriculture.

Answers to Test Your Knowledge

1.	b	6.	a
2.	d	7.	c
3.	b	8.	d
4.	b	9.	c
5.	b	10.	a

CHAPTER 34: PLANT REPRODUCTION

Suggested Answers to Structure Your Knowledge

1.

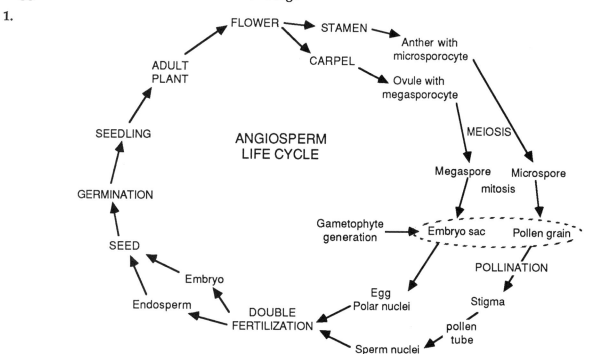

2. *Sexual*: Advantages include increased genetic variability, providing the potential to adapt to changing conditions, and the dispersal and dormancy capabilities provided by seeds. Disadvantages are that the seedling stage is very vunerable, and that genetic recombination may separate adaptive traits.

Asexual: Advantages include the hardiness of vegetative propagation and the maintenance of well-adapted plants in an environment. Disadvantages relate to the advantages of sexual reproduction: there is no genetic variability from which to choose should conditions change, and there are no seeds.

Both of these types of reproduction have been important to the adaptation and success of plants in various habitats.

Answers to Test Your Knowledge

Fill in the Blanks:

1. stamen
2. stigma
3. monoecious
4. embryo sac
5. sepals

6. epicotyl
7. coleoptile
8. parthenocarpic
9. aggregate
10. callus

Multiple Choice:

1. c
2. c
3. b
4. a
5. d

6. d
7. a
8. c
9. b
10. c

CHAPTER 35: CONTROL SYSTEMS IN PLANTS

Suggested Answers To Structure Your Knowledge

1.

Function ↓ Hormone →	Auxin	Cytokinins	Gibberellins	Abscisic Acid	Ethylene
Promote stem elongation	√		√		
Promote root growth	√	√	√		
Promote fruit development	√		√		
Promote fruit ripening					√
Promote seed and bud germination			√		
Promote cell division and growth	√	√	√		
Delay senescence		√			
Contribute to senescence					√
Inhibit growth				√	√
Promote bud and seed dormancy				√	

2.

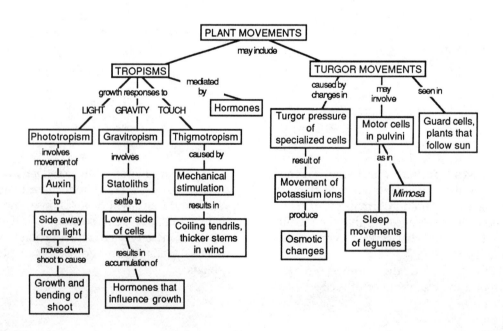

Answers to Test Your Knowledge

True or False:

1. true

2. false: change *gibberellins* to *auxin*

3. true

4. true

5. false: change *bolting* to *fruit ripening;* or change to read *Bolting is gibberellin-induced elongation of a flower stalk.*

6. false: change *ABA* to *ethylene*

7. true

8. false: change *thigmotropism* to *turgor movements;* or change to read *Thigmotropism is the change in the growth of a plant in response to touch.*

9. false: change *circadian rhythm* to *photoperiodism*

10. true

Multiple Choice:

1. b	4. d	7. c	10. a
2. c	5. b	8. d	11. d
3. a	6. a	9. c	12. a

Animals: Form and Function

CHAPTER 36

INTRODUCTION TO
ANIMAL FORM AND FUNCTION

FRAMEWORK

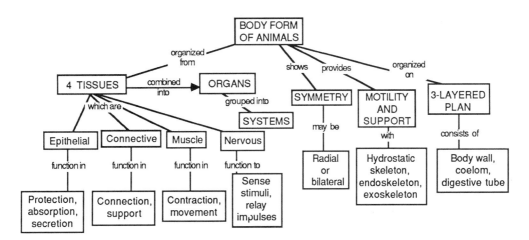

CHAPTER SUMMARY

Almost all plants and animals need an internal transport system, and terrestrial organisms must support themselves against gravity. Networks of tubes and rigid structures have evolved in both these kingdoms. Although fundamentally different in origin, anatomy, and physiology, both groups illustrate the general theme of the correlation of structure and function, and the ability of life to adapt to the environment, both by short-term physiological adjustment and long-term evolution.

Levels of Structural Organization

A hierarchy of structural order characterizes life. Unicellular protozoa are able to carry out all their life functions without integration beyond the cellular level. Multicellular organisms have specialized cells grouped into tissues. In most animals, tissues are combined into structural and functional units called organs, and various organs may cooperate in organ systems.

Tissues are collections of cells with a common structure and function. Histologists—biologists who study tissues—classify tissues into four categories.

Epithelial tissue lines the outer and inner surfaces of the body in sheets of tightly packed cells that function in protection, absorption, and secretion. Cells at the base of an epithelium are attached to a *basement membrane*. Epithelia are classified on the basis of the number of cell layers and cell shape. A *simple epithelium* has one layer, whereas a *stratified epithelium* has multiple layers of cells. The cells at the free surface may be *squamous* (scalelike), *cuboidal* (boxlike), or *columnar* (pillarlike).

Specialized epithelial cells may be organized into *glands* that secrete or excrete their products. Goblet cells are single-celled glands that secrete mucus. They are found embedded in the *mucous membrane* of the gut, where mucus lubricates and protects the gut lining, and in the air passages, where mucus traps particles, which are removed as this ciliated epithelium moves the mucous layer upward.

Most glands are multicellular invaginations of the epithelium. *Exocrine glands* are connected to tubes into which their products are excreted, whereas *endocrine glands* secrete their substances, called hormones, directly into the blood.

Connective tissue connects and supports other tissues, and is characterized by having relatively few cells suspended in an extracellular *matrix* of fibers, which may be embedded in a liquid, jellylike, or solid ground substance.

Loose connective tissue attaches epithelia to underlying tissues and holds organs in place. Its loosely woven fibers are of three types. *Collagenous fibers* are made of collagen, the most abundant animal protein and the longest protein known. Three collagen molecules coil to form a fibril, and several fibrils make up a collagenous fiber, a structure with great tensile strength. *Elastic fibers*, made of the protein elastin, provide loose connective tissue with resilience. Branched *reticular fibers* form a tightly woven connection with adjacent tissues.

The cells scattered in the fibrous mesh of loose connective tissue include *fibroblasts*, which secrete the protein of the extracellular fibers, and *macrophages*, amoeboid cells that engulf bacteria and cellular debris by phagocytosis.

Adipose tissue is a special form of loose connective tissue with adipose cells held in the matrix. Each adipose cell contains a large fat droplet. This tissue pads and insulates the body and stores fuel reserves.

Fibrous connective tissue, with its dense arrangement of collagenous fibers, is found in *tendons*, which attach muscles to bones, and in *ligaments*, which join bones at joints.

Cartilage is composed of collagenous fibers embedded in a rubbery ground substance called *chondrin*, which is secreted by *chondrocytes* found in scattered lacunae, or spaces, in the ground substance. Cartilage is a strong but somewhat flexible support material that makes up the skeleton of sharks and vertebrate embryos. We retain cartilage in the nose, ears, intervertebral discs, trachea, and ends of some bones.

Bone is a mineralized connective tissue formed by *osteocytes* that deposit a matrix of collagen and calcium phosphate. *Haversian systems* consist of concentric layers of matrix deposited around a central canal containing blood vessels and nerves. Osteocytes are located in lacunae within the matrix and are connected to one another by cellular extensions called canaliculi. In long bones, the hard outer region is *compact bone* built of repeating Haversian systems, whereas the interior is a *spongy bone tissue*, called marrow. Red marrow, near the ends of long bones, manufactures blood cells.

Blood is a connective tissue with a liquid extracellular matrix called plasma, which contains water, salts, and dissolved proteins. Erythrocytes (red blood cells) carry oxygen, leukocytes (white blood cells) function in defense, and platelets are involved in the clotting of blood.

Muscle tissue consists of long, excitable cells, packed with microfilaments of actin and myosin and capable of contraction. Three types of muscle tissues are found in vertebrate bodies. *Skeletal muscle*—also called striated muscle because of its striped appearance, which is caused by the arrangement of overlapping filaments—is responsible for the voluntary movements of the body. *Cardiac muscle*, forming the wall of the heart, is also striated; the ends of its branching cells are joined by intercalated discs that rapidly relay electrical impulses. *Visceral muscle*, also called smooth muscle, is composed of spindle-shaped cells lacking striations. Visceral muscle is found in the walls of the digestive tract, arteries, and other internal organs. Skeletal muscle is the only muscle under conscious control.

Nervous tissue senses stimuli and transmits signals. The *neuron*, or nerve cell, consists of a cell body and two or more nerve processes that conduct impulses toward (dendrites) and away from (axons) the cell body.

In all but the sponges and cnidarians, tissues are organized into specialized units of function called *organs*. Many vertebrate organs are suspended by *mesenteries* in fluid-filled body cavities. Mammals have a *thoracic cavity* separated by a muscular diaphragm from an *abdominal cavity*.

Groups of organs are integrated into *systems*, which perform the major functions required for life. Ten organ systems are recognized: integumentary, skeletal, muscular, nervous, endocrine, circulatory, respiratory, digestive, urinary, and reproductive.

Basic Principles of Animal Design

Every cell must be in an aqueous medium to allow for oxygen, nutrient, and waste exchange. Animals with compact bodies must provide extensive, moist internal membranes for exchanging materials with the environment. The surface area of the air chambers of the lung is about 100 m². The internal surface of the kidney is specialized to filter wastes from blood, and the extensive lining of the digestive tract absorbs nutrients. The circulatory system connects these exchange surfaces with the extracellular environment.

Radial symmetry, found in cnidarians and echinoderms, often occurs in sessile or sedentary animals, allowing the animal to encounter food and other stimuli from any direction. In animals with *bilateral symmetry*, a central longitudinal plane divides the body into left and right halves. Bilateral animals have a *dorsal* (top) and *ventral* (bottom) surface and an *anterior* (head) and *posterior* (tail) end. This type of symmetry is often associated with active movement and cephalization.

Most animals are mobile and spend time and energy actively looking for food and mates and escaping from danger. Skeletons function in support for the collection of nonrigid animal cells, for protection of soft tissues, and for movement, by providing rigid structures for muscles to contract against.

Hydrostatic skeletons consist of fluid-filled compartments against which soft-bodied animals can contract their muscles. *Exoskeletons,* such as the calcium carbonate shells of molluscs and chitinous cuticles of arthropods, are hard surface coverings produced by the animal. A hard skeleton encased within the soft tissues of an animal is an *endoskeleton,* of which the spicules of sponges, calcium-containing plates of echinoderms, and vertebrate bony skeletons are examples.

The *axial skeleton* of vertebrates consists of the skull, vertebral column, and rib cage; the *appendicular skeleton* includes the pectoral and pelvic girdles and limb bones. The bones of the skeleton act as levers that attached muscles contract against and move. Antagonistic muscles are needed to move a body part in opposite directions.

Complex animals have a body plan that is basically a tube within a tube. The inner digestive tract is separated from the outer body-wall tube by a coelom in which various organs are suspended. This triple-layered body plan is established by the three embryonic layers: the ectoderm, mesoderm, and endoderm.

A specialized digestive compartment allows an animal to secrete enzymes and acids to break down food molecules without endangering its own cells. Protozoans, with intracellular digestion, utilize food vacuoles to contain digestive enzymes. Cnidarians and flatworms have a digestive sac with a single opening to the outside. More complex animals have *complete digestive tracts* with two openings: a mouth and an anus. This one-way tract allows for specialization of regions for the process of digestion.

The internal environment of vertebrates is the *interstitial fluid* that bathes the cells. The ability of an organism to control the constancy of this environment is called *homeostasis.* Most of the mechanisms by which animals maintain homeostasis are based on *negative feedback loops.* When some factor moves above or below a *set point,* a control mechanism is turned on or off to return the factor to normal. Several negative feedback mechanisms control body temperature in humans. When the hypothalamus monitors a rise in temperature, it sends nervous impulses to sweat glands to increase their activity. When the temperature falls below the set point, the hypothalamus stops sending "sweat" signals.

Positive feedback in a physiological function is a mechanism in which a change in a factor serves to amplify rather than reverse the change. The stimulation of uterine contractions during childbirth is an example. Because they are inherently unstable, positive feedback mechanisms are rare.

STRUCTURE YOUR KNOWLEDGE

1. Fill in the following table on the structure and function of animal tissues.

Tissue	Structural Characteristics	General Functions	Structure and Function of Specific Types

2. What are the functions of a skeletal system? What are the relative advantages of exoskeletons and endoskeletons?
3. Describe the advantages of the progression in complexity of digestive systems found in protozoans to cnidarians and flatworms to vertebrates.
4. How do compact animals deal with the need for gas, nutrient, and waste exchange between their cells and the environment?

TEST YOUR KNOWLEDGE

MULTIPLE CHOICE: *Choose the one best answer.*

1. Which of the following is not an organ system?
   a. skeletal
   b. connective
   c. digestive
   d. nervous

2. A stratified squamous epithelium would be composed of
   a. several layers of flat cells attached to a basement membrane.
   b. ciliated, mucus-secreting flattened cells.
   c. a hierarchical arrangement of boxlike cells.
   d. an irregularly arranged layer of pillarlike cells.

3. Exocrine glands
   a. are usually single-celled glands.
   b. produce hormones.
   c. excrete their products through ducts.
   d. are or do all of the above.

4. Which of the following are incorrectly paired?
   a. fibrous connective tissue—chondrocytes embedded in chondrin
   b. bone—osteocytes connected by canaliculi in Haversian systems
   c. loose connective tissue—collagenous, elastic, reticular fibers
   d. adipose tissue—loose connective tissue with fat-storing cells

5. The best description of visceral muscle is
   a. striated, branching cells, under involuntary control.
   b. spindle-shaped cells, under involuntary control.
   c. spindle-shaped cells connected by intercalated discs.
   d. striated cells with overlapping filaments, under involuntary control.

6. The diaphragm
   a. is a mesentery.
   b. increases the surface area of the lungs.
   c. is part of the mammalian reproductive system.
   d. separates the thoracic and abdominal cavities in mammals.

7. An advantage of bilateral symmetry is that
   a. an organism has a distinct left and right side.
   b. the animal can move in any direction.
   c. sense organs can be concentrated in the anterior region in a motile animal.
   d. the skeleton is adapted for rapid movement.

8. Which of the following is not true of the vertebrate skeletal system?
   a. It contains living cells and is capable of growth.
   b. It does not serve the protective function characteristic of an exoskeleton.
   c. It serves as levers that are moved by the contraction of muscles.
   d. It consists of an axial skeleton, including the skull, vertebrae, and ribs, and the appendicular skeleton, including limb bones and pectoral and pelvic girdles.

9. The interstitial fluid of vertebrates
   a. is the internal environment within cells.
   b. provides for the exchange of nutrients and wastes between cells and blood.
   c. is the site of digestion.
   d. is a result of the triple-layered body plan.

10. Negative feedback loops
    a. are mechanisms that maintain homeostasis.
    b. are activated when a physiological variable deviates from a set point.
    c. are analogous to a thermostat that controls room temperature.
    d. are all of the above.

CHAPTER 37

ANIMAL NUTRITION

FRAMEWORK

Animals eat other organisms in order to obtain fuel for respiration, organic raw materials, and essential nutrients such as essential amino acids, vitamins, and minerals. Digestion is the enzymatic hydrolysis of macromolecules into monomers that can be absorbed across cell membranes.

Gastrovascular cavities, found in cnidarians and flatworms, are digestive sacs in which some extracellular digestion takes place before food particles are phagocytized by cells lining the cavity. All the more complex animal groups have alimentary canals, one-way tracts with specialized regions for mechanical breakdown of food, storage, digestion, absorption of nutrients, and elimination of wastes.

This chapter details the structures, functions, enzymes, and hormones of the human digestive tract.

CHAPTER SUMMARY

Heterotrophs eat other organisms to obtain the organic nutrients they require. *Herbivores* eat plants; *carnivores* eat animals; and *omnivores* consume both plants and animals.

Many aquatic animals are *filter-feeders*, sifting small food particles from the water. *Deposit-feeders* salvage pieces of decaying organic matter in detritus. *Fluid-feeders* suck fluids from a living plant or animal host. *Holotrophs* are predators that ingest relatively large pieces of plants or animals.

The proteins, fats, and polysaccharides that make up food are too large to pass through cell membranes, and these macromolecules must be reorganized to form the macromolecules that make up each heterotroph. In digestion, enzymes break food into its component monomers, which can then be absorbed across the membrane of the digestive compartment and enter the animal's cells.

Nutritional Requirements

Food provides fuel for cellular respiration, organic raw materials for the construction of the animal's molecules, and *essential nutrients*, which the animal cannot synthesize and must obtain in prefabricated form.

The monomers of carbohydrates, fats, and proteins can be used as fuel for cellular respiration, although the first two are used preferentially. Energy content of food is measured in *calories*. The Calorie, as used by nutritionists, is actually a *kilocalorie*, the amount of energy needed to raise the temperature of a kilogram of water 1° C. Fat supplies about two times as many kcal/g as carbohydrate or protein.

Metabolism must supply energy continuously to maintain breathing, heart beat, and stable body temperature. The *basal metabolic rate* (BMR) is the number of kilocalories a resting organism requires for these processes for a given time. Birds and mammals, which are *endothermic* and use metabolic energy to maintain a constant body temperature, have higher basal metabolic rates than do the *ectothermic* fish, reptiles, and amphibians, which absorb environmental energy to maintain body temperatures. In endotherms, body size is inversely related to the number of calories required to maintain each gram of body weight. The smaller the animal, the greater its surface area–to–volume ratio, the greater the loss of heat to the surroundings, and the greater the energy cost of maintaining a stable body temperature.

The BMR for humans averages 1600 to 1800 kcal per day for males and about 1300 to 1500 kcal for females. Any activity increases the caloric requirement above this level.

When an animal consumes more calories than are needed to meet its energy requirements, the excess calories are stored. The liver and muscles store energy as *glycogen*, a polymer of glucose. A human can store about a day's supply of energy in glycogen. When the gylcogen supplies are full, additional calories are stored in adipose tissue as fat.

An *undernourished* person has a diet deficient in calories. If a calorie deficiency continues, the body breaks down its own proteins for energy. Incidents of undernourishment are usually associated with drought or war, although the condition anorexia nervosa can result in undernourishment.

Overnourishment, or obesity, increases the risk of heart attack, diabetes, and other disorders. The determination of "healthy" weights for people often involves the use of insurance companies' statistics on increased health risks.

Animals can fabricate a great variety of organic molecules using enzymes to rearrange the precursors acquired from food. In vertebrates, the conversion of nutrients to organic molecules occurs mainly in the liver.

Molecules that an animal requires but cannot make are called *essential nutrients*. These requirements vary from species to species, depending on biosynthetic capabilities. When the diet is missing one or more essential nutrients, the animal is said to be *malnourished*. Malnutrition is more common than is undernutrition in human populations.

Eight of the 20 amino acids required to make proteins are essential in the human diet. Protein deficiency develops from a diet that lacks one or more essential amino acid. The resulting syndrome of retarded mental and physical development in children is called *kwashiorkor*, which may arise when a child is weaned from mother's milk to a starchy diet.

Meat, eggs, and cheese contain complete proteins with all essential amino acids in roughly balanced proportions. Most plant proteins are incomplete, and diets built on a single staple, such as corn, beans or rice, can result in protein deficiency. The body cannot store amino acids, and a deficiency of a single essential amino acid may prevent protein synthesis and limit the use of other amino acids. A combination of plant foods, complementary in amino acids and consumed at the same meal, can prevent protein deficiencies.

Animals are unable to make certain unsaturated fatty acids. Linoleic acid, used to make some phospholipids found in membranes, is required in the human diet. Deficiencies of essential fatty acids are rare.

Vitamins are organic molecules required in small amounts in the diet. Most vitamins are used as coenzymes or parts of coenzymes, and deficiencies can cause severe syndromes. The first vitamin isolated was thiamine; a deficiency of thiamine results in the disease beriberi.

Thirteen vitamins essential to humans have been identified. Water-soluble vitamins include the B-complex, most of which function as coenzymes in key metabolic processes, and vitamin C, required for production of connective tissue. Excesses of water-soluble vitamins are excreted. The fat-soluble vitamins are A, incorporated into visual pigments; D, aiding in bone formation; E, seeming to protect phospholipids in membranes from oxidation; and K, required for blood clotting. Excesses of fat-soluble vitamins are deposited in body fat, and overdoses may cause toxic accumulations. There is debate between those scientists who believe that the recommended daily allowances (RDAs) are sufficient and those who believe that optimal intakes of certain vitamins are much higher.

A compound that is a vitamin for one species may not be essential for a second species that can synthesize it. Symbiotic microorganisms may produce vitamins that supplement or substitute for vitamins in the diet.

Minerals are inorganic nutrients, and usually are needed in only very small amounts. Requirements vary with species. Vertebrates require relatively large quantities of calcium and phosphorus for bone construction. Calcium is necessary for normal nerve and muscle functioning, and phosphorus is needed in ATP and nucleic acids. Iron is a component of the cytochromes and hemoglobin. Other minerals function as co-factors. Iodine is needed by vertebrates to make thyroxin, a metabolic regulatory hormone. Sodium, potassium, and chlorine are important in nerve function and osmotic balance.

Digestion

Digestion breaks apart food particles and splits macromolecules into monomers: polysaccharides into simple sugars, fats into glycerol and fatty acids, proteins into amino acids, and nucleic acids into nucleotides. Macromolecules are broken down by *hydrolysis*, the enzymatic addition of a water molecule when the bond between monomers is broken. Hydrolysis must occur in a compartment separated from the animal's own protoplasm.

The simplest animals have single-opening sacs called *gastrovascular cavities*, which function in both digestion and transport of nutrients throughout the body. Using tentacles equipped with stinging nematocysts, *Hydra*, a cnidarian, captures prey and stuffs it through its expandable mouth. Cells of the gastrodermis secrete digestive enzymes, which initiate food breakdown, and beating flagella circulate the smaller food particles throughout the cavity. Gastrodermal cells take in food particles by phagocytosis, and hydrolysis of macromolecules occurs by intracellular digestion within food vacuoles. Undigested materials are expelled through the mouth.

Flatworms also have gastrovascular cavities. In a planarian, food is sucked into the mouth, which opens at the end of a muscular *pharynx*. The gastrovascular cavity has three main branches and secondary ramifications that greatly increase the surface area. Digestion is begun extracellularly and continues within cells that take up food particles by phagocytosis.

Nematodes, annelids, molluscs, arthropods, echinoderms, and chordates all have *digestive tracts* or *alimentary canals*, which have two openings: a mouth and an anus. Specialized regions, which may be adapted to specific diets, allow the sequential digestion and absorption of nutrients. Food, ingested through the mouth and pharynx, passes through an esophagus that leads to a crop, a gizzard, or a stomach. These three organs are specialized for storing or grinding food. In the lengthy *intestine*, digestive enzymes hydrolyze macromolecules, and nutrients are absorbed across the tube lining. Undigested wastes are egested through the *anus*.

The mammalian digestive system begins with the teeth, used to bite off and chew chunks of food. Dentition, the types and arrangement of teeth in the mouth, is correlated with diet.

Herbivores have longer alimentary canals because plant material, with its cell walls, is more difficult to digest than meat is. Many herbivorous mammals also have special fermentation chambers filled with symbiotic bacteria and protozoa. These microorganisms both digest cellulose to simple sugars and produce a variety of essential nutrients for the animal. The caecum, a large pouch at the connection of small and large intestines, may house these microrganisms. Since the symbiotic bacteria of rabbits and some rodents live in the large intestine, these animals may ingest their feces so that the nutrients produced by these symbionts can be absorbed as the feces go through the small intestine.

Ruminants have the most elaborate adaptations for a herbivorous diet. The stomach is divided into four chambers, two of which contain symbiotic bacteria that digest cellulose. The cud is regurgitated, rechewed, and swallowed into the other chambers, where both the microbially digested cellulose and the microorganisms themselves are digested and absorbed.

The mammalian digestive tract has a four-layered wall: an inner mucous membrane, a connective-tissue layer, smooth muscle, and a sheath of connective tissue attached to the body-cavity membrane. The rhythmic waves of contraction called *peristalsis* push food through the tract. Ringlike valves called *sphincters* regulate the passage of material between some specialized segments.

Accessory glands, including salivary glands, the pancreas, and the liver with its gall bladder, deliver digestive enzymes to the alimentary canal through ducts.

A Tour Through the Human Digestive Tract

Physical and chemical digestion begins in the mouth, where the relatively unspecialized human dentition chews food to expose a greater surface area to enzyme action. The presence of food in the oral cavity, or learned associations with other stimuli, triggers a nervous reflex that causes the salivary glands to deliver saliva through ducts to the mouth.

Saliva contains several components: *mucin*, a glycoprotein that protects the mouth lining from abrasion and lubricates the food for swallowing; buffers, which help neutralize the cavity-causing acidic environment; antibacterial agents; and *salivary amylase*, which hydrolyzes starch into smaller polysaccharides and prevents a buildup of starch between the teeth.

The tongue is used to taste, to manipulate food, and to push the food ball or *bolus* into the pharynx for swallowing. The pharynx is the intersection leading to both the esophagus and trachea. During swallowing, the esophageal sphincter relaxes and the windpipe moves so that its opening is blocked by the *epiglottis*. The *Heimlich maneuver* is the emergency method used to help dislodge food or a foreign object that has blocked someone's windpipe.

Food moves down through the esophagus to the stomach, squeezed along by a wave of smooth muscle contraction called peristalsis.

The J-shaped stomach, with its elastic wall and folds, or *rugae*, can expand to hold about 2 liters of food and fluid. The epithelium lining the lumen (cavity) secretes *gastric juice*, a digestive fluid containing a high concentration of hydrochloric acid and *pepsin*, an enzyme that hydrolyzes specific peptide bonds in proteins. The low pH value of the stomach contents breaks down food tissues, kills bacteria, and denatures proteins, which increases the exposure of peptide bonds to the action of pepsin.

Pepsin is synthesized and secreted in the inactive form called *pepsinogen*. HCl and pepsin itself activate pepsinogen—an example of positive feedback. Protein-digesting enzymes are usually secreted in inactive forms, generally called *zymogens*. The mucous coating secreted by the epithelium protects the stomach lining from digestion. Lesions called gastric ulcers develop where the lining is eroded faster than it can be regenerated by mitosis.

The sight, smell, or taste of food sends a nervous message from the brain to the stomach that initiates secretion of gastric juice. Substances in the food stimulate the stomach wall to release the hormone *gastrin* into the circulatory system. When gastrin recirculates back to the stomach, it stimulates further secretion of gastric juice. If the stomach contents' pH becomes too low, the release of gastrin is inhibited, which decreases secretion of gastric juice—an example of negative feedback.

Smooth muscles mix the stomach contents about every 20 seconds. The stomach is usually closed off by a *cardiac sphincter*, which prevents backflow into the esophagus, and a *pyloric sphincter*, which regulates passage into the small intestine of the *acid chyme* produced by the action of the stomach and its secretions on the ingested food.

Most enzymatic hydrolysis of macromolecules and absorption of nutrients into the blood take place in the small intestine, the longest section of the alimentary canal.

The *pancreas* produces hydrolytic enzymes and a bicarbonate-rich alkaline solution that offsets the acidity of the chyme. The pancreas is also an endocrine gland, secreting the hormones insulin and glucagon, which regulate glucose concentration in the blood.

The liver produces *bile*, which is stored in the *gall bladder*. Bile aids in the digestion and absorption of fats, and also contains pigments that are by-products of the breakdown of red blood cells in the liver.

Bile, digestive juices from the pancreas, and secretions from gland cells of the intestinal wall are mixed with the chyme in the first section of the small intestine, called the *duodenum*. Food molecules trigger the release of various digestive hormones, which regulate the coordinated release of secretions from the pancreas and liver. The release of *secretin*, a regulatory hormone, is controlled by the pH value of the chyme; its action is to stimulate secretion of bicarbonate ions from the pancreas. *Cholecystokinin* (CCK), produced by lining cells of the duodenum, causes the gall bladder to contract and also triggers the release of pancreatic enzymes. A fat-rich chyme causes the duodenum to release *enterogastrone*, a hormone that slows down entry of chyme into the duodenum by inhibiting peristalsis in the stomach.

The digestion of starch, begun by salivary amylase, is continued by a pancreatic amylase, which hydrolyzes starch into maltose, and the enzyme maltase, which splits maltose into two glucose molecules. *Disaccharidases*, enzymes specific for hydrolysis of different disaccharides, are built into the plasma membranes of intestinal-wall epithelial cells.

Pepsin in the stomach breaks proteins into smaller pieces. Protein digestion is completed in the small intestine by *trypsin* and *chymotrypsin*, enzymes specific for peptide bonds adjacent to certain amino acids; *carboxypeptidase*, which splits amino acids off the free carboxyl end of the polypeptide; and *aminopeptidase*, which works on the peptide bond at the amino end of the chain. Protein-digesting enzymes from the pancreas

are secreted as zymogens, in inactive form. *Enterokinase* is an enzyme that activates these other enzymes in the small intestine.

The digestion of nonpolar, insoluble fats is aided by bile salts. In a process called *emulsification*, bile salts coat tiny fat droplets so the latter do not coalesce. *Lipase* is an enzyme that hydrolyzes fat molecules.

Most digestion is completed while the chyme is still in the duodenum. The *jejunum* and *ileum* are regions of the small intestine specialized for the absorption of nutrients. The huge surface area of the small intestine is created by large folds covered with fingerlike projections called *villi*, on which the epithelial cells have microscopic extensions called *microvilli*. The core of each villus has a net of *capillaries* and a lymph vessel called a *lacteal*. Nutrients are absorbed across the epithelium of the villus and then across the single-celled wall of the capillaries or lacteal. Transport may be passive (the nutrient moving down its concentration gradient) or active (the nutrient pumped against a gradient). The active transport of sodium into the lumen of the intestine and its passive reentry into epithelial cells seem to drive the uptake of certain nutrients.

Amino acids and sugars enter capillaries and are carried to the liver by the bloodstream. Glycerol and fatty acids are absorbed by epithelial cells and then recombined to form fats, which are coated with proteins to make tiny globules called *chylomicrons*. These packages are transported by exocytosis out of the epithelial cells and enter a lacteal. Some fat molecules are bound to carrier proteins and transported as *lipoproteins* into capillaries.

The nutrient-laden blood from the small intestine is carried directly to the liver by the *hepatic portal vein*. The liver converts and stores various molecules and regulates the nutrient content of the blood.

The small intestine leads into the large intestine, or colon, at a junction with a sphincter. A blind pouch, called the *caecum*, attaches at the juncture and has a fingerlike extension, the *appendix*. The colon functions to reabsorb water. An irritation or infection of the colon lining may result in less absorption of water and lead to diarrhea; whereas an excess of reabsorption, occurring when peristalsis moves the intestinal contents too slowly, may result in constipation. The wastes of the digestive tract are called *feces*.

A rich flora of mostly harmless bacteria live on organic material remaining in the feces. These intestinal bacteria produce the gases methane and hydrogen sulfide as a by-product of their metabolism. Some bacteria produce vitamin K, which is absorbed by the host.

The feces contain cellulose, other undigested ingredients of food, bile pigments, salts excreted by the colon, and a large proportion of intestinal bacteria. Feces are stored in the *rectum*. A voluntary and involuntary sphincter between the rectum and anus control the elimination of feces, which is initiated by strong contractions of the colon.

STRUCTURE YOUR KNOWLEDGE

1. Food fills three needs: fuel, organic raw materials, and essential nutrients. Develop a concept map to organize your understanding of these nutritional needs of animals.
2. Fill in the following chart on the processes that occur in the sections or organs of the human digestive tract, and list the glands and digestive fluids associated with these processes.

Organ or Section	Processes Occurring	Associated Glands, Organs and Digestive Fluids
Oral cavity		
Pharynx		
Esophagus		
Stomach		
Small intestine		
Large intestine		

TEST YOUR KNOWLEDGE

MATCHING: *Match the description with the correct enzyme or hormone.*

1. ____ enzyme that hydrolyzes peptide bonds, works in the stomach

2. ____ hormone that stimulates secretion of gastric juice

3. ____ hormone that causes the gall bladder to contract and the pancreas to release enzymes

A. aminopeptidase
B. bile salts
C. cholecystokinin
D. chymotrypsin
E. disaccharidases
F. enterokinase
G. enterogastrone

4. ____ enzyme that begins digestion of starch in mouth

5. ____ enzymes specific for hydrolyzing disaccharides

6. ____ enzyme that hydrolyzes peptide bonds at amino end of polypeptide

7. ____ enzyme that hydrolyzes fats

H. gastrin
I. hydrochloric acid
J. lipase
K. maltase
L. pepsin
M. salivary amylase
N. secretin

8. _____ intestinal enzyme that activates zymogens

9. _____ enzyme specific for peptide bonds adjacent to certain amino acids, works in duodenum

10. _____ hormone that stimulates secretion of bicarbonate ions from pancreas

MULTIPLE CHOICE: *Choose the one best answer.*

1. Deposit-feeders
   a. feed mostly on mineral substrates.
   b. filter small organisms from water.
   c. eat plants.
   d. feed on detritus.

2. The energy content of fats
   a. is released by bile salts.
   b. is inversely related to body size.
   c. is approximately two times that of carbohydrate or protein.
   d. can reverse the effects of malnutrition.

3. BMR
   a. stands for bottom metabolic rate.
   b. is higher for ectotherms than for endotherms.
   c. is higher per gram for smaller animals than for larger ones.
   d. all of the above are correct.

4. Which of the following statements is not true?
   a. The average human can store enough glycogen to supply calories for several weeks.
   b. Eating less and/or exercising more may result in weight loss.
   c. Conversion of glucose and glycogen takes place in the liver.
   d. Excessive calories are stored as fat, regardless of their food source.

5. Kwashiorkor
   a. results from protein deficiency.
   b. is an example of malnutrition.
   c. is a syndrome involving retarded mental and physical development.
   d. all of the above are correct.

6. Incomplete proteins
   a. are lacking in essential vitamins.
   b. are a cause of undernourishment.
   c. are found in proper proportion in meat, eggs, and cheese.
   d. are lacking in one or more essential amino acid.

7. Vitamins
   a. may be produced by intestinal microorganisms.
   b. are constant from one species to the next.
   c. are toxic in excess because they are deposited in body fat.
   d. are inorganic nutrients, needed in small amounts, that usually function as co-factors.

8. Which of the following is most prevalent in animals with gastrovascular cavities?
   a. absorption of predigested nutrients
   b. extracellular digestion
   c. intracellular digestion
   d. fluid feeding

9. Ruminants
   a. have teeth adapted for a herbivorous diet.
   b. must use microorganisms to digest plant starch.
   c. eat their feces to obtain nutrients digested from cellulose by microorganisms.
   d. house symbiotic bacteria and microorganisms in a caecum.

10. The stomach can expand during eating due to its
   a. villi.
   b. rugae.
   c. caecae.
   d. sphincters.

11. The acid pH value of the stomach contents
   a. hydrolyzes proteins.
   b. is regulated by the release of gastrin.
   c. is neutralized by gastric juice.
   d. is produced by pepsin.

12. Acid chyme is the
   a. nutrient broth that leaves the stomach.
   b. nutrient broth that passes through the jejunum and ileum.
   c. gastric secretions of epithelial cells.
   d. mixture of macromolecules passing through the cardiac sphincter.

13. Chylomicrons are
   a. lipoproteins transported by the circulatory system.
   b. small branches of the lymphatic system.
   c. protein-coated fat globules excreted out of epithelial cells.
   d. small peptides acted on by chymotrypsin.

14. Which of the following is not a common component of feces?
    a. intestinal bacteria
    b. cellulose
    c. bile salts
    d. undigested ingredients of food

15. The hepatic portal vein
    a. supplies the capillaries of the intestines.
    b. carries absorbed nutrients to the liver for processing.
    c. carries blood from the liver to the heart.
    d. drains the lacteals of the villi.

CHAPTER 38

CIRCULATION AND GAS EXCHANGE

FRAMEWORK

This chapter surveys the basic approaches to circulation and gas exchange found in the animal kingdom, with special attention to the human systems and the problems of cardiovascular disease. The following concept map organizes some of the chapter's key ideas.

fluid, usually blood, exchanges materials with the environment across the thin and extensive epithelia of organs specialized for gas exchange, nutrient absorption, and waste removal. The blood then exchanges chemicals with the interstitial fluid that bathes the cells.

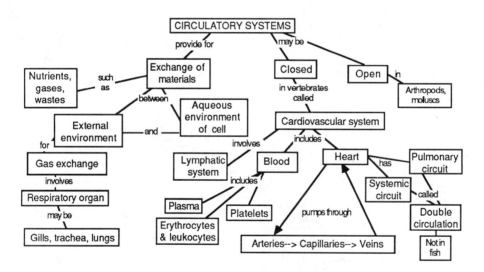

CHAPTER SUMMARY

The exchange of materials is essential for the cells of an organism. Every living cell must be located in an aqueous environment, which provides oxygen and nutrients and permits disposal of carbon dioxide and metabolic wastes.

The diffusion time of a substance is proportional to the square of the distance covered. Diffusion cannot possibly transport chemicals over the macroscopic distances in animals. All but the simplest animals have internal transport systems for body fluids. The body

Internal Transport in Invertebrates

An internal transport system is unnecessary in *Hydra* and other cnidarians. The central gastrovascular cavity inside the two-cell–thick body wall makes exchange between cells and their environment direct. Planarians and other flatworms also have gastrovascular cavities that digest and distribute nutrients throughout the body.

In the *open circulatory system* found in insects and other arthropods, blood bathes the internal tissues directly, and there is no distinction between blood and interstitial fluid. Blood is circulated by movements that squeeze the sinuses or spaces between organs and by the beating of a simple heart that is usually part of a dorsal vessel. Blood is drawn into the vessel through ostia, or pores, when the heart relaxes, and pumped into the interconnected system of sinuses during contraction.

Earthworms and other annelids have *closed circulatory systems*, in which the blood remains in vessels. A dorsal vessel functions as a heart by pumping blood forward. Five pairs of vessels, which loop around the digestive tract and connect the dorsal and ventral main vessels, pulsate and function as auxiliary hearts. Blood circulates more rapidly and efficiently through a closed circulatory system.

Circulation in Vertebrates

The closed circulatory system of vertebrates, also called the *cardiovascular system*, consists of the heart, blood vessels, and blood. The heart has one or two *atria*, which receive blood, and one or two *ventricles*, which pump blood out of the heart. *Arteries*, carrying blood away from the heart, branch into tiny *arterioles* within organs. Arising from arterioles are the microscopic *capillaries* across the thin walls of which exchange between blood and interstitial fluid occurs. Capillaries rejoin to form *venules*, which meet in the *veins* that return blood to the heart.

The ventricle of a fish's two-chambered heart pumps blood first to the capillary beds of the gills. Arteries carry oxygenated blood to the capillary beds in the rest of the body, and veins return the blood to the atrium. The resistance of the first capillary bed in the gills reduces the hydrostatic pressure that pushes blood through vessels, and oxygenated blood leaving the gills flows slowly to other organs.

In the three-chambered heart of amphibians, the single ventricle pumps blood through a forked artery into the *pulmonary circuit*, which leads to lungs and skin and then back to the left atrium, and the *systemic circuit*, which carries blood to the rest of the body and back to the right atrium. This *double circulation* repumps blood after it loses pressure in the capillary beds of the lungs. A ridge in the ventricle diverts most of the oxygenated blood from the left atrium into the systemic circuit and the deoxygenated blood from the right atrium into the pulmonary circuit. The three-chambered reptilian heart has a septum that partially divides the single ventricle. In crocodiles, the septum completely divides the ventricle into two chambers.

Delivery of oxygen for cellular respiration is most efficient in birds and mammals, which, as endotherms, have high oxygen demands. The left side of the four-chambered heart handles only oxygenated blood, whereas the right side receives and pumps deoxygenated blood. In the mammalian circulation, oxygenated blood returning from the lungs in the pulmonary veins flows to the left atrium and flows into the left ventricle, from which it is pumped through the aorta into the systemic circulation. Deoxygenated blood from the capillary beds of the body returns through the vena cavae to the right atrium and flows into the right ventricle, from which it is pumped in the pulmonary artery to the capillary beds of the lungs.

The human heart is enclosed in a two-layered, fluid-filled sac just beneath the sternum. The atria have relatively thin walls, whereas the ventricles have thicker muscular walls. The *heart cycle* lasts about 0.8 sec and consists of the *systole*, during which the heart muscle contracts and the chambers pump blood, and the *diastole*, when the heart is relaxed. Except for the first 0.1 sec of the systole, when the atria contract, they are filling with blood. The slow and powerful contraction of the ventricles lasts 0.3 sec. Blood flows passively into the ventricles during diastole, and atrial contraction finishes filling them.

Atrioventricular valves between each atrium and ventricle are snapped shut when the ventricles contract. Strong fibers prevent the connective-tissue flaps from turning inside out. The *semilunar valves* at the entrance of the aorta and pulmonary artery are forced open by ventricular contraction, and close when the ventricles relax and the elastic walls of the arteries recoil. The heart-beat sounds are caused by the closing of these valves. A *heart murmur* is the detectable hissing sound of blood leaking back through a defective valve.

The average human heart rate, or *pulse*, is 65 to 75 beats per minute. The inverse relationship between body size and pulse is related to the higher metabolic rate per gram of tissue of smaller animals.

The *cardiac output*, or volume of blood pumped per minute into the systemic circuit, depends on the heart rate and the *stroke volume*, or quantity of blood pumped by each contraction of the left ventricle.

Cells of cardiac muscle are self-excitable or *myogenic*; they have an intrinsic ability to contract. The rhythm of contractions is coordinated by the *SA (sinoatrial) node*, or *pacemaker*. When nodal tissue contracts, it generates electrical impulses. The contraction of the SA node, which is located in the wall of the right atrium near the anterior vena cava, initiates a wave of excitation that signals the cells of the two atria to contract. (Remember that cardiac muscle cells are electrically connected by the intercalated discs.) The *AV (atrioventricular) node*, which is located at the base of the two atria, relays the impulse (after a slight delay) to the ventricles. The electrical currents produced during the heart cycle can be detected by electrodes placed on the skin. The SA node is controlled by two sets of nerves with antagonistic signals, and is influenced by hormones, temperature, and stimulation from increased blood flow.

The wall of an artery or vein consists of three layers: an outer connective tissue zone with elastic fibers, a middle layer of smooth muscle and more elastic fibers, and a lining of *endothelium* made of simple squamous epithelium. The middle layer is especially thick in arteries. Capillaries have only the endothelial layer.

The flow of blood decelerates after leaving the heart as a result of the increasing cross-sectional area of the branching vessel system. The enormous number of capillaries creates a total diameter much greater than other parts of the system, and resistance is also greater in the narrow capillaries. Blood flow is very slow in capillaries, improving opportunity for exchange of substances with interstitial fluid.

Blood pressure, the hydrostatic force exerted against the wall of a vessel, is much greater in arteries than in veins and is greatest during systole. *Peripheral resistance*, caused by the narrow openings of the arterioles impeding the exit of blood from arteries, causes the swelling of the arteries during systole. The snapping back of the elastic arteries during diastole maintains a continuous blood flow into arterioles and capillaries.

Both cardiac output and peripheral resistance determine blood pressure. Contraction of smooth muscles in arteriole walls increases resistance and thus increases blood pressure, whereas dilation of arterioles lowers blood pressure. Neural and hormonal signals control these muscles.

Blood pressure drops to almost zero in the capillary beds. The one-way valves in veins and the contraction of skeletal muscles between which veins are embedded force blood to flow back to the heart. Pressure changes during breathing also draw blood into the large veins in the thoracic cavity.

Capillaries branch off *thoroughfare channels*, the connections between arterioles and venules. A sphincter regulates the passage of blood into a capillary. Only about 5% to 10% of the body's capillaries have blood flowing through them at any one time. The brain, kidneys, liver, and heart are usually heavily supplied with blood; the distribution to the other areas of the body varies with need.

The exchange of substances between blood and interstitial fluid may involve endocytosis and exocytosis in the endothelial cells making up the capillary wall, passive diffusion of small substances (water, sugars, salts, oxygen, and urea) both through the cells and between adjoining cells, and the forcing of fluid out of the capillary by hydrostatic pressure. Blood cells and proteins are too large to pass through the endothelium. Blood pressure at the upstream end of a capillary forces fluid out, whereas osmotic pressure at the downstream end tends to draw most of the fluid back in.

The fluid that does not return to the capillary, and any proteins that may have leaked through the capillary wall, are returned to the blood through the *lymphatic system*. The accumulation of interstitial fluid in tissues causes a condition known as *edema*. Fluid diffuses into lymph capillaries intermingled in the capillary net. The fluid, called *lymph*, moves through lymph vessels with one-way valves as a result of the movement of skeletal muscles and the rhythmic contactions of the vessel walls. In *lymph nodes*, the lymph is filtered, and white blood cells attack viruses and bacteria.

Blood

Vertebrate blood is a type of connective tissue with cells in a liquid matrix called *plasma*. When *whole blood* is centrifuged, the *formed elements* or cells form a dense red pellet that makes up about 45% of the volume of blood.

The plasma consists of a large variety of solutes dissolved in water. The collective and individual concentrations of *electrolytes*, or inorganic salts in the form of dissolved ions, are important to osmotic balance between blood and interstitial fluid and to the functioning of muscles and nerves. The kidney is responsible for the homeostatic regulation of ion concentrations. Plasma proteins function as buffers, osmotic components of the blood, antibodies, escorts for lipids, and clotting factors. Blood plasma from which the fibrinogens or clotting factors have been removed is called serum. Nutrients, metabolic wastes, gases, and hormones also are dissolved in the plasma. Except for its higher protein concentration, plasma is very similar in composition to the interstitial fluid.

The numerous (about 25 trillion in the body's 5 liters of blood) *erythrocytes* transport oxygen. Mammalian erythrocytes lack nuclei, and all red blood cells lack mitochondria and generate their ATP by anaerobic metabolism. The small size and biconcave shape of red blood cells create a large surface area of plasma membrane across which oxygen can diffuse. Erythrocytes are packed with *hemoglobin*, an iron-containing protein that binds oxygen.

Erythrocytes are formed in the red marrow of bones, and their production is controlled by a negative feedback mechanism involving the hormone erythropoietin secreted by the kidney in response to low oxygen supply in tissues. Red blood cells circulate for about 3 to 4 months before they are phagocytized by cells in the liver and their components recycled.

Leukocytes fight infections. Some white cells are phagocytes; some are lymphocytes that give rise to cells that produce antibodies. Lymphocytes are produced in lymphoid organs (spleen, thymus, tonsils, lymph nodes), whereas other leukocytes arise from the stem cells in the bone marrow, which also form erythrocytes.

Platelets, pinched-off fragments of large cells in the bone marrow, are involved in the blood-clotting mechanism in which the blood protein *fibrinogen* is converted to *fibrin*, which aggregates into threads or fibers. An inherited defect in any step of the complex clotting process causes *hemophilia*. Anticlotting factors normally prevent clotting of blood in the absence of injury. A *thrombus* is a clot that occurs within a blood vessel and blocks the flow of blood.

Cardiovascular Disease

More than one half of all deaths in the United States are caused by *cardiovascular disease*. Most deaths occur from a heart attack, in which a thrombus blocks a coronary artery, causing the muscle that was served by the artery to die, and the conduction of electrical impulses through the cardiac muscle to be interrupted. A clot that develops elsewhere in the circulatory system and is moved through vessels until it becomes lodged in an artery is called an *embolus*. Strokes are caused when a thrombus or embolus blocks an artery in the brain.

Most heart-attack and stroke victims had been suffering from a chronic disease known as *atherosclerosis*, in which growths called *plaques* develop within arteries and narrow the vessels. Plaques that become hardened by calcium deposits result in *arteriosclerosis*, or "hardening of the arteries." An embolus is more likely to be trapped in narrowed vessels, and plaques are common sites of thrombus formation. Occasional chest pains, known as *angina pectoris*, may be a warning sign that a coronary artery is partially blocked.

Hypertension, or high blood pressure, promotes atherosclerosis and increases the risk of heart attack and stroke. Hypertension is thought to damage the endothelium and initiate plaque formation. This condition can be easily diagnosed and controlled by drugs, diet, and exercise. Hypertension and atherosclerosis tend to be inherited. Smoking, not exercising, and eating a fat-rich diet have been correlated with an increased risk of cardiovascular disease.

Gas Exchange

Gas exchange provides the continuous supply of oxygen needed for cellular respiration and removes carbon dioxide. The *respiratory medium*, which supplies oxygen, is air for a terrestrial animal and water for an aquatic one. The *respiratory surface*, the portion of an animal's body where gas exchange with the respiratory medium occurs, must be moist and large enough to supply the whole body. A localized region of the body surface is usually specialized with a thin, moist epithelium between a rich blood supply and the respiratory medium.

When the entire outer skin serves as a respiratory organ, as in an earthworm, the animal must live in damp or wet places and have a relatively small, long, and thin body to provide a high ratio of surface area to volume. Bulkier animals use an extensively branched localized region of the body surface for gas exchange.

Gills are evaginations of the body surface, ranging from the simple bumps on echinoderms to the complex, finely divided gills of molluscs, crustaceans, fishes, and some amphibians. Delicate gills are usually sheltered by a protective cover. The low oxygen concentration in a water environment usually requires *ventilation*, or movement of the respiratory medium across the respiratory surface. The density of water requires an animal to expend considerable energy to ventilate its gills.

Blood flows through the capillaries in a direction opposite to the flow of water over the gills of a fish. This arrangement sets up a *countercurrent exchange*, in which the diffusion gradient favors the movement of oxygen into the blood throughout the length of the capillary.

The higher concentration of oxygen, faster diffusion rate of O_2 and CO_2 in air, and lower density, make air an easier respiratory medium than water is. To prevent water loss from respiratory surfaces, the respiratory surfaces of most terrestrial organisms are invaginated.

The *trachea* of insects are tiny air tubes that ramify throughout the body to come into contact with nearly every cell. Openings to the tracheal system are *spiracles*. Some insects use rhythmic body movements to ventilate their tracheal systems.

Lungs are respiratory surfaces restricted to one location from which oxygen is transported by the circulatory system. Lungs have evolved in land snails, spiders, and terrestrial vertebrates. Frog lungs are balloon-like, but mammalian lungs have a spongy texture with a much greater surface area of epithelium for gas exchange. The lungs of mammals are located in the thoracic cavity and enclosed in a double-walled sac.

Air, entering through the nostrils, is filtered, warmed, humidified, and smelled in the nasal cavity. Air passes through the pharynx and enters the windpipe through the glottis. When food is being swallowed, the glottis is pushed against the epiglottis. The *larynx* functions as a voicebox in humans and many other mammals; exhaled air vibrates a pair of vocal cords. The trachea, or windpipe, branches into two *bronchi*, which then branch repeatedly into *bronchioles* when they reach the lungs. Ciliated, mucus-coated epithelium lines much of the respiratory tree and removes dust and other particles from the respiratory system.

Multilobed air sacs encased in a web of capillaries are at the tips of the tiniest bronchioles. Gas exchange takes place across the thin epithelium of these *alveoli*.

Vertebrate lungs are ventilated by *breathing*, the alternate inhalation and exhalation of air. A frog uses *positive-pressure breathing* to "blow up" and empty its lungs. Expanding the mouth cavity draws air into the mouth, and raising the jaw forces air into the lungs.

Mammals ventilate their lungs by *negative-pressure breathing*. The volume of the lungs and the thoracic cavity is increased by expansion of the rib cage and contraction of the diaphragm. Air pressure is reduced within this increased volume, and air flows through the nostrils down to the lungs. Relaxation of the muscles between the ribs and of the diaphragm compresses the lungs, increases the pressure, and forces air out.

Tidal volume is the volume of air inhaled and exhaled by a resting animal. The volume that can be inhaled and exhaled by forced breathing is called *vital capacity*. The *residual volume* is the air that remains in the alveoli and lungs after forceful exhalation.

Birds have air sacs that penetrate their abdomen, neck, and wings. The air sacs and lungs are ventilated when the bird breathes, through a circuit that includes a one-way passage through channels in the lungs. A countercurrent exchange is set up in the lungs as the blood in the capillaries flows in the direction opposite to air flow. Air sacs do not participate in gas exchange, but they help to maintain air flow through the lungs, lower the density of the bird, and help to dissipate heat formed by the metabolism of flight muscles.

Breathing is controlled by automatic mechanisms. The *breathing center* in the medulla at the stem of the brain sends nerve impulses to the muscles between the ribs and to the diaphragm to contract. Stretch sensors in the lungs respond to the expansion of the lungs and send nervous impulses that inhibit the breathing center. When CO_2 concentration increases in the blood, more carbonic acid is formed. A drop in the pH value of the blood, sensed by the breathing center, increases the messages to breathe. Oxygen sensors in key arteries react only to severe deficiencies of O_2.

The concentration of gases in air or dissolved in water is measured as *partial pressure*. The partial pressure of oxygen, which makes up 21% of the atmosphere, is 160 mm Hg (0.21 x 760 mm—atmospheric pressure at sea level). The partial presure of CO_2 is 0.23 mm Hg. Partial pressure is proportional to concentration; a gas will diffuse from a region of higher partial pressure to lower partial pressure.

Blood entering the lungs has a lower O_2 partial pressure and a higher CO_2 partial pressure than does the air in the alveoli, and oxygen diffuses into the capillaries and carbon dioxide diffuses out. In the systemic capillaries, pressure differences favor the diffusion of O_2 out of the blood into the interstitial fluid and of CO_2 into the blood.

In most animals, O_2 is carried by *respiratory pigments* in the blood. Hemoglobin, usually located in red blood cells, is the respiratory pigment of almost all vertebrates. Copper is the oxygen-binding component in the respiratory protein *hemocyanin*, common in arthropods.

Hemoglobin is composed of four similar polypeptides, each of which has a prosthetic heme group with iron at its center. The binding of O_2 to the iron atom of one subunit induces a shape change in the other subunits, and their affinity for oxygen increases. Likewise, the unloading of the first O_2 affects a conformational change that lowers the other subunits' affinity for oxygen. The *dissociation curve* for hemoglobin shows the relative amounts of oxygen bound to hemoglobin under varying oxygen concentrations. At low partial pressures of oxygen, little O_2 is bound, but the percent saturation of hemoglobin rises sharply after the concentration rises high enough for the first oxygen to be bound to a hemoglobin molecule. Thus, in the steep part of this S-shaped curve, a slight change in partial pressure will cause hemoglobin to load or to unload a substantial amount of oxygen.

A drop in pH lowers the affinity of hemoglobin for O_2. Since a rapidly metabolizing tissue produces more CO_2, which lowers pH level, hemoglobin will unload more of its O_2 to that tissue.

Most carbon dioxide is transported in blood as *bicarbonate ions*. Carbon dioxide enters into the red blood cells, where it first reacts to form carbonic acid and then dissociates into H^+ and a bicarbonate ion. The hydrogen ions are bound by the hemoglobin, and the pH value of the blood is not greatly lowered during the transport of carbon dioxide. In the lungs, the diffusion of CO_2 out of the blood shifts the equilibrium in favor of the conversion of bicarbonate back to CO_2, and CO_2 is unloaded from the blood.

STRUCTURE YOUR KNOWLEDGE

1. Briefly outline or sketch the circulation of blood through the human heart to the systemic and pulmonary circuits.

2. Develop a brief concept map or table that shows the components of blood and their functions.

3. Create a concept map that shows your understanding of blood pressure, what it does, what causes it, and where and when it is highest.

4. Briefly outline or sketch the exchange of gases in the human body, indicating the major structures and processes involved.

TEST YOUR KNOWLEDGE

MULTIPLE CHOICE: *Choose the one best answer.*

1. A gastrovascular cavity
   a. is found in cnidarians and annelids.
   b. functions to pump fluids throughout the body.
   c. functions in both digestion and distribution of nutrients.
   d. involves all of the above.

2. Which of the following is not a similarity between open and closed circulatory systems?
   a. Some sort of pumping device helps to move blood through the body.
   b. Some of the circulation of blood is a result of movements of the body.
   c. The blood and interstitial fluid are indistinguishable from each other.
   d. All tissues come into close contact with the circulating body fluid so that the exchange of nutrients and wastes can take place.

3. In a system with double circulation,
   a. blood is pumped at two locations as it circulates through the body.
   b. there is a countercurrent exchange within the gills.
   c. there is never any mixing of oxygenated and deoxygenated blood in the heart.
   d. blood is repumped after it returns from the capillary beds of the gas-exchange organ.

4. During diastole,
   a. the atria fill with blood.
   b. blood flows passively into the ventricles.
   c. the elastic recoil of the arteries maintains hydrostatic pressure on the blood.
   d. all of the above occur.

5. An atrioventricular valve
   a. prevents the backflow of blood from a ventricle into an atrium.
   b. prevents the leakage of blood between ventricles.
   c. prevents the backflow of blood from the aorta into the left atrium.
   d. prevents the backflow of blood from the pulmonary artery into the right ventricle.

6. During heavy exercise,
   a. stroke volume increases.
   b. heart rate increases.
   c. cardiac output increases.
   d. All of the above are correct.

7. Heart beat is initiated by
   a. contraction of the SA node.
   b. contraction of the AV node.
   c. contraction of the right atrium.
   d. contraction of the myogenic cardiac muscle cells.

8. Blood flows more slowly in the arterioles than in the arteries because
   a. the arterioles' thoroughfare channels to venules are often closed off.
   b. the arterioles collectively have a larger cross-sectional area than do the arteries.
   c. the arterioles must provide opportunity for exchange with the interstitial fluid.
   d. of all of the above.

9. If all the body's capillaries were open at the same time,
   a. blood pressure would fall dramatically.
   b. peripheral resistance would increase.
   c. blood would move too rapidly through the capillary beds.
   d. the amount of blood returning to the heart would increase.

10. Which of the following is not a factor in the exchange of substances in capillary beds?
    a. endocytosis and exocytosis
    b. passive diffusion through spaces between endothelial cells
    c. hydrostatic pressure
    d. bulk flow through the apoplast

11. Edema is a result of
    a. swollen lymph glands.
    b. an accumulation of interstitial fluid.
    c. swollen feet.
    d. too high a concentration of blood proteins.

12. A function of the kidney is
    a. control of the lymphatic system.
    b. maintenance of electrolyte balance.
    c. production of lymphocytes.
    d. destruction and recycling of red blood cells.

13. Fibrinogen is
    a. a blood protein that escorts lipids through the circulatory system.
    b. a cell fragment involved in the blood-clotting mechanism.
    c. a blood protein that is converted to fibrin to form a blood clot.
    d. one of the formed elements of blood.

14. Angina pectoris
    a. causes atherosclerosis.
    b. is a mild form of stroke.

c. promotes hypertension.
d. may be a warning sign that a coronary artery is partially blocked by plaque.

15. In countercurrent exchange,
  a. the flow of fluids or gases in opposite directions maintains a favorable diffusion gradient along the length of the exchange surface.
  b. oxygen moves from a region of low partial pressure to one of high partial pressure.
  c. oxygen is exchanged for carbon dioxide.
  d. a double circulation keeps oxygenated and deoxygenated blood separate.

16. The tracheae of insects
  a. are stiffened with rings and lead into the lungs.
  b. are filled by positive pressure breathing.
  c. ramify along the circulatory system for gas exchange.
  d. are highly branched so that they come into contact with almost every cell for gas exchange.

17. Which of the following is not involved in speeding up breathing?
  a. a drop in the pH value of the blood
  b. stretch receptors in the lungs
  c. impulses from the breathing center in the medulla
  d. severe deficiences of oxygen

18. The binding of an O_2 to the first iron atom of a hemoglobin molecule
  a. occurs at a very low partial pressure of oxygen.
  b. produces a conformational change that lowers the other subunits' affinity for oxygen.
  c. is an example of cooperativity resulting from a conformational change.
  d. occurs more readily at a lower pH value.

19. Carbon dioxide is transported in the blood
  a. attached to hemoglobin.
  b. as bicarbonate ions.
  c. at a higher partial pressure than oxygen is.
  d. as carbonic acid.

20. According to the dissociation curve for hemoglobin,
  a. the percent saturation of hemoglobin changes with the amount of oxygen bound to hemoglobin.
  b. there is a range in which a slight change in partial pressure of oxygen will cause hemoglobin to load or unload a large amount of oxygen.
  c. the partial pressure of oxygen in the tissues is lower than that in the alveoli.
  d. a drop in pH value lowers the affinity of hemoglobin for O_2.

CHAPTER 39

THE IMMUNE SYSTEM

FRAMEWORK

The *immune system*, a subset of the circulatory and lymphatic systems, functions to protect the body from external and internal threats such as those posed by some bacteria, viruses, and early-stage cancer cells. *Nonspecific defense mechanisms* include the physical and chemical barriers of the skin and mucous membranes and the inflammatory response. The immune system reacts to foreign antigens with *specific defenses*, including antibodies produced by B cells (humoral immunity), and T-cell responses (cell-mediated immunity) as in the production of cytotoxic T cells, helper T cells, and suppressor T cells.

The immune system is characterized by the ability to distinguish self from nonself, high specificity to a huge number of antigenic determinants, coordinated action of various cells and chemical messages (histamines, interferon), and immunological memory. Immune disorders include allergies, immunodeficiency caused by inheritance, disease, or drug treatments, and AIDS.

CHAPTER SUMMARY

Nonspecific Defense Mechanisms

The physical barrier of the skin is reinforced by oil and sweat secretions that create a low pH value and contain lysozyme, an enzyme that attacks bacterial cell walls. Lysozyme is also present in tears and saliva. Gastric juice kills most bacteria that reach the stomach. The tiny hairs of the nostrils filter out particles that may carry microorganisms, and the ciliated mucus-coated epithelial lining of the respiratory tract traps and removes particles.

The *inflammatory response* includes the dilation of small blood vessels to increase blood flow, the leaking of capillaries, and the congregation of phagocytic white blood cells. The pus that may accumulate is a combination of dead cells and body fluids. *Histamine*, released by injured cells, causes blood vessels to dilate and initiates the inflammatory response.

An inflammatory reaction may be systemic, including an increase in the number of circulating white blood cells and a fever, which may stimulate phagocytosis and inhibit growth of microorganisms. Fevers may be triggered by toxins produced by pathogens or by pyrogens released by certain white blood cells.

Specific Defense Mechanisms: The Immune System

The *humoral* component of the immune system involves the production of antibodies. The *cell-mediated* immune system involves specialized cells that circulate in blood and lymphoid tissue. The immune system distinguishes *self* from *nonself* by recognizing the biochemical differences of invading bacteria and viruses as well as of cells of tissue grafts.

The immune system "remembers" foreign substances and can react against them promptly, a property known as *immunity*, which protects an organism after exposure to certain viral and bacterial pathogens. In *vaccination*, the immune system is exposed to an inactive or attenuated pathogen, which initiates the system's long-term capability to respond quickly to the infective agent. In *active* immunity, the body is stimulated to produce antibodies. In *passive* immunity, antibodies are supplied through the placenta to a fetus or by an injection of gamma globulin, and these antibodies provide temporary immunity.

Lymphocytes mediate both humoral and cellular immune responses. *B cells* produce antibodies; *T cells* are responsible for cellular immunity. Both cells develop from stem cells located in the bone marrow. These cells either migrate to the thymus and differentiate into T cells or continue to develop in the bone marrow as B cells. Final maturation of lymphocytes occurs in secondary lymphoid tissues such as the spleen, lymph nodes, or tonsils.

The immune system responds to foreign macromolecules, called *antigens*, by producing *antibodies*, which help combat the antigen. In general, antigens are proteins or polysaccharides that are part of the capsules, cell walls, or coats of bacteria or viruses. Antigens also may be parts of membranes of other cells. Antibodies recognize localized regions, called *antigenic determinants*, of an antigen.

Every B cell acquires a capacity to produce an antigen-specific antibody during its development, but exposure to that antigen is necessary to activate antibody production. Antibodies are a class of proteins called *immunoglobulins*, abbreviated *Ig*.

The *variable (V)* region of an antibody permits it to recognize and bind to a specific antigen; the *constant (C)* region is involved in the destruction and elimination of that antigen. The Y-shaped antibody molecule consists of two pairs of polypeptide chains: two identical short *light (L)* chains and two identical longer *heavy (H)* chains. Both H and L chains have variable sections at the ends of the two arms of the Y, forming the *antigen-binding site*. The amino acid compositions of the V regions create the unique contours and binding potential of the site. Weak bonds that form between antigen and antibody account for the specificity of antibody–antigen binding. The five types of constant regions create five classes of antibodies in mammals: IgM, IgG, IgA, IgD, and IgE. Each class plays a different role in the immune response.

Antibodies label foreign molecules for destruction. In a process called *opsonization*, IgM and IgG antibodies coat the antigen, and phagocytes bind to the constant regions of the antibody and engulf the antigen-bearing invader. IgM and IgG antibodies also activate the *complement system*, a set of blood proteins that produce a pore in the foreign cell's membrane and cause the cell to lyse. An inflammatory response can also activate the complement system.

Antibodies can be used in biological research and clinical testing to detect specific antigens. Techniques for making *monoclonal antibodies*, developed in the late 1970s, are used to supply quantities of identical antibodies specific for any antigen. Cells derived from a tumor line are fused with antibody-producing lymphocytes, obtained from the spleen of an animal exposed to the antigen of interest. These hybrid cells, called *hybridomas*, are cultured to produce a mass of cells, which produce the same antibody.

T cells produce proteins, called *T-cell receptors*, consisting of two polypeptide chains with constant and variable regions. The proteins are anchored in the T-cell membrane, and the variable regions give T cells the ability to recognize many different kinds of antigens. On recognizing foreign antigens, T cells may differentiate into *cytotoxic T cells*, which kill foreign cells and viruses; *helper T cells*, which help other lymphocytes recognize the antigens and activate other blood cells; or *suppressor T cells*, which suppress the response of some B and T lymphocytes. Some helper T cells are necessary for some B cells to make antibodies, and other T cells secrete interleukins, messenger molecules that recruit other lymphocytes and cause phagocytic cells to accumulate.

The ability of the immune system to respond specifically to an enormous variety of different antigens is explained by the *clonal selection theory*. During differentiation, each B and T lymphocyte develops a single type of receptor that can respond to a single antigenic determinant. When an antigen binds to a lymphocyte that has a complementary receptor on its surface, the lymphocyte is activated to mature and proliferate, creating a *clone* of lymphocytes, all having the same antigenic specificity.

The ability to create a large number of identical T and B cells (clonal selection) accounts for the "memory" of the immune system. A *primary immune response* occurs several days after an animal first encounters an antigen. The response rises rapidly and then falls gradually. If the animal reencounters the same antigen, the *secondary immune response* is more rapid, greater in magnitude, and longer in duration. The initial encounter causes B- and T-cell clones to proliferate and differentiate, creating *plasma cells* (derivatives of B cells that produce antibodies), the various kinds of T-cell effectors, and large clones of *memory cells*, which are long-lived cells stored in the spleen and lymph nodes that are able to produce effector cells and more memory cells when stimulated by the same antigen. Memory cells account for the rapid and greater secondary immune response.

Interferons are substances produced by virus-infected cells that help other cells to resist virus infection. Recombinant DNA technology has made it possible to produce large quantities of interferons, which are being tested for their effectiveness in treating viral infections and cancer.

One mechanism of interferon action is believed to involve the "turning on" of interferon genes by contact with viruses. Interferon stimulates the production of other proteins that inhibit the cells from making proteins for the infecting virus. Diffusion of interferon to other cells stimulates the same defense mechanism. The action of interferon in cancer treatment is not completely understood.

Self versus Nonself

The ability of the immune system to distinguish self from nonself develops during fetal growth, when clones of lymphocytes capable of reacting against molecules that tag "self" cells are apparently destroyed by suppressor T cells.

Individuals with type A blood have A-type antigen on the surface of their red blood cells and produce antibodies against the B-type antigen. Likewise, people with type B blood produce anti-A antibodies and have B-type antigens on their blood cells. Animals with AB type blood have both A and B antigens and no antibodies. People with type O blood, who produce A and B antibodies, are called universal donors because their blood cells carry no antigens and will not agglutinate when exposed to the antibodies of a blood recipient.

The *Rh factor* is another blood-group antigen. An Rh-negative mother does not have the Rh determinant in her blood cells. If she is carrying an Rh-positive child and some fetal blood leaks into her circulation, she produces antibodies against the Rh factor. Should she carry a second Rh-positive fetus, her immunological memory may result in the production of antibodies that cross the placenta and agglutinate fetal red blood cells. Treatment with anti-Rh antibodies just after delivery of the first Rh-positive baby destroys any fetal cells with Rh antigens that may have leaked into the mother's circulation and prevents the mother's immunological response to the antigen.

A large set of cell-surface antigens called the *major histocompatibility complex (MHC)* is involved in the immune system's ability to distinguish self from nonself. Transplanted tissues and organs may be rejected because these antigens trigger T-cell responses. Various drugs are used in people who receive transplanted organs to suppress the immune response.

Sometimes the immune system turns against self, leading to *autoimmune diseases*, such as lupus erythematosis and probably rheumatoid arthritis and rheumatic fever. Autoimmune disorders may develop when antibodies or effector T cells that are responding to a foreign invader cross-react with the individual's own tissues. In other cases, the surface components of some cells may change enough that they become recognized as foreign, or body substances that were previously hidden may come into contact with the immune system and trigger a response.

Cancer comprises a variety of diseases, all of which involve changes in normal body cells. Some of these changes to the outer membrane surface of the cells may result in the immune system identifying the cancer cells as foreign and destroying them. Individuals with deficient immune systems are far more susceptible to cancer. The *immune surveillance theory* suggests that the immune system constantly finds and destroys cancer cells in early stages of development. The failure of this surveillance system when tumors develop is being researched.

Immune Disorders

Allergies are overreactions to specific antigens, or allergens. Antibodies of the IgE family trigger allergic reactions by binding to *mast cells* (noncirculating cells found in connective tissue) and then binding to antigens. The mast cells undergo *degranulation*, releasing histamines, which create an inflammatory response. Antihistamines are drugs that interfere with the action of histamines. *Anaphylaxis* is a severe allergic response in which the abrupt dilation of peripheral blood vessels caused by a rapid release of histamines leads to a life-threatening drop in blood pressure and thus to shock. Severe allergic reactions can be counteracted with injections of epinephrine to reconstrict these vessels.

A defect in any of the components of the immune system leads to increased susceptibility to viruses or bacterial infections. In the congenital disease known as *severe combined immunodeficiency (SCID)*, both T and B cells are absent or inactive. Afflicted individuals must receive successful bone-marrow transplants or live in completely sterilized, isolated environments. Certain cancers, such as Hodgkin's disease, affect lymph node cells and can depress the immune system. Certain drug treatments also lead to immune deficiencies.

Acquired immune deficiency syndrome (AIDS) is an immune disorder in which the number of T helper cells that activate other T and B lymphocytes is greatly reduced. The immune system is severely weakened, and most afflicted individuals die within 3 years, usually from opportunistic infections or cancers. A rare form of pneumonia, severe diarrhea, and a skin cancer called Kaposi's sarcoma are among the AIDS-related diseases. The less severe *AIDS-related complex (ARC)* produces swollen lymph nodes, fever, night sweats, and weight loss.

AIDS was first recognized in 1981, when the increasing incidence of Kaposi's sarcoma in homosexual men was investigated by the Centers for Disease Control (CDC) in Atlanta. High-risk groups include sexually active homosexual males, intravenous drug users, recipients of blood products, and sexual partners of members of the high-risk groups.

The infectious agent responsible for AIDS was identified independently in the United States and in France in 1984. The American group named it HTLV III, for human T-cell leukemia virus III. This virus lives inside lymphocytes and generally does not circulate free in the blood. Casual contact does not seem to transmit the virus. Most AIDS patients have circulating antibodies to the virus. A test has been developed that identifies low levels of AIDS antibody and is being used to screen blood donations. Only 10% to 20% of the individuals who have antibodies against the virus display symptoms of AIDS, perhaps partly due to the long incubation time of the disease. Research continues on characterizing the virus, tracing the origins of the disease, and developing effective treatments and vaccines for AIDS.

STRUCTURE YOUR KNOWLEDGE

1. Create a concept map showing the cells involved in the immune system, their origins, and their functions.

2. Describe the structure of an antibody molecule and how this structure relates to its function. Briefly explain the clonal selection theory.

3. Fill in the following chart on some of the molecules involved in the immune system.

Molecule	How or Where Produced	Activity
Lysozyme		
Histamine		
Antibody		
Complement system		
Interleukin		
Interferon		

TEST YOUR KNOWLEDGE

MULTIPLE CHOICE: *Choose the one best answer.*

1. Which of the following is incorrectly paired with its effect?
   a. gastric juice — kills bacteria in the stomach
   b. fever — stimulates phagocytosis and inhibits microbial growth
   c. histamine — causes blood vessels to dilate
   d. vaccination — creates passive immunity

2. Monoclonal antibodies
   a. are used to create active immunity.
   b. explain the ability of the immune system to create a large number of identical T or B cells.
   c. are produced in tissue culture by hybridomas.
   d. initiate the secondary immune response.

3. Antigens are
   a. proteins that consist of two light and two heavy polypeptide chains.
   b. proteins or polysaccharides usually found on the membrane, wall, or capsule of an invading cell or virus.
   c. blood proteins that lyse the membranes of foreign cells.
   d. proteins embedded in T-cell membranes.

4. A secondary immune response is more rapid and greater than a primary immune response because
   a. large clones of memory cells respond to the pathogen and produce more effector cells.
   b. the second response is an active immunity, whereas the primary one was a passive immunity.
   c. helper T cells are available to activate other blood cells.
   d. interleukins cause the rapid accumulation of phagocytic cells.

5. Clones of lymphocytes capable of reacting against "self" molecules
   a. are usually not a problem until a woman's second pregnancy.
   b. are destroyed by suppressor T cells before birth.
   c. are usually kept separate from the immune system
   d. contribute to immunodeficiency diseases.

6. The major histocompatibility complex
   a. is involved in the ability to distinguish self from nonself.
   b. is a large set of cell-surface antigens.
   c. may trigger T-cell responses after transplant operations.
   d. all of the above are correct.

7. In opsonization,
   a. antibodies coating the antigen help phagocytes bind to and engulf the invading cell.
   b. a set of blood proteins lyses a hole in the foreign cell's membrane.
   c. antibodies cause cells to agglutinate.
   d. a flood of histamines is released that leads to an inflammatory response.

8. Severe combined immunodeficiency
   a. is an autoimmune disease.
   b. is a form of cancer in which the membrane surface of the cell has changed.
   c. is a disease in which both T and B cells are absent or inactive.
   d. is an immune disorder is which the number of T helper cells is greatly reduced.

9. A transfusion of B-type blood given to a person who has A-type blood would result in
   a. the recipient's anti-B antibodies clumping the donated red blood cells..
   b. the recipient's B antigens reacting with the donated anti-B antibodies.
   c. the recipient forming both anti-A and anti-B antibodies.
   d. no reaction, because B is a universal donor type of blood.

10. Which of the following are incorrectly paired?
    a. variable region — antibody specificity for an antigenic determinant
    b. helper T cells — production of plasma cells
    c. cytotoxic T cells — destruction of foreign cells and viruses
    d. immunoglobulins — antibodies

CHAPTER 40

CONTROLLING THE INTERNAL ENVIRONMENT

FRAMEWORK

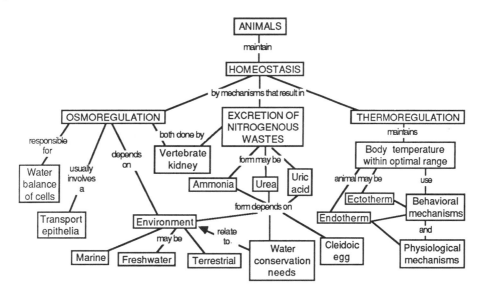

CHAPTER SUMMARY

Animals are able to survive large fluctuations in their external environment by maintaining a relatively constant internal environment, a process known as homeostasis. Changes in the internal body fluid that bathes the cells are tempered by various regulatory devices, usually involving feedback mechanisms.

Osmoregulation

The environment largely determines the problems animals face in regulating the water balance of their cells. Osmosis is the diffusion of water across a selectively permeable membrane separating two solutions that differ in *osmolarity* (moles of solute particles per liter). Isotonic solutions are equal in osmolarity, andthere will be no net osmosis between them. Water flows across a membrane from a hypotonic (more dilute) to a hypertonic (more concentrated) solution.

Osmoconformers are isotonic with their aqueous surroundings. *Osmoregulators* must get rid of excess water if they live in a hypotonic medium, or take in water to offset osmotic loss if they inhabit a hypertonic environment.

Most marine invertebrates are osmoconformers, whereas most marine vertebrates are osmoregulators. Sharks maintain an osmolarity slightly higher than that of seawater by retaining urea, a nitrogenous waste product, within their bodies. Their internal salt concentrations are lower than those of seawater because they use *rectal glands* to pump salt out of their bodies.

Bony fishes (Class Osteichthyes) evolved from freshwater ancestors, and many marine bony fishes are hypotonic to seawater. They must drink large quantities of seawater to replace the water they lose by osmosis. Excess salt is pumped out through salt glands located in the gills. Many marine birds get rid of excess salt through nasal salt glands.

Freshwater animals constantly take in water by osmosis because their internal fluids are hypertonic to their medium. Protoctists use contractile vacuoles to pump out excess water. Freshwater animals excrete large quantities of dilute urine. Salt supplies are replaced from their food, or, in some fishes, by ion pumps in the gills.

Animals that move between freshwater and saltwater environments, or that live in brackish water, often are able to maintain a constant internal osmolarity by using osmoregulatory mechanisms in opposite directions in the different environments.

Terrestrial animals face the threat of desiccation as they lose water by evaporation and waste excretion. Since fluids are replaced by drinking and eating food with high water content, nervous and hormonal control of thirst are osmoregulatory adaptations.

An osmoregulator must balance the water lost through its protective coverings, gas-exchange surfaces, urine, and feces with water gained through food and drink. The osmolarity of the internal environment is usually regulated by the transport of salt, followed by the osmotic movement of water, across a *transport epithelium*. Located at the tissue-environment boundary or *apical surface*, this single sheet of cells, linked by impermeable tight junctions, regulates the passage of solutes between the extracellular fluid and the environment.

The salt glands of fish gills have transport eptihelia that pump salts outward for marine fishes and inward for freshwater fishes. The transport epithelium in vertebrate kidneys functions in metabolic waste excretion as well as in osmoregulation.

Excretory Systems of Invertebrates

The *flame-cell system* of flatworms (phylum Platyhelminthes) consists of a branched system of tubules embedded in the body tissues. Extracellular fluids pass into the bulbous flame cells at the ends of the smallest tubules. Cilia projecting into the tubule propel the fluid along the tubule and into excretory ducts that empty by way of numerous pores. The flame-cell system is primarily osmoregulatory; metabolic wastes are excreted through the gastrovascular cavity.

In insects and other terrestrial arthropods, *Malpighian tubules* remove nitrogenous wastes from the body fluid (blood) and function in osmoregulation.

Transport epithelia lining these blind sacs, which open into the digestive tract at the end of the midgut, pump salts and wastes from the fluid of the body cavity into the tubule. This urine passes through the hindgut into the rectum. The epithelium of the rectum pumps most of the salt back into the blood, and water follows by osmosis. Nitrogenous wastes are eliminated as dry matter along with the feces.

In most animals with closed circulatory systems, the excretory organs have close associations with blood vessels. The *nephridia* of earthworms occur in pairs in each segment of the worm. A ciliated funnel called a *nephridiostome* collects coelomic fluid, which then moves through a folded tubule encased in capillaries. The transport epithelium of the tubule pumps salts out of the tubule, and the salts are reabsorbed into the blood. The dilute urine exits by way of *nephridiopores*.

The Vertebrate Kidney

Nephrons, the excretory tubules of vertebrates, are arranged into compact organs, the *kidneys*. The *vertebrate excretory system* consists of the kidneys, associated blood vessels, and excretory tubes.

Blood enters the kidney through the *renal artery*, and leaves by way of the *renal vein*. Urine exits through the *ureter* and is temporarily stored in the *urinary bladder*. Urine is periodically released from the body through the *urethra*.

Each human kidney contains about a million nephrons, which filter out nitrogenous wastes and adjust the concentrations of various salts in the blood by three processes: filtration, secretion, and absorption.

Filtration occurs when blood pressure forces fluid out of the capillaries and into the nephron tubule. The filtrate contains salts, nitrogenous wastes, glucose, amino acids, and other small molecules. Blood cells and plasma proteins remain in the capillary. *Secretion* of substances from the extracellular fluid into the nephron tubules occurs as the filtrate moves through the tubule. Most of the water, sugar, and other vital nutrients in the filtrate are *absorbed* and returned to the blood.

Modifications of nephron structure and function in various classes of vertebrates are related to the requirements for osmoregulation and excretion of nitrogenous wastes in various habitats.

In the mammalian kidney, the end of the nephron tubule is formed into a cuplike *Bowman's capsule* that encloses a ball of capillaries called the *glomerulus*. Filtrate enters the tubule after being forced out of the glomerulus, and then passes through five specialized regions: a convoluted *proximal tubule*, a *descending limb* of the *loop of Henle*, an *ascending limb*, a convoluted *distal tubule*, and the *collecting duct*. Bowman's capsules and the proximal and distal tubules are located in the outer *cortex* of the kidney; the long loops of Henle extend into the *medulla*.

The capillaries converge as they leave Bowman's capsule and then redivide to form a second capillary network around the tubule. A *portal system* is an arrangement in which the blood passes through two capillary beds before returning to the heart.

The transport epithelium of the proximal tubule selectively secretes K^+ and urea into and actively transports glucose and amino acids out of the filtrate in the tubule. The descending limb of the loop of Henle is freely permeable to water, but not to salt or urea. Water moves into the hypertonic extracellular fluid, leaving a filtrate with high salt and solute concentrations. The ascending limb is not very permeable to water, but is permeable to Na^+ and Cl^-, which diffuse out of the filtrate and add to the high osmolarity of the medulla. Salt is actively transported from the upper portion of the ascending limb.

The distal tubule is specialized for selective secretion and reabsorption, contributing to homeostasis of the body fluids. As the urine moves in the collecting duct back through the increasing osmotic gradient of the medulla, more and more water exits by osmosis. The duct is not permeable to salt, but is to urea. Some urea diffuses out of the duct and eventually diffuses back into the ascending limb of the loop. Urea and salts form the osmotic gradient that enables the kidney to produce urine that is hypertonic to the blood.

Osmoregulation in vertebrates depends on the hormonal control of water and salt reabsorption in the kidneys. *Osmoreceptor cells* in the brain send signals to the pituitary gland when the osmolarity of the blood becomes too high. The pituitary releases *antidiuretic hormone*, or *ADH*, which increases the water permeability of the collecting ducts, resulting in the production of a more concentrated urine. The absence of ADH results in large volumes of dilute urine being produced in a condition called *diuresis*. Osmoreceptor cells can also stimulate nerve cells in the thirst center of the hypothalamus, so an animal drinks.

The adrenal glands respond to a decrease of salt in the body by secreting *aldosterone*, a hormone that stimulates the active absorption of Na^+ ions in the nephrons. This mechanism prevents a loss of total volume of water in the body should the salt concentration (and thus blood osmolarity) drop.

The kidneys of birds also have long loops of Henle and steep osmotic gradients, which result in water conservation and the production of hypertonic urine. Freshwater fishes are hypertonic to their environment and must excrete very dilute urine. Salts are conserved by the efficient absorption of ions from the filtrate. The kidneys of marine bony fishes excrete very little urine, and function mainly to get rid of cations taken in by the drinking of seawater.

Excretion of Nitrogenous Wastes

Ammonia, a small and toxic molecule, is produced when proteins and nucleic acids are broken down for energy or conversion to carbohydrates or fats. Aquatic animals can excrete nitrogenous wastes as ammonia because it is very soluble and easily permeates membranes. Soft-bodied invertebrates lose ammonia across the whole body surface. In freshwater fishes, most of the ammonia is passed across the epithelia of the gills.

Mammals and most adult amphibians excrete *urea*, a much less toxic compound that can be tolerated in more concentrated form. Ammonia and carbon dioxide are combined in the liver to produce urea, which is then carried by the circulatory system to the kidneys.

Land snails, insects, birds, and some reptiles excrete *uric acid*, a compound of low solubility in water that can be excreted as a precipitate after nearly all the water has been reabsorbed from the urine. Animals that produce *cleidoic eggs* excrete uric acid, which can be stored safely within the egg while the embryo develops.

The form of nitrogenous wastes that is excreted is also related to habitat. Terrestrial reptiles excrete uric acid, whereas crocodiles mainly excrete ammonia. Aquatic turtles excrete both urea and ammonia, and are able to shift to uric-acid production when water is less available.

Regulation of Body Temperature

Metabolism is very sensitive to internal changes in temperature. Many animals can maintain their internal temperature within an optimal range even as the environmental temperature fluctuates.

Heat is exchanged between an organism and the environment by the physical processes of *conduction*, the direct transfer of thermal motion between surfaces in contact; *convection*, the mass movement of warmed air or water to or from the organism; *radiation*, the emission of electromagnetic waves by all objects warmer than absolute zero; and *evaporation*, the loss of heat due to the conversion of a liquid to a gas. Convection and radiation account for the bulk of heat loss under relatively mild conditions. Convection and evaporation are the most variable causes of heat loss; wind and the evaporation of sweat can greatly increase the loss of heat.

Animals whose body temperatures conform to the ambient temperature have been called poikilotherms, whereas animals that maintain a constant body temperature have been referred to as homeotherms. However, some poikilotherms in very stable environments have a more constant temperature than do homeotherms.

Thermoregulation can be classified on the basis of

heat source; *ectotherms* absorb most of their heat from their surroundings, whereas *endotherms* use their metabolic heat to supply body heat.

Water temperature fluctuates less than air temperature, both during the day and between seasons. Aquatic ectotherms, such as invertebrates and most fishes, maintain quite stable body temperatures.

In the interview preceding this unit, Bennett pointed out that endothermy solves problems associated with living on land. Endothermy allows animals to maintain a constant body temperature in an environment where severe temperature fluctuations may be common. A warm body temperature requires active metabolism, but high levels of aerobic metabolism facilitate the energetically expensive movement of animals on land. Endotherms have a much higher basal metabolic rate than do ectotherms.

Contraction of muscles (by moving or shivering) produces heat. Certain hormones can increase the metabolic rate and trigger heat production in a process called *nonshivering thermogenesis*. Some mammals have *brown fat* in the neck and between the shoulders that is specialized for this form of rapid heat production.

The physiological and behavioral processes that enable a land mammal to maintain a constant body temperature include: (1) adjustment of the rate of metabolic heat production; (2) adjustment of the amount of blood flowing to the skin—*vasodilation* increases the rate of heat exchange, whereas vasoconstriction decreases it; (3) control of evaporative heat loss through panting or sweating; and (4) changes in behavioral responses including raising the fur, increasing or decreasing movement, and moving to warmer or cooler areas of the environment.

In the hypothalamus of the brain are two thermoregulatory areas: a *heating center* that controls vasoconstriction of superficial vessels, shivering, and nonshivering thermogenesis; and a *cooling center* that controls vasodilation and sweating or panting. Temperature-sensing nerve cells are located in the skin, hypothalamus, and some other areas of the nervous system. Control of body temperature involves feedback mechanisms, with the hypothalamus functioning as the thermostat.

Birds usually have very high body temperatures. A bird uses panting to promote evaporative heat loss when its temperature goes above the optimal range. The arrangement of arteries and veins going to the limbs creates a countercurrent heat exchanger that reduces heat loss. Blood can detour the exchanger should the bird need to lose body heat.

Marine mammals maintain their body temperatures by efficient heat-conserving mechanisms, including a thick layer of insulating blubber and countercurrent heat exchange between arterial and venous blood. When in warmer waters, these animals can lose body heat by dilating superficial blood vessels.

The metabolic rate of the ectothermic reptiles is very low and contributes little to body temperature. Behavioral adaptations, such as body orientation to the sun, allow these animals to regulate their temperature within an optimal range. In diving reptiles, more blood is routed to the body core to conserve heat. Reptiles also release a thyroid hormone that initiates thermogenesis.

Amphibians produce little heat, and easily lose heat by evaporation from their moist skin. Behavioral adaptations enable amphibians to regulate their internal temperature.

The body temperature of fishes is controlled mainly by the water temperature. Some large, active fishes are able to retain the heat produced by their swimming muscles and to maintain an elevated body-core temperature. A countercurrent heat exchanger, called the *rete mirabile* ("wonderful net"), keeps the deep swimming muscles warmer than the surface tissues.

Invertebrates may adjust their temperatures by behavioral or physiological mechanisms. Some large flying insects "warm up" before taking off by contracting their flight muscles. The increase in temperature enables them to sustain the intense activity involved in flight. Honeybees increase their movements and huddle together in the hive to maintain body heat in cold temperatures. They also can cool their hive by bringing in water and fanning it with their wings to promote evaporation and convection.

Many animals are capable of physiological adjustments to seasonal temperature changes. The increased production of certain enzymes may offset the lowered activity of enzymes at nonoptimal temperatures. Enzymes with different temperature optima may also be produced. Changes in the proportions of saturated and unsaturated lipids keep membranes fluid at different temperatures.

Hibernation is a physiological state in which metabolism decreases, the heart and respiratory rates slow down, and body temperature is maintained at a low level. A hibernating animal is able to withstand long periods of cold temperatures and decreased food supplies. *Estivation*, also characterized by slow metabolism and inactivity, allows animals to survive elevated temperatures and diminished water supplies.

A *torpor* is a physiological state that lasts for shorter periods than hibernation or estivation. Many ectotherms enter a state of slowed metabolism and inactivity when their food supply decreases. Some hummingbirds and shrews, which have very high metabolic rates due to their small size, enter a daily torpor during the periods in which they are not feeding. This daily cycle is controlled by the biological clock. Sleep may be a remnant of a more pronounced daily torpor.

Interaction of Regulatory Systems

A major challenge of animal physiology is determining how the various regulatory systems are controlled and how they interact to maintain homeostasis in the internal environment. The feedback circuits of homeostasis involve nervous communication and hormones.

STRUCTURE YOUR KNOWLEDGE

1. Fill in the following table with a brief description of the osmoregulatory, excretory, and thermoregulatory mechanisms for the animals listed.

Animal	Osmoregulation	Excretory System	Thermoregulation
Marine invertebrate			
Marine bony fish			
Freshwater fish			
Flatworm			
Earthworm			
Insect			
Reptile			
Bird			
Mammal			

2. Look at the table you have completed and draw some generalizations relating the control mechanisms of these animals to their environments and their evolutionary history. In particular, think about the problems of osmoregulation presented by marine, freshwater, and terrestrial habitats, the types of nitrogenous waste products suited to these environments, and the thermoregulatory considerations in an aquatic or terrestrial habitat. Now fill in the following table with your generalizations.

Environment	Osmoregulation	Nitrogenous Waste	Thermoregulation
Seawater			
Freshwater			
Terrestrial			

TEST YOUR KNOWLEDGE

MULTIPLE CHOICE: *Choose the one best answer.*

1. There is a net water flow by osmosis through a membrane from
    a. a solution that is hypertonic to one that is hypotonic.
    b. a solution with a lower osmolarity to one with a higher osmolarity.
    c. one isotonic solution to another.
    d. cells in a freshwater environment to the surrounding medium.

2. Sharks maintain urea within their body tissues
    a. because they excrete ammonia.
    b. through the use of their rectal glands.
    c. in order to maintain an osmolarity lower than seawater.
    d. in order to prevent osmotic water loss to their environment.

3. Transport epithelia are responsible for
    a. pumping water across a membrane.
    b. forming an impermeable boundary at the apical surface of an animal.
    c. the exchange of solutes between the extracellular fluid and the environment, followed by the osmotic movement of water
    d. pumping salt into a marine bony fish.

4. Which of the following is incorrectly paired with its excretory system?
    a. earthworm—nephron
    b. flatworm—flame-cell system
    c. insect—Malpighian tubules
    d. amphibian—kidneys

5. A freshwater fish would be expected to
    a. pump salt in through salt glands in the gills.
    b. produce copious quantities of dilute urine.
    c. diffuse ammonia out across the epithelium of the gills.
    d. do all of the above.

6. A portal system
    a. transports salt across an epithelial layer.
    b. transports urine to the exterior of the body.
    c. involves two capillary beds before blood returns to the heart.
    d. connects the proximal and distal convoluted tubules.

7. Which is the correct pathway for the passage of urine in vertebrates?
    a. collecting tubule —> ureter —> bladder —> urethra
    b. renal vein —> renal ureter —> bladder —> urethra
    c. nephron —> urethra —> bladder —> ureter
    d. cortex —> medulla —> bladder —> ureter

8. Aldosterone
    a. is a hormone that stimulates thirst.
    b. is secreted by the adrenal glands in response to a high osmolarity of the blood.
    c. stimulates the active absorption of Na^+ in the nephrons
    d. prevents diuresis.

9. Which of the following statements is incorrect?
    a. Long loops of Henle are associated with steep osmotic gradients and the production of hypertonic urine.
    b. Ammonia is a toxic nitrogenous waste molecule that invertebrates must pump out of their bodies.
    c. The form of nitrogenous waste that requires the least amount of water to excrete is uric acid.
    d. In the mammalian kidney, urea diffuses out of the collecting duct and contributes to the osmotic gradient within the medulla.

10. Which of the following is used by terrestrial animals as a heat-reducing mechanism?
    a. hibernation
    b. countercurrent exchange
    c. evaporation
    d. vasoconstriction

11. Nonshivering thermogenesis
    a. increases the metabolic rate and thus heat production.
    b. produces brown fat.
    c. produces heat by increased muscle contraction.
    d. is a behavioral mechanism of heat production.

12. The heating center of the hypothalamus
    a. responds to messages from temperature-sensing nerve cells in the skin.
    b. controls shivering and vasoconstriction of superficial vessels.
    c. functions along with a cooling center to thermoregulate by feedback mechanisms.
    d. does all of the above.

13. A countercurrent heat exchange
    between arterial and venous blood
    a. warms the blood going to the extremities.
    b. warms the blood returning to the body core.
    c. cools the venous blood.
    d. is used by marine mammals when
        they journey to warmer waters.

14. Physiological adjustments to cooler
    seasonal temperatures may include
    a. production of enzymes with
        higher optimal temperatures.
    b. changes in the lipid components
        of cell membranes.
    c. estivation.
    d. all of the above.

15. Which of the following sections of the mamma-
    lian nephron is incorrectly paired with its function?
    a. Bowman's capsule and
        glomerulus—filtration of blood
    b. proximal convoluted tubule—secretion
        of urea into filtrate and transport of
        glucose and amino acids out of tubule
    c. descending limb of loop of
        Henle—diffusion of urea into filtrate
    d. ascending limb of loop of Henle—diffusion
        and pumping of Na^+ and Cl^- out of filtrate

CHAPTER 41

CHEMICAL COORDINATION

FRAMEWORK

This chapter introduces the intricate system of chemical control and communication within animals. The endocrine system produces hormones, often in response to nervous messages, which regulate homeostasis, growth, and development. Steroid hormones bind with receptors within the nuclei of their target cells and influence gene expression. Peptide hormones bind with cell surface receptors, and second messengers mediate their effects in target cells.

The hypothalamus and pituitary gland play coordinating roles by integrating the nervous and endocrine systems and producing many tropic hormones that control the synthesis and secretion of hormones in other endocrine glands and organs.

CHAPTER SUMMARY

Coordination and communication among the specialized parts of complex animals are achieved by the *nervous system* and the *endocrine system*. The nervous system conveys high-speed messages along neurons, whereas the endocrine system produces chemical messengers called hormones, which are a slower means of communication.

Chemical Messengers of the Body

An animal *hormone* is defined as a molecule that is synthesized by a group of specialized cells, secreted into the circulatory system or body fluids, and elicits a response from *target cells* in another area of the body. Target cells have specific molecular receptors for hormones. *Endocrine glands* are secretory organs that produce hormones.

Endocrinology, the study of hormones, is a rapidly growing area of research. More than 50 different human hormones have been identified. Hormones can be grouped into three general classes: fat-soluble *steroid hormones*, derived from cholesterol; smaller, water-soluble *hormones derived from amino acids*; and *peptide hormones*, which are chains of amino acids of varying lengths. Other chemicals also may function as hormones.

Pheromones are chemical communication signals that function between animals. Serving as mate attractants, territory markers, or alarm substances, these usually small and volatile molecules are active in minute amounts.

Local regulators are chemical regulators that affect target cells near their points of secretion. Neurotransmitters are examples of these local messengers. *Growth factors*, peptides involved in development, are not technically called hormones because they are formed in many different areas.

Prostaglandins (PGs) are a group of modified fatty acids with a wide range of effects on many organ systems. In mammals, PGs stimulate uterine contraction, end the estrus cycle, and influence hormone secretion. PGs function as local regulators and also as hormones.

Mechanisms of Hormone Action

Hormones act at very low concentrations, can have more than one effect on a cell, and can have varying effects on different types of cells and within different species. Two different mechanisms of action have been identified, both involving *receptor* molecules to which the hormones bind: steroid hormones enter the nucleus and influence gene expression, and most nonsteroid hormones attach to the cell surface and stimulate release of intermediaries, which in turn affect cell activity.

Steroids are thought to diffuse through the target cell membrane and into the nucleus, where they bind to receptor proteins. The steroid–receptor complex then attaches to specific regions of the chromatin, initiating transcription of mRNA, which is used for synthesis of new proteins. The chromatin of the target cell appears to influence the action of the hormone–receptor complex.

Peptide hormones, prostaglandins, and most hormones derived from amino acids do not enter their target cells, but instead bind to specific receptors on the outside of the cell. This binding triggers a cascade of biochemical reactions in the cell, involving enzymes that are already present in the cytoplasm.

Through his work with epinephrine and glucose mobilization in liver cells, Sutherland identified *cyclic adenosine monophosphate* (also known as *cyclic AMP* or *cAMP*) and developed the second messenger model of hormone action, for which he received a Nobel prize in 1971. A hormone acts as the *first messenger* when it binds to a receptor molecule in the membrane. Cyclic AMP, as the *second messenger*, relays the message from the membrane to cytoplasmic enzymes.

A great many hormones use cAMP as a second messenger. The binding of a hormone with a specific receptor leads to an increase in the activity of adenylate cyclase, which catalyzes the conversion of ATP to cAMP. Enzymatic degradation of cAMP terminates the response to the hormone.

The effect of cAMP depends on the enzymes that are present in the cell. The enzyme cAMP-dependent kinase catalyzes the addition of phosphate groups to other proteins, which may, in turn, activate other enzymes. The effect of the hormone is amplified when the second messenger sets off an *enzyme cascade*, in which a sequence of enzyme molecules is activated. Through intermediate molecules, the same hormone can have different effects in different cells.

Calcium is also a common second messenger, binding to a protein called *calmodulin* and triggering responses in target cells.

Invertebrate Hormones

Most invertebrate hormones control development and reproduction, although some influence metabolism, pigmentation, and osmoregulation. Arthropods have well-developed endocrine systems.

In insects, hormones control the discrete events of molting and the development of adult characteristics. The steroid hormone *ecdysone*, secreted from a pair of *prothoracic glands*, stimulates the transcription of specific genes and functions to trigger molts and to promote development of adult characteristics. Chromosome puffs, signaling mRNA synthesis, appear in polytene chromosomes when ecdysone is applied.

Brain hormone stimulates the prothoracic glands to secrete ecdysone. *Juvenile hormone* (JH) is secreted by a pair of small glands, the *corpora allata*. This hormone counters the action of ecdysone and promotes retention of larval characteristics.

The Vertebrate Endocrine System

The *hypothalamus* plays a key role in integrating the endocrine and nervous systems. Situated in the lower brain, *neurosecretory cells* of the hypothalamus receive nerve signals and release hormones. One set of neurosecretory cells produces the hormones of the posterior pituitary; another produces *releasing factors* that regulate the anterior pituitary.

The *pituitary gland*, a small appendage at the base of the hypothalamus, consists of two lobes. A *posterior lobe*, or *neurohypophysis*, stores and secretes two small peptide hormones produced by the hypothalamus; the *anterior lobe*, or *adenohypophysis*, produces hormones. Several of these are *tropic* hormones, the targets of which are other endocrine glands.

The hormones stored in the posterior lobe are *oxytocin* and *antidiuretic hormone (ADH)*. Oxytocin induces uterine contractions during labor and milk ejection during nursing. ADH functions in osmoregulation. Nerve cells in the hypothalamus respond to an increase in blood osmolarity by sending impulses to the neurosecretory cells of the hypothalamus, which release ADH. ADH binds to receptors on cells lining the collecting ducts in the kidney and sets off a cAMP second-messenger system that results in an increased permeability of the tubule cells to water. The reabsorption of water from the urine lowers the osmolarity of the blood and causes the hypothalamus to slow the release of ADH, completing a negative feedback loop.

The anterior pituitary produces a variety of protein and peptide hormones. *Growth hormone (GH)* is a polypeptide that promotes growth directly and stimulates other growth factors. Several human growth disorders are related to abnormal GH production, including gigantism and acromegaly, both related to excessive GH, and hypopituitary dwarfism, caused by GH deficiency. GH deficiency can be treated with GH produced by bacteria genetically engineered to contain the human gene for this hormone.

The hormone *prolactin (PRL)* affects a wide variety of target tissues and exhibits effects in different vertebrate species that range from control of mammary gland growth to delay of metamorphosis to regulation of salt and water balance.

Three of the tropic hormones produced by the anterior pituitary are glycoproteins. *Thyroid-stimulating hormone (TSH)* regulates the release of thyroid hormones. *Follicle-stimulating hormone (FSH)* and *luteinizing hormone (LH)*, also called *gonadotropins*, stimulate the activities of the gonads.

Several hormones come from *pro-opiomelanocortin*, a single large protein that is cleaved into several short fragments inside the pituitary cells. *Adrenocorticotropin (ACTH)* is a tropic hormone that stimulates the adrenal cortex to produce and secrete its steroid hormones. *Melanocyte-stimulating hormone (MSH)* regulates the activity of pigment-containing cells in the skin, as seen in the color changes of amphibians. *Endorphins* and *enkephalins*, two other fragments of pro-opiomelanocortin, are also produced by neurons in the brain. These recently discovered hormones have effects similar to those of morphine: they inhibit pain reception.

Releasing factors, which stimulate and inhibit the secretion of hormones by the anterior lobe of the pituitary, are produced at the base of the hypothalamus in an area called the *median eminence*. Veins draining the median eminence lead to capillary beds in the anterior pituitary, creating a portal system that delivers releasing factors directly to the pituitary.

The thyroid gland produces two hormones, triiodothyronine (T_3) and thyroxine (T_4), which contain respectively three and four iodine atoms. Thyroid hormones, critical to the development and maturation of vertebrates, are required for normal functioning of bone-forming cells and for myelination and branching of nerve cells during embryonic development. An inherited deficiency of thyroid hormone in humans results in retarded skeletal growth and mental development, a condition known as cretinism.

The thyroid gland also is involved in the regulation of metabolism. Excess thyroid hormone results in a condition known as hyperthyroidism, the symptoms of which include weight loss, irritability, and high body temperature and blood pressure. Hypothyroidism can cause cretinism or result in weight gain and lethargy in adults. A goiter is an enlarged thyroid gland that produces insufficient thyroid hormone due to a lack of iodine.

A negative feedback loop controls secretion of thyroid hormones. Secretion of TSH-releasing hormone, or TRH, by the hypothalamus stimulates the anterior pituitary to secrete TSH, which binds to receptors on the thyroid gland, generating cAMP and triggering the synthesis and release of thyroid hormone. Thyroid hormone level, monitored by the hypothalamus, influences the secretion of TRH.

The thyroid gland also secretes *calcitonin*, a polypeptide hormone that lowers calcium levels in the blood. The parathyroid glands, located on the thyroid, secrete *parathyroid hormone (PTH)*, which raises blood calcium levels. A balance between PTH and calcitonin results in calcium homeostasis. Vitamin D is needed for proper PTH function.

Insulin is produced by *beta cells* within clusters of pancreatic cells called *islet cells*. The *alpha cells* of the islets secrete the hormone *glucagon*.

Insulin lowers blood sugar levels by promoting the movement of glucose into fat and muscle cells, the formation and storage of glycogen in the liver, the synthesis or protein, and the storage of fat. Glucagon raises glucose concentrations by stimulating conversion of glycogen to glucose in the liver, and by stimulating the breakdown of fats and proteins. The secretion of both these hormones is controlled by the blood sugar level.

Insulin is the only hormone that reduces blood sugar levels. In diabetes, the absence of insulin in the bloodstream means that glucose uptake by cells is restricted, and the body must use fats and body proteins for fuel. Glucose concentration in the blood rises, and its excretion in the urine can result in dehydration as water is excreted with the concentrated urine. Diabetes can be controlled by regular injections of insulin, previously obtained from animal pancreases. Insulin produced from human genes inserted in genetically engineered bacteria is now available.

In mammals, the *adrenal glands*, located on top of the kidneys, consist of two different glands: the outer *cortex* and the central *medulla*. The adrenal medulla synthesizes *epinephrine* (adrenalin) and *norepinephrine* (noradrenalin) from the amino acid tyrosine. Epinephrine, released in response to stress, mobilizes glucose from skeletal muscle and the liver, and stimulates the release of fatty acids from fat cells. Epinephrine and norepinephrine increase the rate and volume of the heart beat, and influence the contraction of smooth muscles such that the blood supply to the heart, brain, and skeletal muscles is increased, while the supply to the skin, gut, and kidneys is reduced. The sympathetic division of the autonomic nervous system stimulates the release of epinephrine. Epinephrine and norepinephrine also function as neurotransmitters in the nervous system.

The adrenal cortex responds to endocrine signals of stress. The pituitary hormone, ACTH, stimulates the adrenal cortex to synthesize and secrete *corticosteroids*, a group of steroid hormones. The hypothalamic regulatory hormone, corticotropin-releasing hormone (CRH), which can be triggered by stress, controls the release of ACTH.

The two main human steroid hormones are the *glucocorticoids*, such as cortisol, and the *mineralocorticoids*, such as aldosterone. Glucocorticoids promote synthesis of glucose from noncarbohydrates, such as proteins, and suppress parts of the body's immune system, including the inflammatory reaction. Cortisone has been used to treat serious inflammatory conditions such as arthritis, but long-term immunosuppression can be dangerous.

The mineralocorticoids have their major effects on salt and water balance. Aldosterone stimulates kidney cells to resorb sodium ions from the urine. Aldosterone secretion is regulated by peptides produced in the liver and kidneys, and by response to plasma ion concentration.

The testes of males and ovaries of females produce steroids that affect growth, development, and reproductive cycles and behaviors. The three major categories of gonadal steroids—androgens, estrogens, and progestins—are found in different proportions in males and females.

The testes primarily synthesize *androgens*, such as *testosterone*, which stimulate development of the male reproductive system and secondary sex characteristics. *Estrogens*, such as *estradiol*, are responsible for the development and maintenance of the female reproductive system and secondary sex characteristics. In mammals, *progestins* are involved in the preparation and maintenance of the uterus for the growth of an embryo.

The gonadotropins from the anterior pituitary gland (follicle-stimulating hormone and luteinizing hormone) control the synthesis of both estrogens and androgens. A hypothalamic regulatory hormone, LH-releasing hormone (LRH), controls secretion of FSH and LH.

Many other organs secrete important hormones; the digestive tract secretes at least eight, including *secretin*. The kidney produces *erythropoetin*, which stimulates production of red blood cells. *Atrial natriuretic hormone*, produced by the heart, helps to regulate salt and water balance.

The *pineal gland*, located near the center of the human brain and closer to the brain surface in other vertebrates, secretes a modified amino-acid hormone, *melatonin*, which regulates functions related to light and changes in daylength. Melatonin is secreted at night, and its production is a link between a biological clock and daily or seasonal activities. The precise roles of the pineal gland and melatonin are not clear.

The *thymus* plays a role in the immune system in young animals. It secretes several factors, including *thymosin*, that stimulate development and differentiation of T lymphocytes.

Endocrine Glands and the Nervous System

The endocrine system and nervous system function together to control chemical communication and coordination in animals. They are *structurally* related in that many endocrine glands are made of nervous tissue, such as the hypothalamus and posterior pituitary, and the insect brain. Other endocrine glands, such as the adrenal medulla, have evolved from the nervous system.

The endocrine and nervous systems are *chemically* related in that several vertebrate hormones, such as epinephrine, are used by both systems. Furthermore, the two systems are *functionally* related. Many physiological processes are coordinated by both nervous and hormonal communications. The output of one system often affects the output of the other. The nervous system controls endocrine glands, and hormones affect the development and functioning of the nervous system.

STRUCTURE YOUR KNOWLEDGE

1. Create a concept map that illustrates your understanding of the mechanisms of action of steroid and peptide hormones.

2. The hypothalamus and pituitary glands produce a number of hormones, tropic hormones, and releasing factors. Fill in the following table with the major products of each of the regions of these glands.

GLAND	HORMONES	MAIN ACTION
HYPOTHALAMUS Neurosecretory cells		
Median eminence		
PITUITARY Posterior lobe (neurohypophysis)		
Anterior lobe (adenohypophysis)		

3. Develop a concept map or diagram of the hormones and mechanisms involved in the body's control of blood sugar levels.

TEST YOUR KNOWLEDGE

MATCHING: *Match the hormone and gland or organ that produces it to the descriptions of hormone action. Not all choices are used.*

Hormones

A. adrenocorticotropin (ACTH)

B. androgens

C. antidiuretic hormone (ADH)

D. corticotropin-releasing hormone (CRH)

E. epinephrine

F. glucagon

G. glucocorticoids

H. insulin

I. melatonin

J. oxytocin

K. thyroxine

Gland or Organ

a. adrenal cortex

b. adrenal medulla

c. hypothalamus

d. pancreas

e. parathyroid

f. pineal

g. pituitary

h. testis

i. thymus

j. thyroid

Hor-mone	Gland	Hormone action
1. ___	___	involved in biological clock and seasonal activities
2. ___	___	involved in synthesis of glucose, suppress inflammatory reaction
3. ___	___	involved in glycogen breakdown in liver, increase blood sugar
4. ___	___	stimulate development of male reproductive system
5. ___	___	stimulate adrenal cortex to synthesize corticosteroids
6. ___	___	increase available energy supplies; increase heart rate and blood supply to skeletal muscles, heart, and brain
7. ___	___	regulate metabolism, development of bone-forming and nerve cells
8. ___	___	control release of ACTH
9. ___	___	control reabsorption of water from collecting ducts of kidney
10. ___	___	stimulate contraction of uterine muscles and mammary gland cells

TEST YOUR KNOWLEDGE

MULTIPLE CHOICE: *Choose the one best answer.*

1. Which of the following is not an accurate statement about hormones?
   a. All hormones are secreted by endocrine glands.
   b. Hormones move through the circulatory system to their destination.
   c. Target cells have specific molecular receptors for hormones.
   d. Hormones are essential to homeostasis.

2. The difference between pheromones and hormones is that
   a. Pheromones are small, volatile molecules, whereas hormones are steroids.
   b. pheromones are involved in reproduction, whereas hormones are not.
   c. pheromones are a form of olfactory communication; hormones are a form of chemical communication.
   d. pheromones are signals that function between animals, whereas hormones communicate among the parts within an animal.

3. Which one of the following examples is incorrectly paired with its class?
   a. local regulators — neurotransmitters
   b. steroid hormone — estrogen
   c. prostaglandin — modified fatty acid
   d. peptide hormone — calmodulin

4. In the second-messenger model of hormone action,
   a. a tropic hormone signals an endocrine gland to release a hormone.
   b. another molecule relays the message from the membrane-bound hormone to the cytoplasmic enzymes.
   c. negative feedback turns off the production of the hormone.
   d. genes are turned on and new proteins are synthesized.

5. Ecdysone
   a. is a steroid hormone found in insects that promotes retention of larval characteristics.
   b. is responsible for the color changes in amphibians.
   c. is produced by chromosome puffs.
   d. is secreted by prothoracic glands in insects and triggers molts and the development of adult characteristics.

6. The role of cAMP in hormone action is to
   a. act as the second messenger and activate other enzymes.
   b. activate adenylate cyclase.
   c. bind a specific hormone to the plasma membrane.
   d. bind to the protein calmodulin and initiate an enzyme cascade.

7. Neurosecretory cells
   a. are found in the hypothalamus.
   b. receive signals from nerve cells and release hormones.
   c. produce releasing factors in the median eminence.
   d. all of the above are correct.

8. Antidiuretic hormone (ADH)
   a. is produced by cells in the kidney and liver.
   b. stimulates the reabsorption of Na^+ from the urine.
   c. acts through a second-messenger system to increase the permeability of kidney collecting tubules to water.
   d. is a steroid hormone produced by the adrenal cortex.

9. The adenohypophysis or anterior lobe of the pituitary
   a. stores two hormones produced by the hypothalamus.
   b. is connected by a portal system to the median eminence.
   c. produces several releasing factors.
   d. does all of the above.

10. Acromegaly is caused by
    a. an excess of thyroxine.
    b. an excess of growth hormone.
    c. an excess of glucocorticoids.
    d. an abnormally high androgen-to-estrogen ratio.

11. Which of the following hormone sequences results in the secretion of estrogens in mammals?
    a. LRH —> FSH —> estrogen
    b. LH —> FSH —> estrogen
    c. CRH —> ACTH —> FSH —> estrogen
    d. gonadotropin —> LH —> estrogen

12. Pro-opiomelanocortin
    a. is produced by the hypothalamus.
    b. is cleaved into several hormones, including ACTH, melanocyte-stimulating hormone, endorphins, and enkephalins.
    c. produces several releasing factors.
    d. all of the above are correct.

13. Hyperthyroidism
    a. is responsible for cretinism.
    b. results in the formation of a goiter.
    c. produces weight loss and a high metabolic rate.
    d. is associated with a deficiency of iodine.

14. Calcium homeostasis is maintained by
    a. a balance between calcitonin, secreted by the thyroid, and PTH, secreted by the parathyroid, which respectively lower and raise blood calcium levels.
    b. a balance between PTH and vitamin D.
    c. a balance between calmodulin, calcitonin, PTH, and vitamin D.
    d. a balance between aldosterone, which stimulates the kidney to resorb calcium, and thymosin, which lowers blood calcium levels.

15. Which of the following is not true of norepinephrine?
    a. It is secreted by the adrenal medulla.
    b. Its action is to increase the heart rate and stroke volume and to constrict selected blood vessels.
    c. Its release is stimulated by ACTH, which is in turn released in response to the release of CRH.
    d. It serves as a neurotransmitter in the nervous system.

CHAPTER 42

ANIMAL REPRODUCTION

FRAMEWORK

This chapter covers the patterns and mechanisms of animal reproduction. In sexual reproduction, fertilization may occur externally or internally, both of which may require behavioral and physical adaptations. Development of the zygote may occur internally within the female, or externally in a moist environment or protected in a resistant egg. Placental mammals provide nourishment as well as shelter for the embryo, and nurse their young.

In the discussion of mamalian reproduction, the human reproductive system is described as an example, including its organs, glands, hormones, gamete formation, and sexual physiology. The chapter also covers pregnancy and birth, contraception, and recent reproductive technology.

CHAPTER SUMMARY

Modes of Reproduction

In *asexual reproduction*, a single individual produces offspring that are genetically identical to itself. These offspring constitute a *clone*. In *sexual reproduction*, two individuals contribute genes to the offspring.

Many invertebrates can reproduce asexually by *budding*, in which a new individual grows attached to the parent's body, or by *fragmentation*, in which the body is broken into several pieces, each of which develops into a complete animal. Specialized groups of cells may be released that grow into new individuals, as in the *gemmules* produced by sponges. *Regeneration* is a means of asexual reproduction in which a piece of an animal can reform a new organism.

The advantages of asexual reproduction are that it is less costly in terms of energy, more rapid, easier for

sessile or widely separated animals, and most effective for maintaining successful genotypes.

In sexual reproduction, the two haploid *gametes*, which are usually a relatively large, nonmotile female *ovum* and a small, flagellated male *spermatozoan*, fuse to form a diploid *zygote*.

The evolution of sex has been attributed to the advantage conferred by genetic variation, which can serve as raw material for natural selection. Chromosome doubling may occur by accident during mitosis, and meiosis may have originated as a mechanism to offset these doublings. The development of sex and the production of diploid organisms may have provided the means to mask or repair genetic damage in one chromosome of a chromosome pair.

Most animals show periodic cycles of reproductive activity. A combination of hormonal and seasonal cues control the timing of these cycles, which may be linked to favorable environmental conditions or energy supplies.

Some animals, such as aphids and rotifers, produce eggs that develop without being fertilized, by *parthenogenesis*. During unfavorable conditions, these organisms may switch to sexual reproduction. In bees, wasps, and ants, males are produced parthogenetically, whereas female workers and reproductive females are produced sexually. In the parthenogenesis used by a few fishes, amphibians, and lizards, doubling of chromosomes creates diploid "zygotes."

In *hermaphroditism*, each individual has functioning male and female reproductive systems, so that mating results in fertilization of both individuals. In *protandry*, an organism may switch sexes during its life.

Mechanisms of Sexual Reproduction

In *external fertilization*, eggs and sperm are shed from the body, and fertilization occurs in the environment.

This type of reproduction occurs almost exclusively in moist habitats, where the gametes and developing zygote are not in danger of desiccation. Environmental or pheromone signals may ensure that gamete release is synchronized. Fishes and amphibians that use external fertilization show mating behaviors, which trigger the release of gametes and provide a means for mate selection.

Internal fertilization involves the placement of sperm in the female reproductive tract, where egg and sperm unite. Behavioral interactions, as well as copulatory organs and sperm receptacles, are required for internal fertilization.

Evolutionary trends of increased protection of the developing embryo and young include resistant eggs, internal development, and parental care. The amniote eggs of birds and reptiles protect the embryo in a terrestrial environment. The young of placental mammals develop within the uterus, nourished from the mother's blood supply through the *placenta*.

Reproductive mechanisms often are more correlated with habitat than with phylogeny. Some invertebrates, such as polychaete annelids, do not have distinct gonads; eggs and sperm develop from cells lining the coelom. Insects have separate sexes and well-developed reproductive systems. Sperm develop in the *testes*, are stored in the *seminal vesicles*, and exit through an *ejaculatory duct* in the *penis*. Accessory glands may add fluid to the semen. Eggs, produced in the *ovaries*, pass through the *oviducts* to the *vagina*. A *spermatheca*, or sperm storing sac, may be present.

Flatworms are hermaphrodites with complex reproductive systems. In addition to ovaries, oviducts, and vagina, the female reproductive system includes yolk and shell glands and a *uterus*, where the eggs are fertilized and begin development. Males have a complex copulatory apparatus, that may be inserted in the vagina or may inject sperm into the female's body through *hypodermic impregnation*.

With the exception of most mammals, vertebrates have a common opening, called the *cloaca*, for the digestive, excretory, and reproductive systems. The uterus of most vertebrates is *bicornate*, having two separate branches. Nonmammalian vertebrates do not have penes, and may evert the cloaca to ejaculate.

Mammalian Reproduction

Human anatomy is used as the example of a mammalian reproductive system.

The external male *genitalia* include the *scrotum*, a fold of skin enclosing the testes, and the *penis*. Sperm are produced in the highly coiled *semiferous tubules* of the *testes*. Interstitial cells produce testosterone and other androgens. In most mammals, a cooler temperature than that of the internal body is needed for sperm production, so the scrotum suspends the testes below the abdominal cavity.

Sperm pass from the testes into the coiled *epididymis*, in which they mature and are stored. During *ejaculation*, sperm are propelled through the *vas deferens*, into a short ejaculatory duct, and out through the *urethra*.

Three glands contribute to the *semen*. The *seminal vesicles* contribute a fluid containing mucus, amino acids, fructose, and clotting proteins. The *prostate gland* produces an alkaline secretion that balances the acidity of the urethra and the vagina. The *bulbourethral glands* produce a small amount of viscous fluid of uncertain function.

The penis is composed of spongy tissue that engorges with blood during sexual arousal, producing an erection that facilitates insertion of the penis into the vagina. Some mammals have a *baculum*, a bone that stiffens the penis. The head of the penis, called the *glans penis*, is covered by a fold of skin called the *prepuce*, or foreskin.

The female gonads, the *ovaries*, consist of many *follicles*, which are sacs of cells that nourish and protect the single egg or ovum contained within each of them. The follicle produces the primary female sex hormones, the estrogens. Following ovulation, the follicle forms a solid mass called the *corpus luteum*, which secretes progesterone and a small amount of estrogen.

The ovum is expelled into the abdominal cavity and swept by cilia into the *oviduct*, or fallopian tube, through which it is transported to the *uterus*. The *endometrium*, or lining of the uterus, provides the maternal part of the placenta. The narrow neck of the uterus, the *cervix*, opens into the vagina, the thin-walled birth canal and repository for sperm during copulation.

The separate openings of the vagina and urethra are enclosed by two pairs of skin folds, the labia minora and the labia majora, which form a *vestibule*. The *glans clitoris* is at the top of the vestibule. Both the labia minora and the clitoris are composed of erectile tissue. The vaginal orifice is initially covered by a thin membrane called the *hymen*.

The *mammary gland*, or breast, is composed of deposits of fatty tissue and a series of *alveoli*, small milk-secreting sacs that drain into ducts that join together and open at the nipple.

The external genitalia of both sexes arise from common *primordia*, or undifferentiated embryonic tissue, which develops into male or female structures depending on the presence or absence of androgens.

Androgens, the most abundant of which is *testosterone*, are responsible for the male primary (associated with reproduction) and secondary (associated with "male" traits of voice, hair growth, muscle growth) sex characteristics. Androgens are also determinants of sexual and other behaviors. Androgen secretion and sperm production are controlled by hormones from the anterior pituitary and hypothalamus.

Female humans and higher primates have *menstrual cycles*, during which the endometrium lining thickens to prepare for the implantation of the embryo, and then is shed if fertilization does not occur. This bleeding, called *menstruation*, occurs on an approximately 28-day cycle in humans. Other mammals have *estrous cycles*, during which the endometrium thickens but, if fertilization does not occur, it is reabsorbed and there is no bleeding. Estrous cycles are often coordinated with season or climate, and females are receptive to sexual activity only during estrus, or heat, the period surrounding ovulation. The length of reproductive cycles is roughly correlated with body size and climate.

Estrogens are responsible for female primary and secondary sex characteristics. *Progesterone* acts on the uterus to prepare it for pregnancy, and on the breasts to stimulate milk-secreting alveoli. FSH and LH from the pituitary, and gonadotropin-releasing hormone (GnRH) from the hypothalamus, in conjunction with estrogen from developing follicles and progesterone produced by the corpus luteum, regulate the menstrual cycle.

Control of *spermatogenesis*, the production of mature sperm cells, is believed to involve FSH and testosterone. The haploid nucleus of a sperm is contained in a thick head, tipped with an *acrosome*, which helps the sperm penetrate the egg. Large numbers of mitochondria provide energy for movement of the flagellum, or tail.

In *oogenesis*, meiosis is unequal, and one large egg is produced along with small haploid *polar bodies*, which disintegrate. Oogenesis occurs in stages; it begins before birth and, in humans, is not completed until after fertilization.

In humans, the onset of reproductive ability, termed *puberty*, is a gradual process that is completed at about the age of 12 to 14 years.

The physiological reactions of humans to sexual stimulation include *vasocongestion* of the penis, testes, labia, vagina, and breasts, and *myotonia*, or increased muscle tension. During the *excitement phase* of the sexual response cycle, vasocongestion occurs. The *plateau phase* continues vasocongestion and myotonia, and breathing rate and heart rate increase. *Orgasm* is characterized by rhythmic, involuntary contractions of the reproductive system in both sexes. In males, emission deposits semen in the urethra, and ejaculation occurs when the urethra contracts and semen is expelled. In the *resolution phase*, vasocongested organs return to normal size, muscles relax, and nonreproductive reactions (breathing and heart rates) return to normal.

The development of an embryo in the uterus, beginning with *conception* and ending with birth, is called *pregnancy* or *gestation*. The duration of pregnancy correlates with the body size and degree of development of the young at birth. Human pregnancy averages 266 days, or 38 weeks.

Human gestation can be divided into three *trimesters*. Following fertilization in the oviduct, the *zygote* travels to the uterus. After about a week of *cleavage* division, in which it develops into a hollow ball of cells called a *blastocyst*, the zygote implants in the endometrium. Tissues grow out of the developing *embryo* to meet with the endometrium and form the *placenta*, in which gas and nutrient exchange and waste removal take place between the maternal and fetal circulations. Differentiation leads to *organogenesis* and, by the eighth week, all the structures of the adult are present, and the embryo is called a *fetus*.

The embryo secretes *human chorionic gonadotropin* (*HCG*), which acts like LH to maintain the corpus luteum's secretion of progesterone and estrogen through the first trimester. High progesterone levels initiate growth of the maternal part of the placenta, enlargement of the uterus, and cessation of ovulation and menstrual cycling.

During the second trimester, the placenta secretes its own progesterone, which maintains the pregnancy. The third trimester is a period of rapid fetal growth. The cervix softens and the pubic symphysis becomes more flexible under the influence of the hormone *relaxin*.

Labor consists of a series of strong contractions of the uterus and results in birth, or *partuition*. During the first stage of labor, the cervix dilates. The second stage consists of the contractions that force the fetus out of the uterus and through the vagina. The placenta is expelled in the final stage of labor.

Lactation is unique to mammals. Prolactin secretion stimulates milk production, and oxytocin controls the release of milk during nursing.

Contraception, the deliberate prevention of pregnancy, can be accomplished by one of several methods: preventing fertilization, preventing implantation, or preventing release of egg or sperm. The *rhythm method* is based on refraining from intercourse during the period that conception is most likely, the 5 to 8 days around the time of ovulation. *Coitus interruptus*, or withdrawal of the penis from the vagina prior to ejaculation, is not a dependable method of contraception. *Condoms* and *diaphragms*, used in conjunction with *spermicidal foam* or *jelly*, present a physical and chemical barrier to fertilization.

The *intrauterine device* (*IUD*) probably prevents implantation by irritating the endometrium. IUDs have been associated with serious side effects.

The release of gametes can be prevented by *chemical contraception*, such as the birth control pill, and *sterilization*, such as *tubal ligation* in women or *vasectomy* in men. Birth-control pills are combinations of synthetic estrogen and progestin (progesteronelike hormone), which act by negative feedback to stop the release of

GnRH by the hypothalamus and FSH and LH by the pituitary. The lack of these hormones results in a cessation of ovulation and follicle development. Cardiovascular problems are a potential side effect of birth-control pills. The development of male chemical contraception is currently under research.

Abortion is the termination of a pregnancy. Spontaneous abortion, or miscarriage, occurs in one-sixth of all pregnancies.

Some genetic diseases and congenital defects can be detected while the fetus is in utero. *Ultrasonography* produces an image of the fetus from the echoes of high-frequency sound waves. In *amniocentesis*, a sample of amniotic fluid is withdrawn through a long needle inserted into the uterus, and fetal cells are cultured and analyzed for chromosomal defects. *Chorionic biopsy* removes a small sample of tissue from the fetal part of the placenta and examines the cells for genetic defects.

A new technique of *in vitro fertilization* involves the removal of ova from a woman whose oviducts are blocked, fertilization within a petri dish, and implantation of the developing embryo in the uterus.

STRUCTURE YOUR KNOWLEDGE

1. Trace the path of a human sperm from the point of production to the point of fertilization, briefly commenting on the functions of the structures it passes through and of the associated glands.

2. Describe the path of a human ovum that does not become fertilized.

3. List the three general classes, along with examples, of birth-control methods. Which are most likely to prevent pregnancy? Which are least likely to do so?

TEST YOUR KNOWLEDGE

FILL IN THE BLANKS

1. _____ type of asexual reproduction in which a new individual grows while attached to the parent's body

2. _____ group of genetically identical offspring

3. _____ development of egg without fertilization

4. _____ individual with functioning male and female reproductive systems

5. _____ common opening of digestive, excretory, and reproductive systems in vertebrates

6. _____ cycle in which thickened endometrium is reabsorbed

7. _____ hormone that prepares the uterus for pregnancy

8. _____ common duct for urine and semen

9. _____ filling of a tissue with blood due to increased blood flow

10. _____ period when reproductive ability begins in humans

MULTIPLE CHOICE: *Choose the one best answer.*

1. Which of the following is an explanation for the periodicity of reproductive cycles in animals?
   a. Reproduction may occur during periods of increased food supply, during which energy can be invested in gamete formation.
   b. Seasonal cycles may allow offspring to be produced during favorable environmental conditions, when chances of survival are highest.
   c. Hormonal control of reproduction may be tied to biological clocks and seasonal cues.
   d. All of the above may contribute to periodic reproductive activity.

2. Which of the following is not required for internal fertilization?
   a. internal development of the embryo
   b. copulatory organ
   c. sperm receptacle
   d. behavioral interaction

3. Insect reproductive systems may include all the following except
   a. seminal vesicles.
   b. an ejaculatory duct.
   c. a spermatheca.
   d. an endometrium.

4. Which of the following is incorrectly paired with its function?
   a. semiferous tubules—add fluid containing mucous, fructose, and amino acids to semen
   b. scrotum—encase testes, hold below abdominal cavity
   c. epididymis—store sperm
   d. prostrate gland—add alkaline secretion to semen

5. The function of the corpus luteum is to
   a. nourish and protect the ovum.
   b. produce prolactin in the alveoli.
   c. produce progesterone and estrogen.
   d. convert into a hormone-producing follicle following ovulation.

6. A difference between estrous and menstrual cycles is that
   a. vertebrates have estrous cycles, whereas only mammals have menstrual cycles.
   b. estrous cycles occur more frequently than menstrual cycles do.
   c. the endometrial lining is reabsorbed in estrous cycles but shed in menstrual cycles.
   d. the endometrium does not thicken in estrous cycles; in menstrual cycles, the thickened lining is shed if fertilization does not occur.

7. Which of the following hormones is incorrectly paired with its function?
   a. GnRH — control release of FSH and LH
   b. oxytocin — stimulate production of milk in alveoli of mammary gland
   c. estrogen — responsible for primary and secondary female sex characteristics
   d. relaxin — loosen pubic symphysis

8. Myotonia is
   a. a congenital birth defect.
   b. the hormone responsible for breast development.
   c. increased muscle tension.
   d. rhythmic contractions of reproductive organs during orgasm.

9. The embryo contributes to the maintenance of pregnancy by
   a. forming the placenta.
   b. secreting human chorionic gonadotropin.
   c. embedding in the endometrium.
   d. secreting high levels of progesterone.

10. Examples of birth-control methods that prevent release of gametes from the gonads are
    a. sterilization and chemical contraception.
    b. birth-control pills and IUDs.
    c. condoms and diaphragms.
    d. abstinence and chemical contraception.

CHAPTER 43

ANIMAL DEVELOPMENT

FRAMEWORK

Fertilization initiates physical and molecular changes, such as the growth of fertilization membranes and increased metabolism and protein synthesis within sea urchin eggs. The early processes of embryonic development include cleavage, gastrulation, and organogenesis. The forms of these cell divisions and morphogenetic movements depend on the amount of yolk in the egg and on the taxonomic position of the animal. This chapter describes the embryology of a cephalochordate and of amphibians, birds, and mammals.

Development may be mosaic, in which the fate of blastomeres is fixed by the first cleavage divisions, or regulative, in which blastomeres remain totipotent for a longer time. Determination and differentiation of cells are the result of the control of gene expression by both cytoplasmic determinants and the positional information a cell receives within a developing structure. Developmental biologists, using transplant experiments and other techniques, are gradually unraveling some of the mechanisms underlying the complex processes of animal development.

CHAPTER SUMMARY

Development involves the irreversible changes that occur within an organism, such as embryonic development, postembryonic growth and maturation, metamorphosis, regeneration, wound healing, and aging. The three key processes of embryonic development are cell division, the production of large numbers of cells; *differentiation*, the formation of specialized cells; and *morphogenesis*, the development of body shape and organization.

Fertilization

The union of egg and sperm, or *fertilization*, activates the egg by initiating metabolic reactions that trigger embryonic development.

In sea urchin fertilization, the acrosome at the tip of the sperm discharges hydrolytic enzymes when in contact with the jelly coat of an egg, allowing the *acrosomal process* to penetrate this layer. Fertilization of egg and sperm of the same species is ensured when a protein called *bindin* on the surface of the sperm attaches to specific receptor molecules on the egg's *vitelline layer*, external to the egg's plasma membrane.

Enzymes digest a hole through the vitelline layer, and the sperm's plasma membrane fuses with that of the egg, allowing the sperm nucleus to enter the egg. Fusion of the membranes opens membrane ion channels, and Na$^+$ ions flow into the egg. The resulting depolarization of the membrane prevents other sperm cells from fusing with the egg and provides a *fast block to polyspermy*.

The depolarization initiates the release of calcium in the egg cytoplasm, causing *cortical granules*, located in the *cortex* (the outer zone of cytoplasm), to fuse with the plasma membrane. These vesicles release their contents by exocytosis and cause the vitelline layer to loosen from the membrane and to swell by the osmotic uptake of water. The elevated vitelline layer, known as the *fertilization membrane*, functions as a *slow block to polyspermy*.

The release of calcium ions into the cytoplasm raises the pH value, which increases the rates of cellular respiration and protein synthesis. Injection of calcium into an egg, or pricking the egg with a needle, can turn on these metabolic responses and initiate parthenogenetic development. Even an enucleated egg can be activated to begin protein synthesis, showing that inactive mRNA had been stockpiled in the egg.

After the sperm nucleus fuses with the egg nucleus, DNA replication begins in preparation for the cleavage division that begins the development of the embryo.

Early Stages of Embryonic Development

The early embryology of amphioxus, a cephalochordate, is used to describe the steps of cleavage, gastrulation, and organogenesis.

Cleavage is a succession of rapid cell divisions during which the embryo becomes partitioned into many small cells, called *blastomeres*. The proteins required for these divisions are synthesized using maternal mRNA that had been stockpiled before fertilization.

The axis of the amphioxus egg is defined by the *animal pole*, which is the point at which the polar body budded from the egg during meiosis. The opposite end is called the *vegetal pole*. The first two cleavage divisions are vertical, or polar; the third division is equatorial. Further cleavage produces a *morula*, a solid ball of cells, followed by the arrangement of cells into a hollow ball, the *blastula*, surrounding a fluid-filled cavity called the *blastocoel*.

In amphioxus, *gastrulation* occurs when one side of the blastula buckles inward by a process known as *invagination*. The *gastrula* is a cup-shaped embryo having two layers of cells that surround the *archenteron*, the cavity that will become the digestive tract. The archenteron opens by way of the *blastopore*, which develops into the anus in deuterostomes and the mouth in protostomes.

A third layer of cells develops from pouches that bud from the archenteron and expand into the space between the outer and inner layers. The triploblastic body plan now consists of the primary germ layers—the outer *ectoderm*, the middle *mesoderm*, and the inner *endoderm*.

Rudiments of organs develop from the primary germ layers in the process known as *organogenesis*. The *notochord* is formed from dorsal mesoderm along the roof of the archenteron. Ectoderm above the developing notochord thickens to form a *neural plate*, which sinks and rolls into a tube, forming the *neural tube*. This tube develops into the central nervous system. The ectoderm also gives rise to the epidermis, inner ear, and eye lens.

The mesoderm forms the notochord, lining of the coelom, muscles, skeleton, gonads, kidneys, and most of the circulatory system. The lining of the digestive tract and organs that form as outpocketings of the archenteron (lungs, liver, pancreas) arise from endoderm.

Comparative Embryology of Vertebrates

Yolk platelets, which store lipids and protein, are concentrated in the vegetal hemisphere of a frog egg, and *pigment granules* are located in the cortex of the animal hemisphere. The first two cleavage divisions are polar. The horizontal third division is displaced toward the top of the embryo. Because the yolk impedes divisions in the vegetal hemisphere, the frog blastula has its blastocoel, enclosed in a wall that is several cells thick, restricted to the animal hemisphere.

A change in the shape of *bottle cells* begins gastrulation, forming a tuck that will become the *dorsal lip* of the blastopore. In a process called *involution*, surface cells roll over the lip and migrate along the roof of the blastocoel. The blastopore is first crescent shaped and then it forms a circle surrounding a *yolk plug* of large, yolk-filled cells. Gastrulation produces an archenteron, surrounded by the three primary germ layers.

In organogenesis, the notochord, derived from dorsal mesoderm, forms first, followed by the neural tube, derived from a plate of ectoderm above the notochord. Mesoderm along the side of the notochord separates into blocks, called *somites*, which give rise to the vertebrae and skeletal muscles. Lateral to the somites, the mesoderm forms the lining of the coelom.

A band of cells called the *neural crest* forms above the neural tube. These cells later migrate to form pigment cells in the skin, skull bones, teeth, adrenal glands, and components of the peripheral nervous system.

The yolk of a bird egg is so dense that cleavage occurs in only a small disc of cytoplasm at the animal pole, producing a cap of cells called the *blastodisc*. *Meroblastic cleavage* is this incomplete division of a yolk-filled egg, whereas *holoblastic cleavage* is the complete division through the egg.

The blastodisc separates into an upper and lower layer, forming the blastocoel in between. A straight invagination, called the *primitive streak*, marks the anterior–posterior axis of the embryo and is the site of gastrulation. Rolling in of the top layer of cells forms the mesoderm between the upper ectoderm and lower endoderm. The borders of this flat sheet fold down and join, forming a triple-layered tube from which organogenesis proceeds.

The primary germ layers also give rise to four extraembryonic membranes, essential to development within the avian egg shell. The *yolk sac* grows to enclose the yolk and develops blood vessels to carry nutrients to the embryo. The *amnion* encloses a fluid-filled sac, which provides an aqueous environment for development and acts as a shock absorber. The *allantois*, growing out of the hindgut, serves as a receptacle for the nitrogenous waste, uric acid, and expands to press the *chorion* against the lining of the eggshell. The vascularized allantois, in conjunction with the chorion, provides gas exchange for the developing embryo.

The egg of placental mammals has little yolk, and cleavage of a mammalian zygote is holoblastic. Gastrulation and early organogenesis, however, are similar in pattern to those of birds and reptiles. Cleavage planes appear to be randomly oriented, and blastomeres are equal in size.

The *blastocyst* consists of an outer epithelium, the *trophoblast*, surrounding a cavity into which protrudes a cluster of cells, called the *inner cell mass*, which will develop into the embryo and some extraembryonic membranes. The trophoblast secretes enzymes that enable the blastocyst to embed in the endometrium and extends projections that develop into the fetal portion of the placenta.

The four extraembryonic membranes are homologous to those of reptiles and birds. The chorion surrounds the embryo and the other membranes. The amnion encloses the embryo in a fluid-filled amniotic cavity. The yolk-sac membrane, enclosing a small fluid-filled cavity, is the site of early formation of blood and development of germ cells that will develop into gametes. The allantois forms blood vessels to connect the embryo with the placenta through the umbilical cord.

The inner cell mass forms a flat *embryonic disc*, which resembles the two-layered blastodisc of birds and reptiles. In gastrulation, cells from the upper layer roll through a primitive streak to form the mesoderm. Organogenesis begins with the formation of the notochord, neural tube, and somites. In the human embryo, all major organs have begun development by the end of the first trimester.

Mechanisms of Development

A bilaterally symmetrical animal has a left and right side, an anterior–posterior axis, and a dorsal–ventral axis. These three polarities may be established at the zygote stage.

In the amphibian egg, an animal–vegetal axis exists due to the yolk platelets concentrated in the vegetal pole and pigment granules concentrated near the animal pole. During fertilization, the cortex of the egg is moved toward the point of sperm entry, and the edge of the pigmented layer opposite that point is pulled toward the animal pole, exposing a lighter-colored cytoplasm. This half-moon–shaped mark is the *gray crescent*, which will be bisected by the first cleavage division and marks the future location of the dorsal lip of the blastopore. The gray crescent determines the anterior–posterior axis, the left–right polarity of the embryo, and the animal–vegetal axis becomes the dorsal–ventral axis of the embryo.

Mammalian eggs, in contrast, do not have an apparent polarity, and early cleavage divisions are randomly oriented.

Cytoplasmic determinants are unevenly distributed cytoplasmic components that fix the developmental fates of cells. In mollusks, a polar lobe that forms on one of the first two blastomeres is necessary for the subsequent formation of mesodermal structures in the embryo. If separated, the blastomere without the polar lobe does not develop normally.

Early blastomeres that can be separated and still produce complete embryos are said to be *totipotent*. Even though the cytoplasmic determinants are asymmetrically distributed in the frog egg, the first cleavage division equally separates these determinants. If an experimentally induced cleavage plane does not bisect the gray crescent, only the blastomere that receives the gray crescent will develop into a normal tadpole.

Development in which the early blastomeres are not totipotent is said to be *mosaic*. In *regulative* development, cells are totipotent longer and their developmental fates can be experimentally altered. Mammalian embryos remain regulative longer than do those of most other animals.

In the 1920s, Vogt developed a *fate map* for the amphibian embryo by labeling regions of the blastula with vital dyes and determining where these dyes showed up in the late gastrula stage.

Extension, contraction, and adhesion are involved in the changes of cell shape and various morphogenetic movements. Reorganization of the cytoskeleton produces the wedge shape of cells that leads to the formation of invaginations and evaginations. Amoeboid movement is also involved in morphogenesis. Cells at the leading edge of migrating tissue may extend pseudopodia and then drag along neighboring cells, attached to them by intercellular junctions. Amoeboid cells wander from the neural crest to various parts of the embryo.

An extracellular matrix of adhesive substances and fibers may help to guide the movement of cells. The orientation of *fibronectin* fibrils corresponds to the arrangement of contractile microfilaments of the cytoskeleton of migrating cells. Substances on the surfaces of cells, called *cell adhesion molecules* (CAMS), function to hold the cells of specific tissues and organs together during morphogenesis.

In *induction*, one group of cells influences the development of another. If the chordamesoderm that forms the notochord is transplanted, a neural plate and tube will form in an abnormal region of ectoderm. Using transplant experiments, Mangold and Spemann established that the dorsal lip is a primary *organizer* of the embryo because of its influence in early organogenesis.

Cells *differentiate*, or develop characteristic structures and functions, because they express different portions of a common genome. Once a cell is committed to a particular path of development, it is said to be *determined*—its developmental potential has become restricted. Transplant experiments are used to show whether tissues at various stages of development have been determined. Transplanted ectoderm from an early gastrula will form a neural plate above the developing notochord, whereas ectoderm from a late-stage gastrula transplanted into an early gastrula will not be induced and no neural plate will form.

Determination appears to develop progressively; the developmental options of cells become more and more narrow. Differentiation is a product of the developmental history of the cell. The cytoplasmic environment of the cell apparently controls the expression of the genome. Morphogenetic movements also contribute to determination and differentiation by exposing cells to differing chemical and physical environments.

Pattern formation is the ordering of cells and tissues into three-dimensional structures that make up the various parts of an animal. The mechanism of this organization appears to be that the cells of a rudi-mentary organ are initially similar, and they differentiate based on their position within the field of cells. Gradients of chemical signals moving along three planes would provide the *positional information* needed for a cell to determine its location in the three-dimensional developing organ. These hypothetical substances producing the chemical gradients are called *morphogens*.

Wolpert has studied positional information in chick limb development by doing transplant experiments. Undifferentiated mesoderm from the thigh region developed into a toe when transplanted to the tip of a wing bud. The mesoderm had already been determined to form a part of the leg, but it still responded to its location at the tip of a developing limb. Installments of positional information are provided to cells throughout their development, and determination becomes more and more specific.

STRUCTURE YOUR KNOWLEDGE

1. Create a flow chart that shows the sequence of events in the fertilization of a sea urchin egg. Include the functions of key events.

2. Fill in the following table, briefly describing the early stages of development for amphioxus, frog, bird, and mammalian embryos.

Stage	Amphioxus	Amphibian	Bird	Mammal
Cleavage				
Blastula				
Gastrulation				

3. Use the following terms to create a concept map that illustrates the relationships among these concepts. Add additional concepts as needed.

development   determination
differentiation   cytoplasmic determinants
pattern formation   positional information

TEST YOUR KNOWLEDGE

MULTIPLE CHOICE: *Choose the one best answer.*

1. The vitelline layer
   a. forms the fertilization membrane.
   b. releases calcium that initiates the cortical reaction.
   c. has bindin proteins on the surface that attach to receptor molecules on the acrosomal process of sperm.
   d. all of the above are correct.

2. The fast block to polyspermy
   a. prevents sperm from other species from fertilizing the egg.
   b. is produced by the depolarization of the membrane.
   c. is a result of the formation of the fertilization membrane.
   d. is caused by the exocytosis of cortical granules.

3. The activation of an enucleated egg by pricking shows that
   a. calcium ions initiate activation.
   b. inactive mRNA had been stockpiled in the egg cell.
   c. the release of calcium ions raises the pH level of the cell.
   d. the nucleus is not needed for the development of a sea urchin embryo.

4. Which of the following groups does not have eggs with definite animal and vegetal poles?
   a. amphioxus
   b. frog
   c. bird
   d. mammal

5. The archenteron develops into
   a. the anus in deuterostomes.
   b. the blastocoel.
   c. the endoderm.
   d. the digestive tract.

6. Organogenesis
   a. produces the gastrula by invagination.
   b. is the movements of cells that produce body structures.
   c. is the rudimentary development of organs from the primary germ layers.
   d. begins with the formation of the neural plate.

7. Which of the following is incorrectly paired with its primary germ layer?
   a. muscles — mesoderm
   b. central nervous system — ectoderm
   c. lens of the eye — mesoderm
   d. liver — endoderm

8. In a frog embryo, the blastocoel
   a. is completely obliterated by yolk platelets.
   b. becomes lined with endoderm during gastrulation.
   c. is created by the change in shape of bottle cells.
   d. is displaced into the animal hemisphere.

9. The embryonic disc of mammalian embryos
   a. develops from the inner cell mass.
   b. resembles the blastodisc of birds.
   c. develops into the embryo and some extraembryonic membranes.
   d. does all of the above.

10. The function of the allantois in birds is to
    a. provide for nutrient exchange between the embryo and yolk sac.
    b. store nitrogenous wastes in the form of urea.
    c. form a respiratory organ in conjunction with the chorion.
    d. produce blood vessels in the umbilical cord.

11. Which of the following is a correct statement of a difference between avian and mammalian early embryology?
    a. The bird egg has meroblastic cleavage, whereas the mammalian egg has holoblastic cleavage.
    b. The bird embryo is surrounded by an amnion membrane, whereas the mammalian embryo is enclosed by the placenta.
    c. Gastrulation occurs through the primitive streak in birds, but by way of the blastopore in mammals.
    d. The bird egg has four extraembryonic membranes, whereas the mammalian embryo develops only two.

12. Somites are
    a. blocks of mesoderm circling the archenteron.
    b. cells from which the notochord arises.
    c. serially arranged mesoderm blocks that develop into vertebrae and skeletal muscles.
    d. mesodermal derivatives of the notochord that line the coelom.

13. A band of cells called the neural crest
    a. rolls up to form the neural tube.
    b. produces amoeboid cells, some of which migrate to form teeth, skull bones, and parts of the peripheral nervous system.
    c. induces the formation of the optic vesicle.
    d. develops into the brain.

14. The trophoblast
    a. produces the fetal portion of the placenta.
    b. encloses a fluid-filled cavity in which blood cells and the germ cells that will develop into gametes are produced.
    c. is the two-layered blastodisc found in mammalian embryos.
    d. secretes enzymes that hydrolyze through the egg's vitelline layer.

15. The gray crescent
    a. is the future location of the dorsal lip of the blastopore.
    b. is created when the pigmented layer is pulled toward the animal pole at fertilization.
    c. is bisected by the first cleavage division, creating an equal division of cytoplasmic determinants.
    d. is all of the above.

16. In regulative development,
    a. cells remain totipotent until the gastrula stage.
    b. early blastomeres that are separated produce complete embryos.
    c. blastomeres without the polar lobe do not develop normally.
    d. none of the above is correct.

17. A fate map for amphibian embryos
    a. was developed by Spemann and Mangold in the 1920s.
    b. indicates which parts of the embryo will be derived from different regions of the blastula.
    c. reveals that the extraembryonic membranes derive from the dorsal lip.
    d. indicates that early cleavage and development are mosaic.

18. Cytoplamic determinants
    a. are unevenly distributed cytoplasmic components that fix the developmental fates of cells.
    b. are involved in the regulation of gene expression.
    c. are involved in the determination of cells and tissues.
    d. are all of the above.

19. Cells of specific tissues and organs are held together during morphogenesis by
    a. fibronectin fibrils.
    b. cell adhesion molecules.
    c. induction.
    d. morphogens.

20. Pattern formation appears to be determined by
    a. positional information a cell receives from gradients of chemical signals.
    b. differentiation of cells that then migrate into developing organs.
    c. when in development the transplant experiment is performed.
    d. the induction of cells from chordamesoderm cells found in the center of organs.

CHAPTER 44

NERVOUS SYSTEMS

FRAMEWORK

This chapter describes the structural components of nervous systems and their functional transmission of information as they integrate and coordinate information from and responses to the internal and external environments. Work toward an understanding of the following clusters of concepts as you study this chapter.

Structures	Functions	Vertebrate Nervous System	Parts of human brain
Neurons--cell body, axons, dendrites; glia--myelin sheath, blood-brain barrier; nerves--sensory, motor, mixed; ganglion	Resting potential, action potential, depolarization, voltage-sensitive ion channels, electrical synapse, chemical synapse, neurotransmitter	Peripheral nervous system-- sensory and motor: somatic and autonomic: sympathetic and parasympathetic; central nervous system-- spinal cord and brain	Medulla, pons, cerebellum, midbrain, reticular formation, thalamus, hypothalamus, corpus collosum, cerebral cortex

CHAPTER SUMMARY

The *nervous system* uses specialized cells and a combination of electrical and chemical signals to receive and transmit information from the internal and external environments, to integrate this information, and to relay response instructions to muscles, glands, and other target organs. Three characteristics that distinguish the nervous system from the endocrine system are rapid communication, specific control with messages traveling directly to individual target cells, and structural complexity that enables the integration of more kinds of information and responses.

Cells of the Nervous System

The cells of the nervous system include *neurons*, which transmit signals by making use of differences in electri-cal charge across their cell membranes, and *glial cells*, which support, insulate, and protect the neurons.

A neuron consists of a *cell body*, which contains the nucleus and most of the organelles and cytoplasm, and long fiberlike extensions of the cell, *axons* and *dentrites*, which conduct impulses. A neuron usually has one axon, which transmits signals away from the cell body. The more numerous, shorter, and branched dendrites carry signals toward the cell body. Chemical or electrical signals transmit messages at a *synapse* between the axon of one neuron and the dendrite of the next. *Nerves* are bundles of these fibers and may be sensory, containing only dendrites; motor, consisting only of axons; or mixed.

Sensory neurons transmit information from the internal and external environments toward the brain or spinal cord. *Motor neurons* carry information from the brain or spinal cord to effector organs. *Interneurons* transmit signals from one neuron to another. Interneurons are confined to the *central nervous system*—the brain and spinal cord. The *peripheral nervous system* consists of the sensory and motor neurons.

Among the types of glia found in the human nervous system are *astrocytes*, lining the capillaries in the brain to form a *blood-brain barrier*, which restricts the passage of most substances into the brain, and *oligodendrocytes* in the central nervous system and *Schwann cells* in the peripheral system, which wrap and insulate axons in a coat called the *myelin sheath*.

Transmission Along Neurons

The *electrical* or *membrane potential*, caused by the charge difference between the cytoplasm and extracellular fluid, is a fundamental feature of all cells. Excitable cells, such as neurons, use this potential difference to conduct an electrical impulse.

The *resting potential* of a neuron is about –70 mV (millivolts). The resting potential depends on the different abilities of ions to cross the membrane, the balance of diffusion and electrical forces acting on the ions, and the sodium–potassium pump.

The sodium–potassium pump moves three sodium ions out of the cell for every two potassium ions moved in. This pump creates steep concentration gradients for sodium and potassium ions across the membrane and accumulates more positive ions on the outside, making the inside of the cell membrane negatively charged relative to the outside.

The concentration gradients drive the ions back across the membrane by facilitated diffusion through specific channels. Electrical interactions with other ions may counteract diffusion forces—positively charged potassium ions moving back out of the cell encounter the high concentration of positively charged sodium ions. The electrical charge or potential that balances the chemical diffusion gradient for a particular ion is the *equilibrium potential* for that ion. The equilibrium potential for sodium is about +65 mV; that for potassium is about –90 mV. At the –70 mV resting potential of the cell membrane, energy is stored in the gradients of Na^+ and K^+ ions. The resting potential is closer to potassium's equilibrium potential because the membrane is about ten times more permeable to potassium than to sodium.

An *action potential* is a rapid change in the membrane's electrical potential, which creates a nerve impulse. A *depolarization* of the membrane means that the inside of the cell becomes less negative relative to the outside and the voltage across the membrane approaches zero. This electrical change opens *voltage-sensitive channels* that permit the rapid diffusion of sodium back into the cell, bringing the membrane potential to a positive value.

The action potential operates on the *all-or-none principle*; the action potential always brings the membrane potential to the same level. The short duration of the action potential is due to the rapid closing of the sodium gates, called *sodium inactivation*, which will not reopen until the membrane has returned to its original resting potential.

The *refractory period* is the short time after an action potential when the neuron first cannot respond to another stimulus, and then requires a much greater stimulus to begin a second action potential. The opening of voltage-sensitive potassium channels helps to restore the resting potential, and the outflux of potassium may temporarily hyperpolarize the membrane and produce a voltage more negative than the resting potential.

The initial depolarization must reach a *threshold* before an action potential is generated; the stimulus must be intense enough to induce the sodium gates to open.

As the positive ions moving into the cell during the action potential spread out, they depolarize adjacent sections of the membrane to the threshold, open new voltage-sensitive channels, and propagate the action potential along the membrane. The greater the axon diameter, the faster the rate of conduction of the action potential. Some invertebrates have giant axons, which conduct impulses rapidly.

In vertebrates, the Schwann cells form a thick, myelin sheath with small gaps, called *nodes of Ranvier*, where Schwann cells abut. The lipid layer insulates the neuron membrane so that the action potential travels from node to node, conducting the nerve impulse much faster.

The Synapse: Transmission Between Cells

The action potential propagates from the point of stimulation to the tip of the axon. Passage of the impulse to the next cell may occur across an electrical synapse, in which cytoplasmic connections called *gap junctions* allow ions to flow between two cells and transmit the action potential across the synapse. In a *chemical synapse*, a chemical message travels from the presynaptic cell across a cleft to the postsynaptic cell.

Electrical synapses conduct uniform impulses much faster than chemical synapses do, providing for speed and reliability in synchronizing the activity of many neurons. Chemical synapses are slower and are capable of more variations. Chemical synapses are most prevalent in vertebrate nervous systems.

In chemical synapses, *synaptic knobs* at the tips of the finely divided ends of axons contain large numbers of *synaptic vesicles*, in which the chemical messenger, called a *neurotransmitter*, is stored. The depolarization of the presynaptic membrane of an axon opens calcium channels in the membrane, allowing calcium to diffuse into the cell. The increase in calcium concentration causes the synaptic vesicles to fuse with the membrane and release their neurotransmitter into the cleft.

The *postsynaptic membrane* contains two types of proteins: receptor proteins for the neurotransmitter molecules, and degrading enzymes that break down the signal after it is received.

The binding of neurotransmitter to receptors on the postsynaptic membrane opens ion channels. If sodium channels are opened, sodium moves into the cell, the membrane depolarizes, and an *excitatory postsynaptic potential (EPSP)* is created, bringing the membrane potential closer to an action potential threshold. If the transmitter opens potassium channels, the outflow of potassium hyperpolarizes the membrane and produces an *inhibitory postsynaptic potential (IPSP)*, making the triggering of an action potential more difficult.

The number of neurotransmitter molecules binding to the postsynaptic membrane determines the strength of the postsynaptic potential. *Summation* of EPSPs and IPSPs usually determines whether the postsynaptic cell fires. Temporal summation occurs when two action potentials travel down the same presynaptic cell in rapid succession. Spatial summation occurs when two different presynaptic nerve cells release neurotransmitter simultaneously. Two EPSPs arriving simultaneously will bring the membrane potential closer to the threshold than a single EPSP will, whereas a concurrent EPSP and IPSP will make the generation of an action potential less likely. Different neurotransmitters are usually responsible for the EPSP and IPSP.

The membrane ion channels in the postsynaptic cell will remain open as long as the neurotransmitter is bound to the receptor. Enzymes in the postsynaptic membrane, such as acetylcholinesterase, quickly degrade the neurotransmitter so that the resting condition is reformed and a new signal can be processed.

The importance of the various molecules to synaptic transmission can be seen by the effects of curare, which outcompetes acetylcholine for receptor binding sites and prevents neuromuscular transmission; by diseases in which receptors are destroyed and transmission is prevented; and by chemicals, such as diazepam in Valium, that enhance receptor binding.

The nervous system produces a great variety of neurotransmitters. *Acetylcholine* is one of the most common in invertebrates and vertebrates. In vertebrate neuromuscular junctions, acetylcholine produces EPSPs in skeletal muscle cells. It also can be inhibitory, slowing down the heart rate of vertebrates and molluscs.

Norepeniphrine and epinephrine are biogenic amines, derived from the amino acid tyrosine, which usually transmit signals in the brain and spinal cord. Dopamine and serotonin are biogenic amines that are involved in sleep, mood, attention, and learning. Amino acids also may be neurotransmitters.

Many peptides are being recognized either as neurotransmitters or as neuromodulators that bind to receptor proteins and influence the response of the cell to neurotransmitters. The endorphins and enkephalins are neuropeptides that have effects like those of opium derivatives.

Organization of Nervous Systems

Groups of neurons that interact and carry information along specific pathways are called circuits. In *convergent circuits*, several neurons come together along a pathway and feed information into a few cells. *Divergent circuits* spread out information from one pathway in several directions; and *reverberating circuits* are circular paths in which signals return to their source.

Nerve cell bodies are arranged in groups called *ganglia*. A ganglion in the brain is often called a nucleus.

The cnidarian *nerve net* is a loosely organized system of nerves, connected by electrical synapses, in which impulses are conducted in both directions. Some centralization is seen in jellyfish, in which clusters of nerve cells around the margin of the bell coordinate swimming movements. Modified nerve nets are also found in echinoderms.

Bilateral animals with more active life styles show *cephalization*, the concentration of sense organs and nerves in the head. A brain and one or more nerve trunks produce a central nervous system.

Flatworms have a simple brain and two or more nerve trunks, with ladderlike transverse nerves. Annelids and arthropods have a prominent brain and a ventral nerve cord, which may have ganglia within each segment of the body. Sessile molluscs have little cephalization and only simple sense organs. Cephalopods, such as the octopus, have large brains, image-forming eyes, and giant axons.

Vertebrates have a dorsal central nervous system, protected by the vertebrae and skull. Three regions, the *rhombencephalon* or *hindbrain*, the *mesencephalon* or *midbrain*, and the *prosencephalon* or *forebrain*, are present in all vertebrates, although these regions may be subdivided to provide a greater capacity for integration of complex activities.

Three evolutionary trends in the vertebrate brain include an increase in relative size of the brain with phylogenetic position, in compartmentalization of function, and in complexity of the forebrain. The folding or convolution of the cerebral cortex of the forebrain increases the surface area of this layer.

The Vertebrate Nervous System

The peripheral nervous system consists of the *sensory* or *afferent nervous system*, which brings information to the *central nervous system (CNS)*, and the *motor* or *efferent nervous system*, which carries signals away from the CNS to effector organs. The human peripheral nervous system contains 12 pairs of cranial nerves and 31 pairs of spinal nerves. Most spinal nerves branch and processes from adjacent nerves mingle to form collections of fibers, called *plexuses*, that travel to target organs.

The sensory nervous system brings in messages from both the external and internal environments. The motor nervous system has two divisions, one of which governs responses to the external environment, while the other coordinates functions of the internal organs. The *somatic nervous system* carries signals to skeletal muscles, the movement of which is largely under conscious control. Many skeletal muscle movements, however, are determined by *reflexes*, which are automatic, subconscious reactions to stimuli mediated by the spinal cord or lower brain. The *autonomic (visceral) nervous system* controls smooth and cardiac muscles and various organs.

The autonomic nervous system is subdivided into the *sympathetic* and the *parasympathetic* nervous systems. In general, the parasympathetic division slows the heart rate, stimulates digestion, and enhances activities that gain and conserve energy. The sympathetic division accelerates the heart rate and increases the metabolic rate, preparing an organism for action.

The central nervous system, which bridges the sensory and motor components of the peripheral nervous system, consists of the *spinal cord* and the *brain*. These bilaterally symmetrical organs are covered by protective layers of connective tissue called the *meninges*. Axons and dendrites are located in tracts; the *white matter* is named for the white color of their myelin sheaths. The cell bodies of the neurons make up the *gray matter*, located on the outer layer in the brain but on the inside of the white matter in the spinal cord.

Spaces called *ventricles* in the brain, which are continuous with the narrow *central canal* of the spinal cord, are filled with cerebrospinal fluid. This fluid, which is formed by filtration of the blood, cushions the brain and carries out circulatory functions.

The spinal cord integrates simple responses to some stimuli (in the form of reflexes) and carries information to and from the brain. The knee-jerk reflex involves a stretch receptor, a sensory neuron, and a motor neuron. Most reflexes have interneurons between sensory and motor neurons.

The human brain develops from the three primary vesicles (the hindbrain, midbrain, and forebrain), which differentiate into specialized regions. The *brainstem*, which includes the hind- and midbrain, extends from the spinal cord to the middle of the brain.

The hindbrain includes the *medulla oblongata*, which contains control centers for homeostatic functions including respiration, heart and blood–vessel actions, and digestion; the *pons*, which functions with the medulla in some of these activities and in conducting information between the rest of the brain and the spinal cord; and the *cerebellum*, which functions in

coordination of movement. The pyramidal tracts, carrying motor neurons from the mid- and forebrain, cross in the medulla, so that the right side of the brain controls much of the movement of the left side of the body and vice versa. The cerebellum integrates information from the auditory and visual systems with input from motor pathways from the cerebrum to provide unconscious coordination of movements and balance.

The upper portion of the brainstem, or *midbrain*, receives and integrates sensory information and sends this information to specific regions of the forebrain. Fibers involved in hearing pass through or terminate in the inferior colliculi. The superior colliculi are important visual centers. The *reticular formation*, another major group of nuclei in the midbrain, regulates states of arousal.

The *forebrain* is the region of the most intricate neural processing. The lower *diencephalon* contains the thalamus and hypothalamus, and the *telencephalon* contains the cerebrum.

The thalamus is a major projection area; most of the input to the cerebral cortex comes from neurons with cell bodies in the thalamus. The thalamus also contains nuclei of the reticular formation, which filters the information sent to the cerebral cortex.

The hypothalamus is the major site for regulation of homeostasis. It produces the posterior pituitary hormones and the releasing factors that control much of the hormone secretion of the anterior pituitary. The hypothalamus contains the regulating centers for many autonomic functions. The region also regulates sexual response and mating behaviors, the alarm response, and pleasure.

The *basal ganglia*, a cluster of nuclei below the cortex, are important in relaying motor impulses and coordinating motor responses. Parkinson's disease is associated with degeneration of cells entering the basal ganglia. Other basal ganglia are involved in the limbic system, which is associated with emotions.

The *cerebral cortex* has changed most during vertebrate evolution. The human cortex is divided into five lobes. The two hemispheres of the cortex are connected by the *corpus callosum*, a thick band of fibers. The cortex contains both primary sensory and motor areas, which directly process information, and association areas, which integrate information from several sources.

The proportion of the primary sensory and motor areas devoted to controlling each part of the body is correlated with the importance of that area. A map called a homunculus shows the location and size of brain area associated with each body part.

Arousal is a state in which the animal is aware of the external world, whereas sleep is a state in which the animal is not conscious of external stimuli. Different patterns in the electrical activity of the brain, which can be recorded in an electroencephalograph, are produced by wakefulness and sleep. Slow synchronous alpha waves are produced by a person lying quietly with eyes closed. Faster beta waves are associated with having opened eyes or with thinking about a complex problem. Quite slow and highly synchronized delta waves occur during S sleep. During D sleep, a desynchronized EEG is produced, and rapid eye movement (REM) behind the lids is observed. Dreaming occurs during D, or REM, sleep.

Almost all fibers reaching the cerebral cortex pass through the reticular formation, which filters the sensory information reaching the cortex. Arousal is related to the amount of input the cortex receives. Sleep-producing centers are located in the pons and medulla, and serotonin may be the neurotransmitter involved in these areas.

Human emotions have been tied to interactions between the cerebral cortex and a group of nuclei in the lower forebrain called the *limbic system*. Destruction of this portion of the brain may result in docility or in the inability to control emotions appropriately.

The association areas of the cerebral cortex are not bilaterally symmetrical. The left hemisphere controls speech, language, and calculation, whereas the right hemisphere controls artistic concepts and spatial perception. Much of the information on this *lateralization* of the brain comes from Sperry's work with "split-brain" patients—people whose corpus callosums have been severed.

The number and size of synapses in the brain increase with learning. It has been proposed that chemical changes in the brain are associated with learning and memory. Short-term memory involves fairly immediate sensory perceptions or ideas. In long-term memory, these perceptions or ideas can be recalled after the passage of time.

Specific areas of the brain appear to be responsible for storage of specific types of memory. For example, Wernicke's area stores information required for speech content, whereas Broca's area involves information required for speech production. Different types of *aphasia*, the inability to speak coherently, occur if one or the other of these regions is damaged.

Neural biologists study simpler organisms to try to understand the mechanisms involved in memory. The habituation of the sea slug *Aplysia* to mild touch stimuli is accompanied by long inactivation times for the voltage-sensitive channels that trigger neurotransmitter release. During sensitization to harmful stimuli, ion channels open more readily and a postsynaptic action potential is more likely to occur. If membrane permeability changes are permanent, memory may be established. Membrane changes of the axon, and release of proteins that stimulate growth of synapses, have also been suggested as mechanisms for memory formation.

STRUCTURE YOUR KNOWLEDGE

1. Develop a flow chart, diagram, or description of the sequence of events in the creation and propagation of an action potential, and in the transmission of this potential across a chemical synapse.

2. The vertebrate nervous system can be separated into a sequence of divisions and subdivisions.

Create a concept map that shows these subsets and their functions.

3. Fill in the following table by briefly describing the functions of the parts of the human brain and identifying the section (hind-, mid-, or forebrain) in which they are found.

Part of brain	Section	Function
Medulla		
Pons		
Cerebellum		
Inferior, superior colliculi		
Reticular formation		
Thalamus		
Hypothalamus		
Cerebral cortex		

TEST YOUR KNOWLEDGE

MULTIPLE CHOICE: *Choose the one best answer.*

1. Which of the following is not a difference between the nervous system and the endocrine system?
   a. Nervous communication is more rapid.
   b. Endocrine system uses chemical communication; nervous system uses electrical communication.
   c. Nervous system messages are delivered directly to target cells or organs.
   d. The structural complexity of the nervous system allows for integration of more information and responses.

2. Interneurons
   a. may connect sensory and motor neurons.
   b. are confined to the peripheral nervous system.
   c. do not have cell bodies.
   d. are carried in mixed nerves.

3. The myelin sheath
   a. is formed by Schwann cells wrapping themselves around axons.
   b. forms a blood-brain barrier.
   c. creates a greater axon diameter so that impulse conduction is faster.
   d. all of the above are correct.

4. Which of the following is not true of the resting potential of a neuron?
   a. The inside of the cell is more negative than the outside is.
   b. There are concentration gradients with more sodium outside the cell and a higher potassium concentration inside the cell.
   c. At –70 mV, it is closer to the equilibrium potential for potassium than for sodium.
   d. It is formed by the sodium–potassium pump and the opening of voltage-sensitive channels.

5. Voltage-sensitive potassium channels
   a. open more slowly than sodium channels do.
   b. help to restore the resting potential.
   c. add to the refractory period by hyperpolarizing the membrane.
   d. do all of the above.

6. A depolarization of the membrane
   a. means that the action potential of the membrane approaches zero.
   b. opens voltage-sensitive channels and permits the rapid outflow of sodium ions.

   c. must reach a threshold value before an action potential is generated.
   d. operates on the all-or-none principle.

7. Which of the following is not true of chemical synapses?
   a. Synaptic knobs at the ends of branching dendrites contain synaptic vesicles, which enclose the neurotransmitter.
   b. The influx of calcium when an action potential reaches the presynaptic membrane causes synaptic vesicles to release their neurotransmitter into the cleft.
   c. The binding of neurotransmitter to receptors on the postsynaptic membrane changes the membrane's permeability to certain ions.
   d. An excitatory postsynaptic potential forms when sodium channels open and the membrane potential moves closer to an action potential threshhold.

8. In spatial summation,
   a. the sum of simultaneously arriving neurotransmitter from different presynaptic nerve cells determines whether the postsynaptic cell fires.
   b. two action potentials arrive from the same presynaptic cell.
   c. several IPSPs arrive concurrently, bringing the presynaptic cell closer to its threshold.
   d. different neurotransmitters are used.

9. Enzymes rapidly degrade the neurotransmitter because
   a. ion channels remain open as long as neurotransmitter is bound to degrading enzymes.
   b. summation would result in too high a neurotransmitter concentration.
   c. the components of the neurotransmitter must be recycled.
   d. the resting condition must be reformed so that a new signal can be received.

10. Which of the following would not function as a neurotransmitter?
    a. acetylcholine
    b. neuropeptides
    c. biogenic amines
    d. steroids

11. Groups of neurons that interact and carry information along pathways are called
    a. ganglia.
    b. nuclei, if they occur in the brain.
    c. circuits.
    d. nerve nets.

12. Which of the following animals is mismatched with its nervous system?
    a. sea star — modified nerve net, central nerve ring with radial nerves
    b. hydra (cnidarian) — ring of ganglia, paired ventral nerve cords
    c. annelid worm — brain, ventral nerve cord with segmental ganglia
    d. vertebrates — dorsal central nervous system of brain and spinal cord

13. Which of the following is not true of the autonomic nervous system?
    a. It is a subdivision of the somatic nervous system.
    b. It consists of the sympathetic and parasympathetic divisions.
    c. It is part of the peripheral nervous system.
    d. It controls smooth and cardiac muscles.

14. Gray matter
    a. is made up of three protective layers called meninges.
    b. is located on the outside in the spinal cord.
    c. makes up the brain.
    d. is made up of the cell bodies of neurons.

15. Cerebrospinal fluid
    a. cushions the brain.
    b. supplies the brain with nutrients and oxygen, and removes waste.
    c. is a filtrate of blood.
    d. all of the above are correct.

16. Which of the following structures is incorrectly paired with its function?
    a. pons — conducts information between spinal cord and brain
    b. cerebellum — contains pyramidal tracts that cross motor neurons from one side of the brain to the other side of the body
    c. thalamus — major projection area, screen and relay incoming impulses
    d. corpus callosum — band of fibers connecting left and right hemispheres

17. During REM sleep,
    a. a desynchronized EEG is produced.
    b. dreaming occurs.
    c. rapid eye movement occurs.
    d. all of the above are correct.

18. The reticular formation
    a. filters sensory information and produces arousal.
    b. relays motor impulses and coordinates motor responses.
    c. is responsible for sleep and uses serotonin as its neurotransmitter.
    d. is severed in "split-brain" patients.

19. The limbic system
    a. controls speech patterns and prevents aphasia.
    b. is responsible for lateralization of the brain.
    c. is a group of nuclei in the lower forebrain associated with emotions.
    d. regulates sexual response, mating behaviors, and pleasure.

20. Memory
    a. is controlled by the left hemisphere.
    b. may involve changes in membrane permeability in axons or dendrites.
    c. is controlled by Wernicke's area.
    d. is a function of age.

CHAPTER 45

RECEPTORS AND EFFECTORS

FRAMEWORK

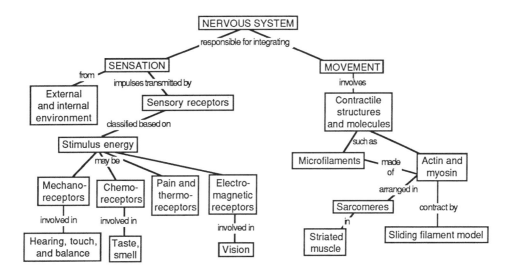

CHAPTER SUMMARY

Sensory Receptors

Information is transmitted as action potentials within the nervous system. Impulses, or *sensations*, are routed to different parts of the brain, which interpret them into *perceptions*. The destination of the impulses determines what is perceived.

Receptors collect and transmit information from environmental stimuli. *Reception* is the absorption of the energy of a stimulus. The latter's conversion into the electrochemical energy of changes in membrane potentials and action potentials is called *transduction*. The stimulus energy may need to undergo *amplification*, either by accessory structures of sense organs or as part of the transduction process.

Transmission of the changes in the cell membrane's electrical potential to the nervous system may take two forms. When the depolarization of a *primary receptor* has reached a threshold level, called a *generator potential*, an action potential is formed and a fixed amount of neurotransmitter is released at a synapse with a sensory neuron. *Secondary receptors* respond to a stimulus with a graded change in membrane potential called a *receptor potential*, which produces a graded change in neurotransmitter release. In either case, the frequency of spontaneous firing of sensory cells is correlated with both the presence and the intensity of the stimulus.

The *integration* of information from a stimulus begins on the cellular level with summation of signals, *adaptation* of the receptor cell to continued stimulation, and varying *sensitivity* of receptors to different conditions. *Phasic receptors* are specialized for detecting change and adapt very rapidly, whereas *tonic receptors* adapt much more slowly and thus continue to send signals to the central nervous system over a longer period of stimulation.

Sensory receptors may be *exteroreceptors*, which receive information from the outside environment, or *interoreceptors*, which provide information from the inside of the body. Receptors can be categorized on the basis of the type of energy stimulus to which they respond.

Mechanoreceptors detect physical deformations in the environment. Bending or stretching of the cell membrane increases its permeability to sodium and potassium ions and generates an action potential. The human *pacinian corpuscles* occur in deep skin layers and respond to strong pressure, whereas the *Meissner's corpuscles* and *Merkel's discs* are closer to the surface and detect light touch. The position of body parts is monitored by *muscle spindles*, or stretch receptors, which respond to stretching of the muscles. Specialized cilia in parallel rows project upward from *hair cells*. When motion produces bending in the cilia, the hair cell membrane stretches and ion permeabilities change.

Chemoreceptors may be general receptors that monitor the total solute concentration in a solution, or specific receptors that respond to individual kinds of molecules. Stimulus molecules bind to membrane sites on the receptor cell and initiate changes in membrane permeability. *Gustatory* and *olfactory* receptors have intermediate specificity, responding to groups of related chemicals.

Thermoreceptors respond to heat or cold and help to regulate body temperature. Ruffini's end organs are naked nerve endings sensitive to heat, and end-bulbs of Kranse, which respond to cold, are dendrites encapsulated by connective tissue.

Naked dendrites called *nociceptors* detect pain. Different groups of receptors respond to excess heat, pressure, or chemicals released from damaged tissues. Histamines and acids trigger pain receptors, and prostaglandins increase perception of pain by sensitizing receptors.

Photoreceptors detect the electromagnetic radiation of visible light, and are often organized into eyes. Some animals have receptors that detect ultraviolet light, infrared rays, and electric currents. Evidence also shows that some homing or migrating animals use magnetic field lines of Earth to orient.

Vision

Receptors containing light-absorbing pigments are used by most invertebrates to detect light. The *eye cup* of planaria detects light intensity and direction. Image-forming eyes include the *compound eye* of insects and crustaceans, made up of thousands of light detectors called *ommatidia*, and the *camera eye* of cephalopods, consisting of a single lens that focuses light onto the retina, a layer of photosensitive cells.

The vertebrate eye is also a camera-type eye. The eyeball consists of the tough, outer-connective tissue layer called the *sclera*, and the thin, pigmented inner layer called the *choroid*. At the front of the eye, the sclera becomes the transparent *cornea*, and the choroid forms the *iris*, which regulates light entering the *pupil*, the hole in the center of the iris. The retina, layered on the choroid, contains the photoreceptor cells. The optic nerve attaches to the eye at the optic disc.

The transparent *lens* focuses an image onto the retina. Most vertebrates focus by *accommodation*, in which muscles of the *ciliary body* change the shape of the lens. The ciliary body produces the *aqueous humor* that fills the anterior eye cavity. *Glaucoma* is a condition in which pressure in the eye is increased by an accumulation of aqueous humor. The jellylike *vitreous humor* fills the posterior cavity of the eye.

Vision involves the formation of a light image on the retina, the transduction of the image into signals of action potentials, and the interpretation of those signals by the brain.

Rods and *cones*, the photoreceptor cells, contain the photopigment *rhodopsin*, which isomerizes when struck by light. This shape change decreases the membrane's permeability to sodium ions, hyperpolarizes the cells, and reduces the release of neurotransmitter.

Rods, concentrated toward the edge of the retina, respond to low levels of light. The center of the visual field, the *fovea*, is filled with cones, which are responsible for color vision. Cones are absent in species that do not have color vision.

Rods and cones are connected to *bipolar cells* in the retina. Some bipolar cells depolarize and others hyperpolarize when stimulated by the receptor cells. Bipolar cells stimulate action potentials in *ganglion cells*.

In the mammalian visual system, signals from rods and cones may follow the vertical pathway directly from the receptor cells to bipolar cells to ganglion cells. Lateral integration of visual signals involves both *horizontal cells*, which carry signals from one receptor cell to others and to several bipolar cells, and *amacrine cells*, which relay information from one bipolar cell to several ganglion cells. The *lateral inhibition* of nonilluminated receptors and bipolar cells by the horizontal cells, and of ganglion cells by amacrine cells, enhances the contrast between a spot of light and its surroundings.

The receptive fields of the ganglion cells (the rods and cones supplying information to the cell) are arranged in two different patterns, one of which responds to light spots surrounded by darkness (on-center receptive fields), and the other to spots of darkness surrounded by light (offcenter fields).

The ganglion cells are the sensory neurons that form the optic nerve. The left and right optic nerves meet at the *optic chiasma* at the base of the cerebral cortex. Nerves from the left sides of both eyes go to the left side of the brain; nerves from the right sides go to the right side of the brain. Most axons of the ganglion cells go to the *lateral geniculate* nuclei of the thalamus, where cells lead to the *primary visual cortex* in the occipital lobe. Other interneurons carry information to other visual centers in the cortex.

The receptive fields in the lateral geniculate nuclei have oncenter or offcenter organizations. The receptive fields in the cortex respond to straight edges and movement. Somehow the coded spots, lines, and movements are integrated into our perception and recognition of objects.

Hearing and Balance

In mammals and most terrestrial vertebrates, the mechanoreceptors for hearing and balance are located within the ear. The ear consists of three regions. The external *pinna* and the *auditory canal* make up the outer ear. The *auditory (eustachian) tube*, connecting the pharynx and the middle ear, equalizes pressure within the ear. The *tympanic membrane* (eardrum) of the *middle ear* transmits sound waves to three small bones—the *malleus* (hammer), *incus* (anvil), and *stapes* (stirrup), which conduct the waves to the *inner ear* by way of a membrane beneath the stapes, called the *oval window*. The coiled, fluid-filled *cochlea* of the inner ear has two large chambers separated by a smaller chamber into which hair cells from the *organ of Corti* project. The ends of the hair cells are embedded in the tectorial membrane.

Sound waves—which were transmitted and amplified by the tympanic membrane, the three bones of the middle ear, the oval window, and pressure waves in the fluid—are transduced into action potentials in the cochlea. Pressure waves, traveling from the vestibular canal through the tympanic canal and dissipating when they strike the *round window*, vibrate the basilar membrane on which the organ of Corti lies. The hair cells alternately press into the tectorial membrane and draw back, bending the hairs and stretching the cell membranes, which makes the cell membranes more permeable to sodium. The resulting depolarization increases neurotransmitter release and the frequency of action potentials carried by the sensory neuron through the auditory nerve to the brain.

Volume is a result of the *amplitude*, or height, of the sound wave; a stronger wave bends the hair cells more and results in an increase in action potentials. *Pitch* is related to the *frequency* of sound waves. Different parts of the basilar membrane vibrate at different frequencies and transmit their impulses to specific regions of the cerebral cortex, where the sensation is perceived as a particular pitch.

Within the inner ear, two chambers, the *utricle* and *saccule*, and three *semicircular canals* make up the vestibular apparatus responsible for balance and equilibrium in humans and most mammals. The cilia of the hair cells in the utricle and saccule project into a gelatinous material containing calcium carbonate particles. Changes in head position change the pull on the hair cells, increasing or decreasing their output of action potentials.

A cluster of hair cells called an *ampulla*, with cilia projecting into a flat, gelatinous sheet, called the *cupula*, is located at the base of each semicircular canal. The inertia of the endolymph, or fluid, in the canals when the head rotates causes the hair cells to bend and increase the frequency of action potentials.

Mechanoreceptors with sensory hairs embedded in a gelatinous cap are contained in the lateral lines of fishes. Water moving through a tube past these mechanoreceptors stimulates the hair cells and enables the fish to perceive its own movement, water currents, and pressure waves generated by other moving objects.

Invertebrates have mechanoreceptors, called *statocysts*, which often consist of a layer of hair cells around a chamber containing statoliths, dense granules or grains of sand. The stimulation of the hair cells under the statocysts provides positional information to the animal.

Taste and Smell

The human chemical senses of gustation and olfaction are produced when a molecule binds to a receptor protein in a receptor cell membrane, and triggers a membrane depolarization and the release of neurotransmitter. Taste sensory systems involve separate receptor cells and sensory neurons, whereas olfactory systems have only sensory neurons.

Taste buds, which contain groups of receptor cells, are scattered on the tongue and in the mouth. The four primary taste sensations—sweet, sour, salt, and bitter—are detected in distinct regions of the tongue.

Olfactory receptor cells line the upper part of the nasal cavity and send their axons to the olfactory bulb of the brain. Cilia from the receptor cells extend into the mucous layer of the nasal cavity. Chemicals diffuse into this layer and bind to specific receptor molecules.

Movement

In eukaryotic cells, *microtubules* are responsible for movements such as separation of chromosomes and beating of cilia and flagella. *Microfilaments* are involved in amoeboid movement and muscle contraction.

Amoeboid movement involves the streaming of the fluid central cytoplasm, called *endoplasm*, into the stiff *ectoplasm* in the *pseudopodia*, extending movement of the cell into these projections. The contraction of the ectoplasm may push or pull the endoplasm forward and involves interactions of actin filaments and myosin.

Vertebrate *skeletal muscle* consists of a bundle of multinucleated muscle cells, called fibers, running the length of the muscle. Each fiber is a bundle of *myofibrils*, each composed of two kinds of *myofilaments*: *thin filaments*, consisting of two strands of actin coiled with a strand of regulatory protein, and *thick filaments*, made of myosin molecules.

The regular arrangement of myofilaments produces repeating light and dark bands; thus skeletal muscle is also called *striated muscle*. A *sarcomere* is delineated by Z *lines*, to which thin filaments attach and project toward the center of the sarcomere. Thick filaments lie free in the center, overlapping the thin filaments in an *A band*, which is broken by an *H zone* in the center where the thin filaments do not overlap the thick filaments when the muscle is at rest. At the edges of the sarcomere, where the thick filaments do not extend, are *I bands* of only thin filaments.

According to the *sliding-filament model* of muscle contraction, the thick filaments ratchet the thin filaments toward the center of the sarcomere, pulling the Z lines together, shortening the I bands, and eliminating the H zone. The synchronous shortening of sarcomeres in the myofibrils of the fibers results in the contraction of the muscle.

When the muscle is at rest, the myosin-binding sites of the actin molecules are blocked by the *tropomyosin* strand in the thin filament and by a set of regulatory proteins, called the *troponin complex*, at each binding site. When troponin binds calcium ions, the tropomyosin–troponin complex changes shape and the actin filament is exposed.

The heads of myosin molecules spontaneously bind to the exposed active sites on the actin filament, forming *cross-bridges* between the thick and thin filaments, and then bend, pulling the thin filaments toward the center of the sarcomere. The myosin molecule hydrolyzes an ATP molecule, releasing energy to break the cross-bridge, and straightens out. The "attach, bend, detach, straighten" cycle continues as long as the binding sites are exposed and ATP is available.

Calcium ions are actively transported into the *sarcoplasmic reticulum*. When an action potential of a motor neuron releases acetylcholine into the neuromuscular junction (and the resulting EPSP is strong enough), an action potential spreads across the cell membrane and into *transverse (T) tubules*, extensions of the membrane that project into the muscle cell. The action potential carried by the T tubules depolarizes the sarcoplasmic reticulum membrane, and calcium ions are released into the cell. Calcium binding with troponin exposes the actin sites and contraction begins. The sarcoplasmic reticulum pumps calcium back out of the cytoplasm when the action potential passes, and the tropomyosin–troponin complex again blocks the actin sites.

A *twitch*, or contraction of a single muscle fiber, is an all-or-none action. Graded muscular contractions result from the innervation of additional fibers. The strength of muscle contraction can be increased by the stimulation of additional motor neurons, called multiple-motor-unit summation, or by waveform summation, in which action potentials occur so rapidly that the muscle does not relax between them, producing a state of increased and sustained contraction called *tetanus*.

Slow muscle fibers have less sarcoplasmic reticulum, and thus calcium remains in the cytoplasm longer and twitches last longer. Slow fibers are common in muscles that must sustain long contractions. They have many mitochondria, a good blood supply, and *myoglobin*, which extracts and stores oxygen from the blood. Fast muscle fibers are used for rapid and powerful contractions.

Most of the energy for muscle contraction comes from phosphagens, such as creatine phosphate in vertebrates, which transfer high-energy phosphate bonds to ADP to replenish ATP supplies.

Muscles pull against and move the bones of the skeleton in antagonistic pairs.

Vertebrate *cardiac muscle* cells are branched and connected by *intercalated discs*, across which action potentials spread to all the muscle cells of the heart. Cardiac muscle cells have high sodium permeability; their resting potentials are more positive than an action potential threshold; and they can generate action potentials without nervous input.

The lack of an orderly arrangement of actin and myosin filaments in *smooth muscle* accounts for its nonstriated appearance. Smooth muscle has smaller and slower contractions.

Invertebrates have muscles similar to the skeletal and smooth muscles of vertebrates. The electric organs of many fishes are modified muscle cells that depolarize and generate an electric current.

An animal's behavior is a function of the interpretation and integration of inputs from sensory receptors and organs by a nervous system, which then signals muscles to move and respond.

STRUCTURE YOUR KNOWLEDGE

1. Label the parts of the eye.

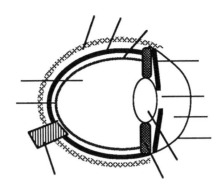

Trace the path of a light stimulus from where it enters the eye to its transmission as an impulse through the optic nerve.

2. Arrange the following structures in an order that enables you to trace the passage of sound waves from the external environment to the brain.

round window	cochlea	vestibular and tympanic canal
oval window	auditory canal	tectorial membrane
organ of Corti	basilar membrane	tympanic membrane
auditory nerve	cerebral cortex	malleus, incus, stapes

3. Trace the sequence of events in muscle contraction from an action potential in a motor neuron to the relaxation of the muscle.

TEST YOUR KNOWLEDGE

MATCHING: *Match the term with its description.*

1. _____ oxygen-storing compound in muscles

2. _____ receptor on muscle, monitors position

3. _____ conversion of stimulus energy into electrochemical energy

4. _____ deep pressure receptors in humans

5. _____ naked dendrites that detect pain

6. _____ cluster of hair cells in semicircular canals

7. _____ invertebrate mechanoreceptors for position; chamber of hair cells

8. _____ energy-storing compound in muscles

9. _____ focusing by changing shape of lens

10. _____ numerous light detectors making up compound eye

11. _____ decrease in sensitivity of receptor during constant stimulation

12. _____ fibers in muscle cells

A. accommodation
B. adaptation
C. ampulla
D. creatine phosphate
E. cupula
F. ganglion cells
G. muscle spindle
H. myofibrils
I. myoglobin
J. nociceptors
K. ommatidia
L. pacinian corpuscles
M. statocyst
N. transduction
O. transmission
P. tropomyosin

MULTIPLE CHOICE: *Choose the one best answer.*

1. Which of the following receptors is incorrectly paired with its category?
   a. hair cell — mechanoreceptor
   b. Ruffini's end organ — pain receptor
   c. gustatory receptor — chemoreceptor
   d. rod — photoreceptor

2. Prostaglandins
   a. increase perception of pain by sensitizing pain receptors.
   b. act like histamines to trigger pain receptors.
   c. sensitize phasic receptors to changes in pain levels.
   d. are stimulated by aspirin and decrease perception of pain.

3. Which of the following statements is not true?
   a. The iris regulates the amount of light entering the pupil.
   b. The transparent cornea focuses an image on the retina.
   c. The ciliary body changes the shape of the lens.
   d. The thin aqueous humor fills the anterior eye cavity.

4. Which of the following is incorrectly paired with its function?
   a. cones—respond to different light wavelengths, producing color vision
   b. horizontal cells—carry signals between receptor cells and bipolar cells
   c. bipolar cells—stimulate ganglion cells
   d. ganglion cells—produce lateral inhibition of receptor and bipolar cells

5. The axons of the ganglion cells
   a. lead to the lateral geniculate nuclei of the thalamus.
   b. make up the optic nerve.
   c. meet with axons from the other eye at the optic chiasma.
   d. do all of the above.

6. The fovea is
   a. the blind spot.
   b. the center of the visual field, containing only cones.
   c. the layer that contains photoreceptor cells.
   d. the optic disc, where the optic nerve attaches to the eye.

7. Oncenter and offcenter receptive fields
   a. respond to bright or dark spots.
   b. are involved in the detection of movement.
   c. are stimulated by straight lines.
   d. all of the above are correct.

8. The utricle and saccule
   a. are involved in lateral integration of visual signals.
   b. are visual centers in the cortex.
   c. are part of the vestibular apparatus responsible for positional information about the head.
   d. are semicircular canals that respond to head rotation.

9. The tympanic membrane
   a. conducts sound waves to the fluid in the vestibular canal.
   b. sets up vibrations in the basilar membrane.
   c. transmits sound waves to the malleus, incus, and stapes.
   d. equalizes pressure between the middle ear and the outside.

10. The transduction of sound waves to action potentials takes place
    a. within the hair cells as they are bent against the tectorial membrane.
    b. within the tectorial membrane as it is stimulated by the hair cells.
    c. as the tectorial membrane becomes more permeable to sodium.
    d. as the basilar membrane vibrates at different frequencies.

11. The binding of a molecule to a receptor protein on a receptor cell membrane
    a. triggers a membrane depolarization and release of neurotransmitter.
    b. is the mechanism of reception and transduction in chemoreceptors.
    c. is involved in both gustatory and olfactory sensation.
    d. all of the above are correct.

12. Amoeboid movement
    a. is caused by the action of microtubules.
    b. involves the streaming of the ectoplasm into the stiffened endoplasm of pseudopods.
    c. involves interactions of actin filaments and myosin.
    d. all of the above are correct.

13. The role of calcium in muscle contraction is
    a. to break the cross-bridges as a co-factor in the ATP hydrolysis.
    b. to bind with troponin, changing its shape so that the actin filament is exposed.
    c. to transmit the action potential across the neuromuscular junction.
    d. to spread the action potential through the T tubules.

14. Tetanus
    a. is the result of puncture wounds.
    b. is the all-or-none contraction of a single muscle fiber shape so that the actin filament is exposed.
    c. is the result of stimulation of additional motor neurons, called multiple-motor-unit summation.
    d. is the result of waveform summation, which produces increased and sustained contraction of a muscle.

15. Cardiac muscle cells
    a. do not have an orderly arrangement of actin and myosin filaments.
    b. are connected by intercalated discs, through which action potentials spread to all muscle cells in the heart.
    c. have less extensive sarcoplasmic reticulums and thus smaller and slower contractions than do skeletal muscle cells.
    d. have a low sodium permeability and resting potentials more positive than an action potential threshold.

FILL IN THE BLANKS: *Use this diagram of a section of a skeletal muscle fiber to fill in the following list.*

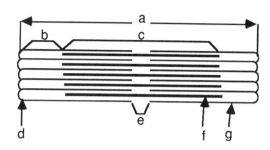

1. _____ Identify a.
2. _____ Identify b.
3. _____ Identify c.
4. _____ Identify d.
5. _____ Identify e.
6. _____ Identify f.
7. _____ Identify g.
8. _____ In which of these structures is troponin found?
9. _____ Which band stays the same length during contraction?
10. _____ Which structure forms cross-bridges?

ANSWER SECTION

CHAPTER 36: INTRODUCTION TO ANIMAL FORM AND FUNCTION

Suggested Answers to Structure Your Knowledge

1.

Tissue	Structural Characteristics	General Functions	Structure and Function of Specific Types
Epithelial	Packed cells, basement membrane, cuboidal, columnar, squamous, simple or stratified	Protection, absorption, secretion, lines body surfaces	Mucous membrane: goblet cells secrete mucus, creating lubricated surface, may have cilia. Glands: exocrine excretes into ducts, endocrine secrete hormones into blood.
Connective	Few cells that secrete extracellular matrix with fibers in liquid, gel or solid ground substance	Connect and support other tissues	Loose connective: loose weave of collagenous,elastic fibers, hold organs in place. Adipose: adipose cells store fat. Fibrous connective:collagenous fibers, form tendons, ligaments.Cartilage: chondrocytes secrete chondrin,rubbery matrix, flexible support. Bone: osteocytes in Haversian system , mineralized matrix, compact and spongy. Blood: erythrocytes, leukocytes, platelets in liquid plasma.
Muscle	Long cells with large numbers of micro-filaments of actin and myosin	Contraction, movement	Skeletal: striated, voluntary movement. Visceral: smooth, spindle-shaped cells, in organs, involuntary. Cardiac: striated, branched cells, intercalated discs, in heart.
Nervous	Neurons with cell body, axons, dendrites	Sense stimuli, conduct impulses	

2. A skeletal system serves to support nonrigid animal cells, provide for movement, and protect soft tissues. An exoskeleton is a protective covering that also provides anchorage for muscles. It is made of nonliving material, and arthropods must shed their cuticle periodically in order to grow. An endoskeleton also provides muscle attachments, and is less bulky than an exoskeleton. Vertebrate skeletons, containing living cells, are capable of growth and remodeling. The skull and rib cage provide protection for vital organs.

3. Protozoans use intracellular digestion within food vacuoles to protect the cell from digestive enzymes. The sac of cnidarians and flatworms allows for processing of larger quantities of food, extracellular digestion, and a large surface area for absorption of nutrients. In a complete digestive tract, the animal's cells are also protected from digestive enzymes, and the one-way tube allows for specialization of regions for various functions—storage, digestion, absorption, and so on.

4. A compact animal must provide gas exchange, nutrients, and waste removal for each cell. Large internal surface areas are specialized for obtaining oxygen, for digestion and absorption, and for removal of waste from the blood. The circulatory system usually provides these resources to all the cells of the body through the interstitial fluid.

Answers to Test Your Knowledge

Multiple Choice:

1.	b	6.	d
2.	a	7.	c
3.	c	8.	b
4.	a	9.	a
5.	b	10.	d

CHAPTER 37: ANIMAL NUTRITION

Suggested Answers To Structure Your Knowledge

1.

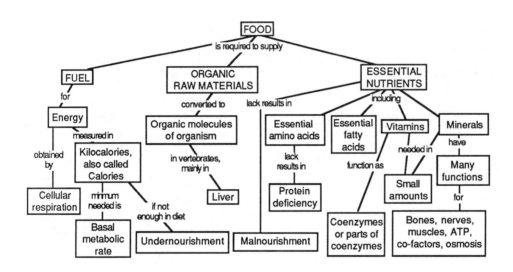

2.

Organ or Section	Processes Occurring	Associated Glands, Organs and Digestive Fluids
Oral cavity	Mechanical breakup of food, mix with saliva, begin breakdown of starch, mucin to lubricate.	Salivary glands secrete salivary amylase, mucin, buffers.
Pharynx	Swallowing of bolus, epiglottis closes windpipe.	
Esophagus	Bolus through esophageal sphincter, peristalsis moves bolus to stomach.	
Stomach	Food tissues broken down by low pH, beginning of protein hydrolysis, churning creates acid chyme, exit by pyloric sphincter.	Epithelial glands secrete gastric juice with pepsin, HCl. Hormone gastrin regulates gastric-juice release.
Small intestine	Complete digestion of starch and proteins, fats emulsified and hydrolyzed, absorption of nutrients across epithelial lining into capillaries and lacteals, large surface area due to villi and microvilli.	Pancreatic enzymes: amylase, trypsin, and other protein-hydrolyzing enzymes; bicarbonate neutralizes. Liver produces bile salts, stored in gall bladder, emulsify fats. Lipase hydrolyzes fats. Hormones: secretin, CCK, enterogastrone control secretions and peristalsis.
Large intestine	Reabsorption of water, storage of feces in rectum. Some microorganisms may produce vitamin K.	

Answers to Test Your Knowledge

Matching:

1. L	**6.** A
2. H	**7.** J
3. C	**8.** F
4. M	**9.** D
5. E	**10.** N

Multiple Choice:

1. d	**6.** d	**11.** b
2. c	**7.** a	**12.** a
3. c	**8.** c	**13.** c
4. a	**9.** a	**14.** c
5. d	**10.** b	**15.** b

CHAPTER 38: CIRCULATION AND GAS EXCHANGE

Suggested Answers To Structure Your Knowledge

1.

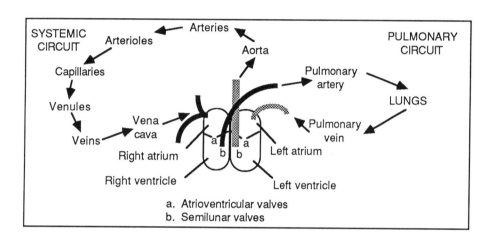

2.

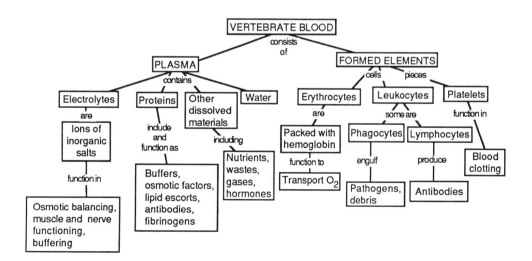

3.

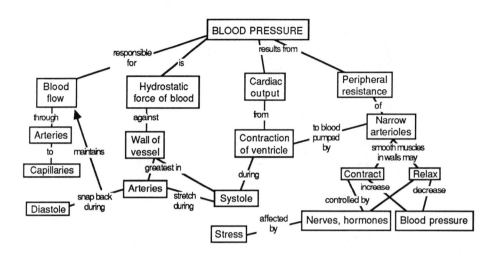

4.

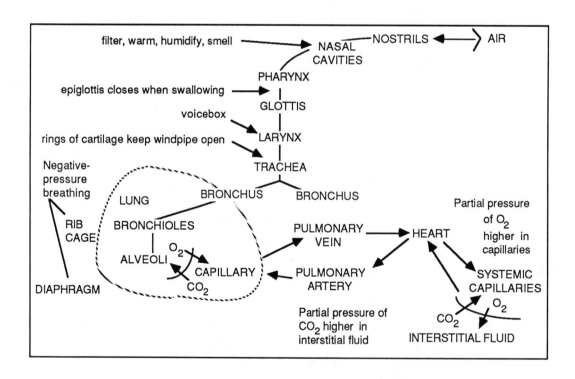

Answers to Test Your Knowledge

Multiple Choice:

1. c	**6.** d	**11.** b	**16.** d
2. c	**7.** a	**12.** b	**17.** b
3. d	**8.** b	**13.** c	**18.** c
4. d	**9.** a	**14.** d	**19.** b
5. a	**10.** d	**15.** a	**20.** b

CHAPTER 39: THE IMMUNE SYSTEM

Suggested Answers to Structure Your Knowledge

1.

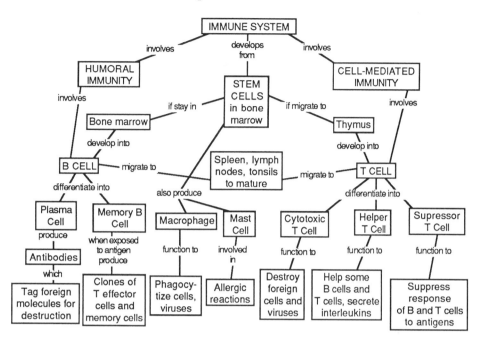

2. An antibody is a protein that typically consists of two identical light polypeptide chains and two identical heavy chains, held together by disulfide bonds in a Y-shaped molecule. The amino-acid sequences in the variable sections of the L and H chains in the arm portion of the Y account for the specificity in binding between antibody and antigen. The constant region of the antibody determines its effector function: five types of constant regions correspond to five classes of antibodies. As an example, phagocytes have receptor sites for the constant regions of IgM and IgG antibodies.

According to the clonal selection theory, T and B cells become fixed early in their development to produce a specific antigen receptor or antibody. The body's ability to respond to a huge variety of antigens depends on a lymphocyte population with a great diversity of receptor specificities. When a T or B cell encounters its antigen, it is selectively activated to proliferate and to produce a clone of lymphocytes, all of which have the same antigenic specificity.

Answers to Test Your Knowledge

1.	d	6.	d
2.	c	7.	a
3.	b	8.	c
4.	a	9.	a
5.	b	10.	b

3.

Molecule	How or Where Produced	Activity
Lysozyme	Enzyme in perspiration, tears, and saliva	Attack cell walls of many bacteria
Histamine	Released by injured cells and by degranulation of mast cells	Cause dilation and leakiness of small blood vessels, create inflammatory response
Antibody	Protein (Ig) made by plasma cells derived from B cells, has V and C regions	Antigen specificity, bind to antigen, mark it for destruction, initiate complement system
Complement system	Set of blood proteins, activated by IgM and IgG and inflammatory response	Produce pore in membrane of foreign cell marked by antibodies, cause cell to lyse
Interleukin	Messenger molecule secreted by helper T cells	Recruit other lymphocytes, accumulate phagocytic cells
Interferon	Production initiatied by virus infection, messenger that diffuses to other cells	Stimulate production of proteins that inhibit cells from making viral proteins, may recruit T cells and macrophages

CHAPTER 40: CONTROLLING THE INTERNAL ENVIRONMENT

Suggested Answers to Structure Your Knowledge

1.

Animal	Osmoregulation	Excretory System	Thermoregulation
Marine invertebrate	Isotonic to seawater, osmoconformer	Diffusion of ammonia through body wall	Ectothermic
Marine bony fish	Drink water to compensate for loss to hypertonic seawater, salt glands in gills pump out salt	Little urine excreted, ammonia diffuses out	Ectothermic, may have heat exchanger
Freshwater fish	Dilute urine to compensate for osmotic gain, may pump salts in through gills	Kidneys produce copius, dilute urine, ammonia lost across epithelium of gills	Ectothermic
Flatworm	Flame-cell system, dilute fluid excreted	Most wastes excreted into gastrovascular cavity	Ectothermic
Earthworm	Nephridia, excrete dilute urine to offset osmosis from habitat	Nephridia	Ectothermic, behavior responses
Insect	Malpighian tubules, salt pumped back and most water reabsorbed	Malpighian tubules, uric acid conserves water	Ectothermic, behavior responses
Reptile	Nephron in kidney, uric acid conserves water	Nephron in kidney, uric acid in most, related to cleidoic egg	Ectothermic, behavior responses
Bird	Nephron in kidney, nasal salt glands in sea birds, uric acid conserves water	Nephron in kidney, uric acid, related to land habitat, cleidoic egg	Endothermic, high metabolism, pant, heat exchanger
Mammal	Nephron in kidney, concentration of urine related to habitat, hormonal, feedback control	Nephron in kidney, urea	Endothermic, control of heat production, vasodilation, sweat, behavior responses

2.

Environment	Osmoregulation	Nitrogenous Waste	Thermoregulation
Seawater	Most are osmoconformers, hypotonic fishes lose water	Ammonia diffuses out easily into water	Stable environment, mostly ectotherms, mammals have blubber, countercurrent
Freshwater	Animals are hypertonic, gain water, excrete large quantities of urine	Ammonia diffuses out easily into water	Fairly stable environment, mostly ectotherms
Terrestrial	Lose water by evaporation, drink, regulate water lost through excretion	Urea less toxic, needs less water to excrete, uric acid most concentrated, cleidoic egg	Variable environment, behavioral adaptations if ectotherms, endotherms can maintain temperature and move rapidly

Answers to Test Your Knowledge

1. b	6. c	11. a
2. d	7. a	12. d
3. c	8. c	13. b
4. a	9. b	14. b
5. d	10. c	15. c

CHAPTER 41: CHEMICAL COORDINATION

Suggested Answers to Structure Your Knowledge

1.

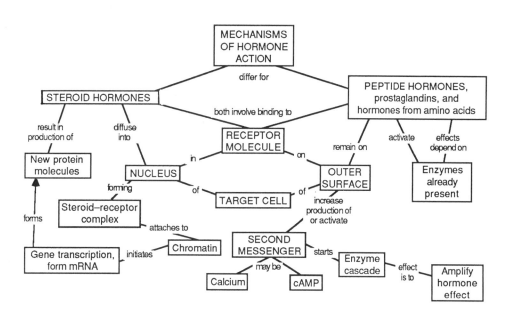

2.

GLAND	HORMONES	MAIN ACTION
HYPOTHALAMUS Neurosecretory cells	Oxytocin	Stimulate uterine muscle contraction, mammary glands
	Antidiuretic hormone (ADH)	Increase water permeability of collecting ducts, promote reabsorption of water from urine
Median eminence	Releasing factors	Carried by portal system to anterior pituitary, stimulate and inhibit release of hormones
	TSH-releasing hormone (TRH)	Stimulate anterior pituitary to secrete TSH
	LH-releasing hormone (LRH)	Stimulate anterior pituitary to secrete FSH and LH
PITUITARY Posterior lobe (neurohypophysis)	Stores and secretes oxytocin and ADH	See actions listed for oxytocin and ADH.
Anterior lobe (adenohypophysis)	Growth hormone (GH)	Promote growth, stimulate other growth factors
	Prolactin (PRL)	Various effects, depending on species—mammary gland growth, osmoregulation, metamorphosis
	Thyroid-stimulating hormone (TSH)	Stimulate thyroid gland to secrete hormones
	Follicle-stimulating. lutenizing hormones, (FSH and LH)	Stimulate activities of gonads, ovulation, and sperm production
	Adrenocorticotropin (ACTH)	Stimulate adrenal cortex to produce and secrete glucocorticoids
	Melanocyte-stimulat- ing hormone (MSH)	Regulate activity of pigment-containing cells in skin
	Endorphins and enkephalins	Effects similar to morphine, reduce pain

3.

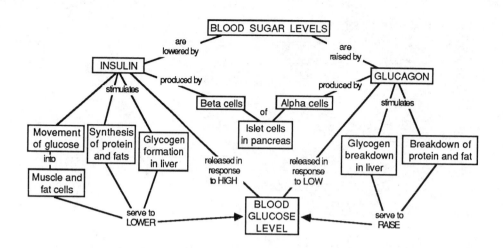

Answers to Test Your Knowledge

Matching:

1.	I f	6.	E b
2.	G a	7.	K j
3.	F d	8.	D c
4.	B h	9.	C c, stored in g
5.	A g	10.	J c, stored in g

Multiple Choice:

1.	a	6.	a	11.	a
2.	d	7.	d	12.	b
3.	d	8.	c	13.	c
4.	b	9.	b	14.	a
5.	d	10.	b	15.	c

CHAPTER 42: ANIMAL REPRODUCTION

Suggested Answers to Structure Your Knowledge

1.

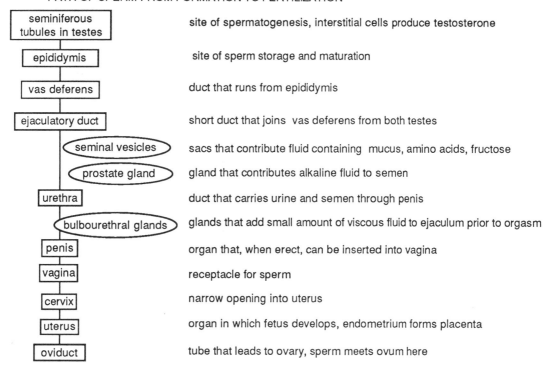

PATH OF SPERM FROM FORMATION TO FERTILIZATION

seminiferous tubules in testes	site of spermatogenesis, interstitial cells produce testosterone
epididymis	site of sperm storage and maturation
vas deferens	duct that runs from epididymis
ejaculatory duct	short duct that joins vas deferens from both testes
seminal vesicles	sacs that contribute fluid containing mucus, amino acids, fructose
prostate gland	gland that contributes alkaline fluid to semen
urethra	duct that carries urine and semen through penis
bulbourethral glands	glands that add small amount of viscous fluid to ejaculum prior to orgasm
penis	organ that, when erect, can be inserted into vagina
vagina	receptacle for sperm
cervix	narrow opening into uterus
uterus	organ in which fetus develops, endometrium forms placenta
oviduct	tube that leads to ovary, sperm meets ovum here

2. A mature ovum within a follicle is released during ovulation and swept into the fallopian tube or oviduct. It travels to the uterus, where it flows out through the vagina with the bleeding associated with menstruation.

3. Birth-control methods (a) prevent fertilization—abstinence, rhythm method, condom, diaphragm, coitus interruptus; (b) prevent implantation of embryo—IUD; (c) prevent release of gametes from gonads—sterilization, chemical contraception such as birth-control pills. The most effective methods are abstinence, sterilization, and chemical contraception. Least effective are the rhythm method and coitus interruptus.

Answers to Test Your Knowledge

Fill in the Blanks:

1. budding
2. clone
3. parthenogenesis
4. hermaphrodite
5. cloaca
6. estrous cycle
7. progesterone
8. urethra
9. vasocongestion
10. puberty

Multiple Choice:

1. d
2. a
3. d
4. a
5. c
6. c
7. b
8. c
9. b
10. a

CHAPTER 43: ANIMAL DEVELOPMENT

Suggested Answers to Structure Your Knowledge

1.

Step	Description
Sperm contacts egg, acrosomal reaction, discharge of hydrolytic enzymes	allows acrosomal process to penetrate jelly coat
Bindin on acrosomal process attaches to receptor on vitelline layer	ensures that egg is fertilized by sperm of same species
Enzymes digest through vitelline layer	brings plasma membranes in contact
Sperm and egg plasma membranes fuse, sperm nucleus can enter cytoplasm of egg	forms two haploid nuclei in same egg, membrane fusion causes ion channels to open
Ion channels open, Na^+ flows into egg, membrane depolarizes	creates fast block to polyspermy through depolarization
Calcium released in egg cytoplasm	causes cortical reaction, triggered by depolarization
Exocytosis by cortical granules, fertilization membrane forms	loosens vitelline layer with enzymes; layer swells by osmotic water uptake, forms slow block to polyspermy
pH rises, rates of cellular respiration and protein synthesis increase	changes pH with calcium; activation of egg, synthesis of proteins from mRNA stockpiled in egg
Sperm and egg nucleus fuse, DNA replication begins	prepares for first cell division, ready for development

2.

Stage	Amphioxus	Amphibian	Bird	Mammal
Cleavage	Holoblastic, animal pole at polar body buds	Holoblastic, but yolk slows divisions, creates uneven cells	Meroblastic, only in cytoplasmic disc on top of yolk	Holoblastic, no polarity to egg, blastomeres equal
Blastula	Hollow ball of cells, one cell thick	Blastocoel displaced to animal hemisphere, wall several cells thick	Cavity between two layers of blastodisc	Embryonic disc, resembles blastodisc of birds and reptiles
Gastrulation	Invagination to form cup-shaped gastrula, mesoderm forms from pouches off archenteron	Invagination by bottle cells, involution of cells at dorsal lip of blastopore, located at gray crescent, 3 germ layers formed	Cells roll in at primitive streak, forming mesoderm, flat, 3-cell layered embryonic disc folds under at edges	Flat embryonic disc, gastrulation through primitive streak, similar to birds

3.

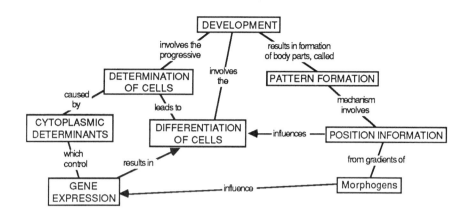

Answers to Test Your Knowledge

Multiple Choice:

1. a	**6.** c	**11.** a	**16.** b
2. b	**7.** c	**12.** c	**17.** b
3. b	**8.** d	**13.** b	**18.** d
4. d	**9.** d	**14.** a	**19.** b
5. d	**10.** c	**15.** d	**20.** a

CHAPTER 44: NERVOUS SYSTEMS

Suggested Answers to Structure Your Knowledge

1.

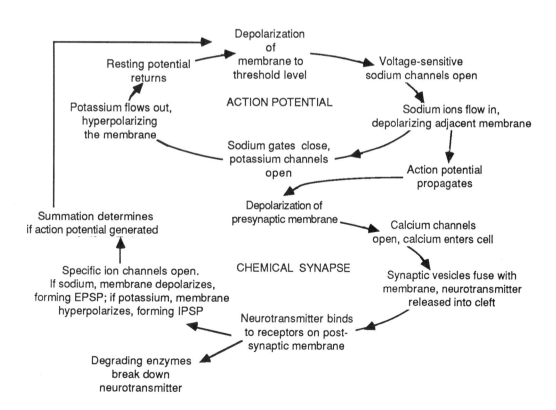

2.

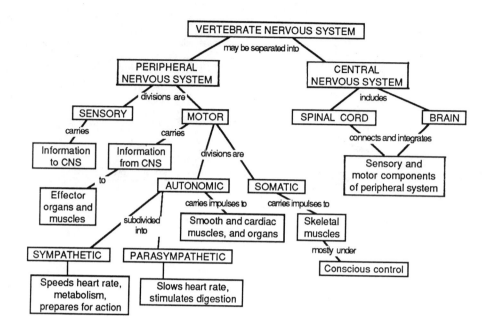

3.

Part of brain	Section	Function
Medulla	Hindbrain	Control centers for homeostatic functions--respiration, heart and blood vessel actions, digestion; region where pyramidal tracts cross
Pons	Hindbrain	Aids in some of functions of medulla; conducts information between brain and spinal cord
Cerebellum	Hindbrain	Controls unconscious coordination of movement and balance
Inferior, superior colliculi	Midbrain	Inferior--involved with hearing; superior--visual center
Reticular formation	Midbrain	Regulates state of arousal; filters all sensory information going to cerebral cortex
Thalamus	Forebrain	Major projection area; relays input to cerebral cortex; contains nuclei of reticular formation
Hypothalamus	Forebrain	Produces hormones of posterior pituitary; produces releasing factors that control anterior pituitary; regulates pleasure, alarm response, sexual response and mating behaviors, autonomic functions
Cerebral cortex	Forebrain	Processes sensory and motor information; integrates information; responsible for thinking, memory, emotions, speech

Answers to Test Your Knowledge

Multiple Choice:

1. b	**6.** c	**11.** c	**16.** b
2. a	**7.** a	**12.** b	**17.** d
3. a	**8.** a	**13.** a	**18.** a
4. d	**9.** d	**14.** d	**19.** c
5. d	**10.** d	**15.** d	**20.** b

CHAPTER 45: RECEPTORS AND EFFECTORS

Suggested Answers To Structure Your Knowledge

1.

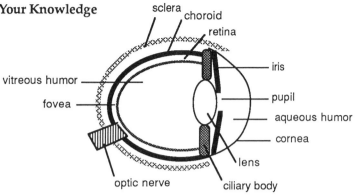

Light enters the eye through the pupil and is focused by the lens onto the retina. When rhodopsin, the photopigment in rods and cones, absorbs light energy, it isomerizes and changes the receptor cell's permeability to sodium. The reduction in the release of neuro- transmitter by rods and cones serves to depolarize or hyperpolarize connected bipolar cells, which in turn stimulate action potentials in ganglion cells—the sensory neurons that form the optic nerve.

2. Sound waves travel down the *auditory canal* to the *tympanic membrane*, from which they are amplified and transmitted to the *malleus, incus,* and *stapes.* Vibration of the stapes against the *oval window* sets up pressure waves in the fluid in the *vestibular* and *tympanic canal* within the *cochlea.* These pressure waves are dissipated when they strike the *round window.* The pressure waves vibrate the *basilar membrane,* on which the *organ of Corti* is located; and the tips of the hair cells, embedded in the *tectorial membrane,* are bent, triggering depolarization, release of neurotransmitter, and initiation of action potentials in the sensory neurons of the *auditory nerve,* which leads to the *cerebral cortex.*

3. A motor neuron releases acteylcholine into the neuromuscular junction, initiating an action potential within the muscle fiber, which spreads into the cell through the T tubules. The action potential depolarizes the sarcoplasmic reticulum membrane and calcium ions are released. The calcium binds with troponin, changes the shape of the tropomyosin–troponin complex, and exposes the myosin-binding sites of the actin molecules. Heads of myosin molecules bind to these sites, bend, and pull the thin filament toward the center of the sarcomere. The myosin molecule hydrolyzes an ATP molecule to break the cross-bridge, straightens out, and forms another cross-bridge, continuing this sequence as long as there is ATP and until calcium is pumped back into the sarcoplasmic reticulum and the tropomyosin–troponin complex again blocks the actin sites.

Answers to Test Your Knowledge

Matching:

1.	I	**5.**	J	**9.**	A
2.	G	**6.**	C	**10.**	K
3.	N	**7.**	O	**11.**	B
4.	L	**8.**	D	**12.**	H

Multiple Choice:

1.	b	**6.**	b	**11.**	d
2.	a	**7.**	a	**12.**	c
3.	b	**8.**	c	**13.**	b
4.	d	**9.**	c	**14.**	d
5.	d	**10.**	a	**15.**	b

Fill in the Blanks:

1.	sarcomere	**6.**	thick filament, myosin
2.	I band	**7.**	thin filament, actin and troponin
3.	A band	**8.**	in g, thin filament
4.	Z line	**9.**	part c, A band
5.	H zone	**10.**	part f, myosin of thick filament, attaches to g

Unit VIII

Ecology

CHAPTER 46

THE PHYSICAL ENVIRONMENT

FRAMEWORK

Ecology deals with the distribution and abundance of organisms and the interrelationships between organisms and the abiotic and biotic factors in their environments. This chapter describes the organizational levels at which ecological questions are asked, the abiotic factors to which organisms have adapted in both an ecological and an evolutionary time frame, and the major world communities, or biomes, in which adaptations to climate and abiotic factors have produced similar and characteristic life forms.

CHAPTER SUMMARY

The Development and Scope of Ecology

Ecology is the study of the interactions of organisms with the *abiotic* and *biotic* factors of their environments. Ecology encompasses the study of the distribution and abundance of organisms, and considers the factors that determine these phenomena. Historically, ecology was a descriptive science; currently, however, an experimental approach, in both the laboratory and the field, is being used to investigate ecological questions.

The levels of ecology are as follows: the individual or *autecology*, which includes the *physiological ecology* of an organism's responses to its physical environment; the *population*, which is concerned with the factors that control the size of groups of individuals of the same species in an area; the *community*, which considers interactions such as predation and competition between different populations in an area; and the *ecosystem*, which includes the abiotic factors as well as the communities that exist in an area, and the flow of energy and chemical cycling.

Interactions between organisms and their environments occur within *ecological time*, whereas the cumulative effects of these interactions may be looked at on the scale of *evolutionary time*. The distribution and abundance of organisms are the result of both past history and present interactions with the environment.

Varying Environments of the Biosphere

The portion of Earth that is inhabited by life is called the *biosphere*. The patchiness of the biosphere is a result of abiotic factors, such as temperature, rainfall, and light, which create a variety of habitats. Natural selection has resulted in diverse adaptations to the wide spectrum of abiotic factors. The ranges of environmental variables that organisms can tolerate is a major factor determining the distribution of species.

Temperature is an important environmental factor because of its effect on metabolism. Most ectotherms cannot maintain body temperatures that vary appreciably from the environmental temperature. Even endotherms, who use metabolic processes to maintain internal temperatures, function best within certain temperature ranges.

The availability of *water* is reflected in the adaptations organisms have to regulate their osmolarity, to obtain water, and to reduce water loss. These adaptations then affect the distribution of organisms.

Light energy drives almost all ecosystems. Within forests, the availability of light influences the distribution of species. The intensity and quality of light are important abiotic factors in aquatic environments. Most photosynthesis occurs relatively near the surface of the water; deeper-dwelling photosynthetic organisms have accessory pigments that can absorb the penetrating wavelengths of light. Many plants and animals are sensitive to photoperiod, which serves as an indicator of seasonal changes.

Variations in the physical structure, pH, and mineral composition of *soil* affect the distribution of plants, which, in turn, affects the distribution of animals. *Wind* increases the rates of heat loss, evaporation in animals, and transpiration in plants.

Fire and other disturbances can drastically affect biological communities. Many plants have evolved adaptations to periodic fires.

Organisms must cope with the entire set of environmental variables within their habitats. The *principle of allocation* is used to analyze the amount of energy that an organism uses for its various processes, such as reproduction, eating, growing, escaping from predators, and dealing with environmental fluctuations. The distribution of organisms is related to different systems of energy allocation. In stable environments, organisms can allot more energy for growth and reproduction, but they are then restricted to stable environments. Organisms that put more of their energy into dealing with environmental changes can successfully live in a wider range of habitats.

Responses of Organisms to Environmental Change

Individuals may use behavioral, physiological, or morphological mechanisms to respond to changes in their environments. Behavioral responses may include a relocation to a more favorable environment, either close by or at a great distance, as in seasonal migrations. Cooperative social behavior, such as huddling, can help organisms respond to unfavorable conditions.

Physiological responses to environmental change are usually slower than behavioral responses are. *Regulators* are organisms whose physiological responses allow them to maintain constant internal conditions, whereas *conformers* cannot regulate their internal environment. Organisms living in stable environments are more likely to be conformers.

The optimal environmental conditions and tolerance limits for an organism can be determined experimentally by varying abiotic factors. Tolerance limits help to determine the spatial distribution of organisms. Physiological adjustment to environmental changes, which can extend tolerance limits, is called *acclimation.*

Morphological responses to environmental changes, such as changes in growth or development, are most common in plants and provide these rooted organisms with a means of adapting to their environments.

Behavioral, physiological, and morphological responses occur within an ecological time scale, but are based on adaptations that have developed by natural selection acting over an evolutionary time span. The adaptation of organisms to localized environments restricts their geographic distribution and often their

ability to respond to changes within their environments. The presence or absence of a species within a location may relate to its ability to tolerate the abiotic factors present or simply to its history of dispersal.

Terrestrial Biomes

Similar physical environments are generally covered by similar types of vegetation. The geographic distribution of the world's major communities, known as *biomes,* is related to the prevailing climate—to the temperature, rainfall, and seasonal fluctuations of the various altitudes and latitudes.

Tropical forests, found near the equator, have little variation in temperature (averaging around 25° C) or daylength. The variation in the amount of rainfall results in the occurrence of tropical thorn forests, where rainfall is scarce; tropical deciduous forests, in which trees and shrubs drop their leaves during the long dry season that alternates with the monsoon; or tropical rain forests, where rainfall is abundant.

The *tropical rain forest,* with its dense canopy formed by tall trees, vines, or *lianas,* and epiphytes, has a large diversity of species and is the most complex of all communities. The soil is generally poor and thin, because the high temperatures and rainfall lead to rapid decomposition and recycling of nutrients back into plant material. The animals are typically tree-dwellers. Human destruction of the tropical rain forest by mining, lumbering, and farming may greatly reduce species diversity and cause large-scale climate changes.

Savannas, tropical grasslands with only scattered trees, are typical of areas with low precipitation, such as the interior of continents. Grasslands have growth forms restricted to grasses and *forbs,* or small broad-leaf plants, but are often rich in the number of species. Frequent fires prevent the invasion of trees. Herbivores and burrowing animals are most common.

Deserts are characterized by low precipitation (less than 30 cm/yr). Temperatures may be hot or cold, depending on location. Where perennial vegetation occurs, it consists of widely scattered shrubs, cacti, or succulents. Rainy periods are marked by rapid blooms of annual plants. Desert plants have physiological and morphological adaptations to dry conditions and extreme temperatures. Seed-eaters, and their reptilian predators, are common animals. Desert animals have physiological and behavioral adaptations to their harsh environment.

The *chaparral,* or shrubland, is common along coast lines in midlatitudes. The shrub and annual vegetation is adapted to periodic fires, which are related to seasonal dryness; species may have deep root systems that permit quick regeneration, or seeds that germinate only after a fire. Animals are typically browsers, fruit-eating birds, seed-eating rodents, and lizards and snakes.

Temperate grasslands are found in relatively cold regions. Occasional fires and drought prevent the invasion of woody shrubs and trees. Large grazing mammals, along with their large carnivore predators, are typical animals.

Temperate forests, characterized by broad-leaved deciduous trees, grow in midlatitude regions, which have adequate moisture to support the growth of large trees. Temperatures range from cold winters to hot summers. Growth and decomposition rates are moderately high, and the soil has a thick humus layer. Herbs, shrubs, and one or two strata of trees are typical, and a rich diversity of animal life is associated with the variety of food and habitat.

The *taiga*, also known as the coniferous or boreal forest, is characterized by harsh winters and short summers, and is found in northern areas and higher elevations. The soil is thin and acidic. Low temperatures and the waxy covering of needles lead to slow decomposition rates. The growth of coniferous trees is usually so thick that little undergrowth is present. Animals include seed-eaters, larger browsers, and predators such as bears, wolves, and lynxes.

The *tundra* is characterized by low shrubby or mat-like vegetation. The arctic tundra, circling the North Pole, has long periods of very little light, interspersed with a brief warm summer marked by long days and rapid plant growth. The *permafrost* of the arctic tundra prevents roots from penetrating very far into the continually saturated soil. In the alpine tundra, found above the tree line on high mountains, daylength is more even and plant growth is slow but steady.

Animals of the tundra adapt to the cold by living in burrows or having large, well-insulated bodies. Some large herbivores are found, and many animals are migratory.

Aquatic Communities

As in terrestrial biomes, organisms living in similar aquatic environments show similar adaptations and body forms as a result of convergent evolution.

Rivers and *streams* are freshwater, flowing habitats, whose physical and chemical characteristics vary from their source to point of entry into ocean or lake. The biotic communities reflect these abiotic factors. Fishes, some invertebrates, algae, and attached mosses are more common upstream, whereas algae, cyanobacteria, bacteria, fungi, and invertebrates occur in the more turbid, slower-flowing, and nutrient-rich downstream areas.

Rivers and streams may flow into *lakes* and *ponds*. The settling of suspended particles and resulting clarity of the water determine the depths to which photosynthesis can occur. Phytoplankton and zooplankton float in the upper regions. Nutrients such as phosphorous

and nitrogen are often limiting, except when mixing of the waters brings these elements up from the sediments. Massive *algal blooms* can occur when nutrients are available. Fishes, aquatic insects, snakes, amphibians, and fish-eating birds are common.

Estuaries, where a river or stream meets the ocean, usually are highly productive areas with diverse animal and plant populations. Estuaries serve as spawning grounds for many marine organisms, and as feeding and breeding areas for waterfowl, amphibians, reptiles, and mammals.

Marine communities are found in the oceans, where algae produce almost 70% of the world's oxygen. The marine environment is classified into a *photic* zone, in which photosynthesis occurs, and an *aphotic* zone, where light does not penetrate.

Marine communities also may be divided on the basis of depth. The *intertidal*, or *littoral*, zone is the shallow area where the water meets the land. A rocky intertidal zone is inhabited by organisms that compete for attachment sites to avoid being washed away by the tides. A sandy or mud intertidal zone is home to burrowing worms and clams. The *neritic* zone is the shallow area over the continental shelves; the *pelagic* zone includes areas of open water; and the bottom surface of all these zones is considered to be the *benthic* zone. The *abyssal* zone is the benthic region where light does not penetrate. Organisms in these areas are adapted to darkness, cold, and high pressures. They may feed on the *detritus*, or organic matter, that sifts down from above, or, in the case of many bacteria, they may be chemosynthetic.

STRUCTURE YOUR KNOWLEDGE

1. Describe ecology. What types of questions are considered? What methods are used to answer those questions? What theory guides the interpretation of data?

2. What are biomes? What are the key abiotic factors that determine them? What accounts for the similarities in life forms throughout a biome?

TEST YOUR KNOWLEDGE

MULTIPLE CHOICE: *Choose the one best answer.*

1. Which level of ecology considers energy flow and chemical cycling?
   a. community
   b. ecosystem
   c. autecology
   d. population

2. Evolutionary history
   a. influences the distribution of organisms.
   b. involves the cumulative effects of the interactions between organisms and their environments.
   c. has produced the adaptations of organisms to different environments.
   d. involves all of the above.

3. Which abiotic factor has the least direct influence on the distribution of animals?
   a. temperature
   b. water
   c. wind
   d. light

4. According to the principle of allocation,
   a. the number of organisms an area can support, called its carrying capacity, is determined by its energy supply.
   b. physiological adjustments to environmental changes can extend the tolerance limits of organisms.
   c. the total amount of energy available to an organism is partitioned into such processes as reproduction, eating, growing, and coping with the environment.
   d. organisms that allot more energy to growth and reproduction are conformers.

5. Tolerance limits
   a. determine whether organisms can live in particular environments.
   b. can be extended by acclimation.
   c. are likely to be greater in regulators.
   d. involve all of the above.

6. Which of the following is incorrectly paired with its description?
   a. neritic zone—shallow area over continental shelves
   b. abyssal zone—benthic region where light does not penetrate
   c. littoral zone—area of open water
   d. intertidal zone—shallow area at edge of water

7. Estuaries
   a. are characterized by high productivity and species diversity.
   b. are areas where rivers enter lakes.
   c. are responsible for 70% of the world's oxygen production.
   d. are all of the above

8. Which of the following is incorrectly paired?
   a. savanna — warm temperature, low precipitation, continental interiors
   b. chaparral — coast lines, mild and wet winters, hot and dry summers
   c. taiga — extreme cold, permafrost, brief summer
   d. temperate grasslands — relatively cold, occasional drought and fires

9. In which area are algal blooms, phytoplankton, and zooplankton most likely to be found?
   a. headwaters in river or stream
   b. downstream in river or stream
   c. lake or pond
   d. littoral area

10. The characteristic animal forms of each biome
    a. are adapted to the various abiotic factors of the area.
    b. are determined by the type of vegetation available for food and shelter.
    c. may be the result of convergent evolution.
    d. are all of the above.

MATCHING: *Match the biotic descriptions with the biomes.*

Biome	Biotic Description
1. _____ chaparral	A. broad-leaved deciduous trees
2. _____ desert	B. lush growth, trees, lianas, epiphytes
3. _____ savanna	C. shrubs, fire-adapted vegetation
4. _____ taiga	D. tropical grasslands, grasses and forbs
5. _____ temperate forest	E. coniferous forests
6. _____ temperate grassland	F. low shrubby or matlike vegetation
7. _____ tropical rain forest	G. grasslands in relatively cold regions
8. _____ tundra	H. widely scattered shrubs, cacti, succulents

CHAPTER 47

POPULATION ECOLOGY

FRAMEWORK

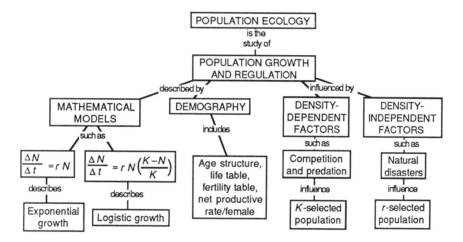

CHAPTER SUMMARY

Population ecology is the study of the fluctuations in and regulation of population size. This branch of ecology overlaps autecology and community ecology; it considers the individual's success in the environment and the interactions of various populations within a community.

Density and Dispersion

Ecologists determine the geographic boundaries of the population they are studying in order to find the means to answer the questions they ask. The number of individuals per unit area or volume is the population *density*; the spacing of those individuals within the boundaries of the population is referred to as *dispersion*.

Measurement of density is done most often by estimates based on a number of sample areas, or *quadrats*.

Indirect indicators, such as burrows or nests, also may be used. In a *mark–recapture* technique, animals are trapped, marked, and released. The proportion of marked and unmarked animals captured in a second trapping is used to estimate population density.

The density of individuals in the entire population's *range*, or geographic area, is called the *absolute density*. Because a range may contain a limited number of suitable habitats, the ecological density of inhabited areas is a useful measure. Dispersion may be *clumped, uniform,* or *random*.

Clumping may indicate a heterogeneous environment, with organisms showing *habitat selection* by congregating in suitable microenvironments. Clumping also may be related to social interactions between individuals. Uniform distribution is usually related to competition for resources, resulting in antagonistic interactions between organisms or the establishment of territories. Random spacing may occur in the absence of strong attractions or repulsions between individuals, or it may reflect a random distribution of a resource.

Models of Population Growth

Natality, or birth, and *immigration* add individuals to a population, whereas *mortality*, or death, and *emigration*, remove population members. Mathematical models have been developed to describe rates of population growth and to attempt to predict population sizes. Models may represent fairly accurately the growth of populations of simple organisms within a laboratory setting, but can serve only as comparisons for the actual growth of populations observed in the field. Models may help to identify the factors affecting the fluctuations in the growth of a particular population.

Unlimited, constantly accelerating growth is called *exponential* or *geometric growth*. The change in a population's size during a specific time period, signified as $\Delta N / \Delta t$, depends on an *intrinsic rate of increase (r)*, or the difference between natality and mortality rates, multiplied by the starting number (N) in the population. The formula and graph for exponential growth are:

$$\frac{\Delta N}{\Delta t} = r N$$

The larger the population (N), the faster the population grows.

Each species has a characteristic r, which partly depends on the *generation time*, or the average age when females begin reproducing. The shorter the generation time, the larger the r value and the greater the potential for rapid growth.

Due to environmental limitations, which result in increased mortality, decreased natality, or both, exponential growth is rarely maintained. Actual growth curves tend to be S-shaped, with population growth leveling off at a *carrying capacity (K)* for a population within a particular environment. The *logistic equation* includes a term that reflects the impact of the increasing population size as it approaches the carrying capacity. The formula and graph for logistic growth are:

$$\frac{\Delta N}{\Delta t} = r N \left(\frac{K - N}{K} \right)$$

When N is small, the $(K - N)/K$ term is close to 1, and growth is approximately exponential (rN). As N approaches K, the carrying capacity, the $(K - N/K)$ term

becomes a small fraction, and population growth is slower than exponential growth. When N reaches K, the term is 0, and population growth ($\Delta N / \Delta t$) is 0.

In terms of the logistic equation, one could think of r realized as the rate at which the population actually grows: r realized = r intrinsic $(K - N/K)$. Maximum increase in population numbers occurs when N is equal to one-half K, when the population is at one-half the carrying capacity of the environment.

The logistic model predicts that no population can grow indefinitely. The growth of some laboratory populations of small organisms and a few populations sampled in the field show an S-shaped growth curve with the population stabilizing at a carrying capacity. Most populations, however, respond to environmental and biological factors and fluctuate to a greater or lesser extent around a stable population number.

Regulation of Populations

Population size is controlled by *density-dependent* and *density-independent factors*. As population density increases, the density-dependent factors of *intraspecific competition*, predation, parasites, and pathogens have greater effects on further population growth. The ultimate effect of density-dependent factors is to stabilize populations around the carrying capacity. Density-independent factors, such as natural disasters or habitat disruption, reduce the population by the same proportion whether the population is large or small. Density-independent factors may have large effects on population size, but they do not help to stabilize populations around the carrying capacity.

It is difficult to determine which factors are actually regulating population sizes in the field. Both density-dependent and density-independent factors may be acting concurrently, and correlations between factors and population sizes may not indicate cause-and-effect connections. Limited success has followed attempts to predict the results of population-regulation programs for agricultural pests.

Some populations of birds, mammals, and insects show regular density fluctuations called *population cycles*. The effects of crowding, perhaps on the endocrine systems of organisms, may regulate these cycles. Cycles also may be caused by the effects of predators or parasites on the increasing density of prey or host populations. Alternately, the effects of an increasing herbivore population on its food supply may regulate population size more than increasing predator pressure does.

Some Important Demographic Statistics

Demography is the study of vital statistics that affect population growth, such as births and deaths, the ratio of males to females, and the age structure. Each age group in a population has a characteristic fertility and death rate. The *age structure* of a population is the relative number of individuals of each age.

Life tables summarizing mortality rates for each age group can be constructed by following a *cohort*, or group, of newborn organisms throughout their lives. Often only data for females are tabulated, since the reproductive rates for females determine the growth of most populations.

Age-specific mortality also can be represented on a *survivorship curve*, which shows the number of members of a cohort that are still alive at each age. The three types of survivorship curves are shown below.

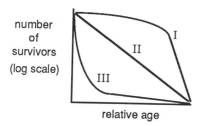

Type I, typical of modern human populations, shows that mortality is low during early and middle age and increases rapidly with old age. In a Type II curve, death rates are relatively constant throughout the life span. A Type III curve is typical of populations that produce many offspring, most of which die off rapidly. The few that survive are likely to reach adulthood.

The variation of birth rates with age is shown in a *fertility table*. Data from life tables and fertility tables can be used to determine the *net reproductive rate*, the number of female offspring that a female member of the population is expected to produce. The growth rate of a population is dependent on the net reproductive rate and the generation time.

Human Population Growth

The human population has been growing exponentially for centuries. In the 80 years between 1850 and 1930, it doubled to 2 billion; it doubled again in the next 45 years; and it is projected to reach 8 billion by the year 2017. In 1986, the world population stood at about 5 billion. The growth since the Industrial Revolution has resulted mainly from a drop in death rates, especially among infants.

Human population growth is unique in that it can be consciously controlled by voluntary contraception or government sanctions. In most developed countries, women are delaying marriage and reproduction. Different countries of the world have different growth rates. The age structure of a population determines

growth; a large proportion of individuals of reproductive age results in relatively rapid growth.

Technology has increased the carrying capacity of Earth for humans, but it is hard to predict what this carrying capacity is, and how and when we will approach it.

Evolution of Life Histories

The *life histories* of organisms—their pattern of birth, reproduction, and death—affect population growth in ecological time, but evolved as a result of natural selection operating in evolutionary time. Three factors of life histories influence the intrinsic rate of increase: (1) number of reproductive episodes; (2) *clutch size*; and (3) age at first reproduction. The shorter the generation time, the greater is *r*. *Semelparous* organisms reproduce only once, whereas *iteroparous* individuals reproduce more than once. Clutch size usually is inversely related to the relative size of offspring.

Populations that have a high *r* value are said to be *r-selected*; individuals reproduce at an early age, are usually semelparous, and have large clutch sizes of relatively small offspring. This type of life history is most common in variable environments, where density-independent factors regulate population size. Reproductive success is ensured by producing large numbers of offspring.

K-selected populations are found in stable environments with limited resources, where density-dependent factors tend to maintain population sizes around the carrying capacity. Offspring are fewer, and reproductive success may depend on the competitive capabilities of the offspring.

Most populations have life histories somewhere in between an *r-* and a *K-*selected scheme.

STRUCTURE YOUR KNOWLEDGE

1. Create a concept map to organize your understanding of the exponential and logistic equations—the mathematical models of population growth.

2. Describe the demographic factors that influence population growth rates.

TEST YOUR KNOWLEDGE

MATCHING: *Match the term with its description.*

1. _____ area from which sample is taken
2. _____ table with mortality rates for each age group
3. _____ birth
4. _____ group followed from birth to death
5. _____ relative number in each age group
6. _____ number of female offspring each female is expected to produce
7. _____ having more than one reproductive event per lifetime
8. _____ graph of age-specific mortality
9. _____ number of offspring produced at one time
10. _____ birth rates for each age group

A. age structure
B. clutch size
C. cohort
D. fertility table
E. generation time
F. iteroparous
G. life table
H. natality
I. net reproductive rate
J. quadrat
K. semelparous
L. survivorship curve

MULTIPLE CHOICE: *Choose the one best answer.*

1. In a range with a heterogeneous distribution of suitable habitats,
   a. the ecological density and absolute density would be the same.
   b. the ecological density would be greater than the absolute density.
   c. the ecological density would be less than the absolute density.
   d. one could not tell which density was greater without sampling.

2. In a mark–recapture study,
   a. all the organisms within several quadrats are counted.
   b. only the organisms that are recaptured are counted.
   c. the proportion of marked individuals that are recaptured is used to estimate the population size.
   d. 10% of all the organisms in an area are captured and marked.

3. A random dispersion pattern may indicate that
   a. antagonistic interactions between individuals do not occur.
   b. resources are randomly distributed.
   c. social interaction between individuals is not a factor in dispersion.
   d. all of the above may apply.

4. Exponential growth
   a. may occur during the early growth of a population.
   b. never occurs.
   c. is possible only in populations with short generation times.
   d. is associated with populations that have large r values.

5. The logistic growth model
   a. predicts that no population can grow indefinitely.
   b. best describes human population growth since 1850.
   c. shows that reproduction ceases when the population size reaches K.
   d. does all of the above.

6. The shorter the generation time,
   a. the smaller the carrying capacity.
   b. the greater the value of r.
   c. the greater the effect of density-independent factors on growth.
   d. the larger the K value possible.

7. The term $(K - N)/K$ influences $\Delta N/\Delta t$ such that
   a. the increase in population numbers ($\Delta N/\Delta t$) is greatest when N is small.
   b. as N approaches K, r (the intrinsic rate of increase) becomes smaller.
   c. when N equals K, population growth, $\Delta N/\Delta t$, is zero.
   d. all of the above are correct.

8. The carrying capacity for a population
   a. is the number of individuals in that population.
   b. is determined by the resources of a particular environment.
   c. decreases as N approaches K.
   d. is set at 8 billion for the human population.

9. Density-dependent factors
   a. tend to maintain a population around the carrying capacity.
   b. may be involved in the population cycles seen in some mammals.
   c. include intraspecific competition, predation, parasites, and pathogens.
   d. involve all of the above.

10. Which of the following is not true?
    a. A population with a large r value probably has a short generation time.
    b. A population with a large r value probably has large clutch sizes.
    c. Populations with large r values often are found in variable environments.
    d. A population with a large r value is most likely to be regulated by density-dependent factors.

11. The age structure of a population
    a. can be inferred from a survivorship curve.
    b. is a result of age-specific fertility.
    c. determines the net reproductive rate.
    d. is most uniform in populations with high growth rates.

12. The clutch size
    a. is inversely related to the number of offspring produced.
    b. influences the intrinsic rate of increase of a population.
    c. is directly proportional to the size of offspring.
    d. is greatest for iteroparous individuals.

CHAPTER 48

COMMUNITIES

FRAMEWORK

Communities are composed of populations of various species that may interact through competition, predation, and symbiosis. The structure of a community—its species diversity, relative abundance, and species composition—is determined by these interactions, the past history of the community, and its successional stage. Disturbances and chance events contribute to the dynamic equilibrium (or nonequilibrium) of a community.

Biogeography studies the distribution of species throughout the world. Island biogeography permits the study of community structure and equilibrium, especially in relationship to size of habitat and migration rate, within a more simple system.

CHAPTER SUMMARY

A collection of populations living close enough to allow for interaction is called a *community*. Community ecology studies the factors that are involved in structuring a community—in determining the diversity and relative abundance of species in a community.

Two Views of Communities

Gleason, in 1926, described communities as chance groupings of species found in the same area because of their similar requirements for environmental factors. Clements, in 1936, claimed that a community was a kind of superorganism organized around prescribed interactions that caused it to function as an integrated unit.

According to Gleason's views, there should be no distinct boundaries between communities, because each species would have its own distribution along the environmental gradients. Clements's hypothesis predicts that species would be clustered into groups, with relatively discrete boundaries between communities.

Most communities appear to vary on a continuum, with each species independently distributed. Abrupt changes between communities are usually correlated with abrupt changes in an abiotic factor. Ecologists define the boundaries of a community in order to study the interactions that help to determine the diversity and relative abundances of species.

Properties of Communities

The community level of organization has several unique properties: (1) *species diversity*, the number of species found in a community; (2) *relative abundance and dominance*, the relative numbers of individuals in each species and the most abundant or dominant species; (3) *prevalent form of vegetation*, the vertical profile or the most common vegetative growth forms; (4) *trophic structure*, the feeding relationships, including plant–herbivore and predator–prey interactions; and (5) *stability*, the ability of a community to return to its original makeup following a disturbance.

Co-Evolution in Communities

Both the abiotic and biotic components of the environment act as selective factors for evolution. *Co-evolution* refers to reciprocal adaptations, in which a change in one species acts as a selective agent on another species, whose adaptations in turn act as selective forces on the first species. The adaptations of flowers and their exclusive pollinators are examples of co-evolution.

Another example of co-evolution is the passion-flower vines and the butterfly *Heliconius*. The toxic compounds of the vines protect them from most herbivorous insects. The *Heliconius* larvae, however, apparently have enzymes that destroy these compounds, and have become specialized feeders on these vines. The female butterflies tend to avoid depositing their bright yellow eggs on leaves that already have eggs. Some species of *Passiflora* have leaves with yellow dots that mimic these eggs and protect the vine from a heavy infestation of *Helioconius* larvae.

In the interview for this unit, Lubchenco cautioned against assuming that every adaptation that affects community interactions is a result of co-evolution. Co-evolution is marked by reciprocity and specificity. Even when species have intertwined adaptations, it is difficult to determine whether they have served as selective factors for each other. It is generally accepted, however, that interactions in ecological time can translate into co-evolution over evolutionary time.

Community Interactions

Intraspecific competition (between individuals of the same species) for limited resources limits population growth. If two species use the same resource, *interspecific competition* should affect the density of both species. Competition for shared resources is believed to be important in determining species diversity.

As a result of his laboratory experiments with *Paramecium*, Gause formulated the *competitive exclusion principle*: two species that compete for the same limiting resource cannot co-exist in the same habitat. Whether competitive exclusion occurs in natural communities, and whether extinction, emigration, and/or co-evolution result from these competitions, are areas of research by community ecologists.

An organism's *niche* is described as its "role" in an ecosystem—its habitat and its usage of the biotic and abiotic resources. The *fundamental niche* includes the resources an organism theoretically could use; the *realized niche* is the resources that it actually can use as determined by the biological constraints of competition or predation.

The competitive exclusion principle holds that two species with the same niche cannot co-exist in a community. Slight variations in niche allow closely related species to co-exist. Competition for niches apparently must be resolved by localized extinction of the less successful species or by character displacement so that the species assume somewhat different niches.

Many ecologists do not accept the generalization that competition is the major factor that structures communities. Niches are so complex that it is difficult to determine whether two species actually are competing, and even more difficult to know what has occurred in evolutionary history.

Predators are species that eat other organisms; *prey*, whether plants or animals, are the species that are eaten. The evolution of defensive adaptations will increase the reproductive success of prey. Plants may defend against animal predators by camouflage, mechanical devices, or chemical compounds. Distasteful or toxic chemicals, called *secondary compounds*, include such well-known compounds as strychnine, morphine, digitalis, nicotine, mescaline, and mustard oils. Counteradaptations may enable certain herbivores to overcome these plant defenses.

Animals can defend against predation by hiding, escaping, or defending. A prey may blend in with its background by shape, behavior, or *cryptic coloration*. Deceptive markings, such as fake eyes or false heads, may discourage or confuse predators. The porcupine's mechanical defenses and the skunk's chemical defenses discourage predation. Some animals passively accumulate toxins from their plant diets, and then use *aposematic coloration*, or bright warning colors, to help predators learn not to eat them.

Mimicry, in which one species resembles another, may serve as a defense. In *Batesian mimicry*, a harmless species resembles a poisonous or distasteful species. *Mullerian mimicry* is mutual imitation by two bad-tasting species.

Adaptations to increase success in predation include acute senses, speed and agility, and physical structures such as claws, fangs, stingers, and pincers.

The role of predation in structuring communities may be to stabilize species diversity. As one prey species dwindles in numbers, most predators will feed on another species. This *switching behavior* may serve to maintain a variety of species within a community.

Symbiotic interactions influence community structure. In *parasitism*, the host is harmed; in *commensalism*, one individual is benefitted and the other unharmed; and in *mutualism*, both symbionts benefit from the relationship.

Both parasites best adapted to find and feed on their hosts, and hosts best able to defend against parasites, would be favored by natural selection. Some secondary plant products and the immune systems of vertebrates are examples of host defenses. Many parasites have evolved adaptations that allow them to feed without killing their specific host. Co-evolution may stabilize host–parasite relationships.

It is difficult to establish that one member of a commensal relationship is completely unaffected by the other. Cattle egrets and the cattle that flush the insects on which the egrets feed illustrate commensalism. Co-evolution cannot be involved in commensal relationships, because the fitness of the one species is not affected by the species that benefits from the association.

Co-evolution may be involved in mutualistic relationships, as changes in one species may affect the fitness of the other. Examples of mutualism include the nitrogen-fixing bacteria within legumes, cellulose-digesting microorganisms within the guts of termites and ruminants, mychorrhizae, flowering plants and their pollinators, and the ants on acacia trees. Some mutualistic interactions may have evolved from predation or parasitism, when organisms became able to derive some benefit from their predator or parasite.

A variety of community interactions are considered to be important to the determination of species diversity within a community. Field experiments, such as Lubchenco's periwinkle study in intertidal communities, often show that predators increase the diversity of species by preventing competitive exclusion by the most successful species. Plant–herbivore and predator–prey relations, as well as symbiotic relationships, may be more important than competition in structuring some communities. Multiple interactions between organisms and their biotic and abiotic environment are involved in determining community structure.

Succession

A transition in species composition in a community, usually following some disturbance, is known as *succession*. If no previous organisms were present, the process is called *primary succession*. *Secondary succession* occurs when an existing community has been disrupted by a fire, logging, or farming.

Clements believed that succession follows a definite sequence, leading to a *climax community*, which was the final, stable stage of succession that would occur within each particular abiotic environment. This unidirectional succession was said to occur because of *facilitation*; each stage of vegetation altered the environment to make it more suitable for species in the next stage.

Facilitation is observed in some cases of succession. The progression from barren ground, left by retreat of a glacier, to a stable spruce–hemlock forest is associated with changes in the soil brought about by the accumulation of humus.

Succession does not always progress predictably toward a climax community. A contrasting view of succession is that the early pioneers, which arrive first on the scene by chance, may inhibit the growth of other plants. Transitions occur when latecomers are able to outcompete the pioneers. An example of *inhibition* may be a rapidly growing plant that shades the growth of most other plants.

Succession also may represent the gradual success of the species with the greatest *tolerance* for the specific abiotic environment of an area. Early successional stages are dominated by colonizing plants, which disperse readily and grow rapidly. Later stages may be dominated by longer-living species that can withstand shading and other forms of competition. Facilitation, inhibition, and tolerance all may contribute to succession in some communities.

Disturbance may destroy a community and begin secondary succession, or it may, in certain situations, serve to stabilize communities by preventing succession. Grasslands may be maintained because periodic fires prevent the invasion of trees.

Human activities have altered the structure and succession of communities all over the world. Logging and agriculture disrupt mature communities, and restart successional growth. The disappearance of the tropical rain forests and development of vast barren areas have resulted from human disturbances.

Species diversity generally increases during succession, partly as a result of more extensive and complex community interactions. According to the equilibrium model, succession reaches a climax community when interactions are so intricate that no new niches are available for additional species.

The nonequilibrium model views communities as being in continual nonequilibrium, with the number of species changing even in the so-called climax stage. Chance events such as dispersal and disturbance are given major roles in the process of succession. Succession depends both on the species that happen to colonize the area and on environmental disturbances that prevent species diversity from becoming constant. Severe and frequent disturbances may restrict the community to good colonizers, whereas infrequent disturbances may allow late-successional, highly competitive species to become dominant. Species diversity may be greatest when disturbances are intermediate in severity and frequency, allowing organisms from different successional stages to be present.

Geographic Aspects of Diversity

Biogeography, the study of past and present distributions of species, is concerned with both the number and the identity of species in an area. Biogeographic realms, roughly correlated with the patterns of continental drift after the breakup of Pangaea and separated in places by deserts or mountain ranges, have their own taxonomic groups. Thus, species distribution reflects past history as well as present interactions.

A species may be limited to a particular range because it has never dispersed beyond that range or because pioneers to areas outside the range have failed to survive. Transplant experiments can be used to determine which of these two reasons has limited the species's range. If the transplanted species fails to survive, the reason may be that it could not tolerate the physical environment or could not compete with the resident species.

Successful transplants indicate that historical separation has kept the species from a particular range. The rabbits in Australia and African honey-bees of South America are examples of successful transplants that have presented ecological problems.

Islands, or any habitat surrounded by an inhospitable habitat, allow ecologists to study the role of dispersal in determining species diversity. In the 1960s, MacArthur and Wilson developed a general *theory of island biogeography*, stating that the size of the island and its distance from the mainland (or source of dispersing species) are the two most important factors determining species diversity. Following a period of initial immigration and colonization, fewer new species can become established, and the extinction rate of present species increases. An equilibrium in species diversity develops, although species composition may continue to change. Species diversity will be greater both for larger islands in which more niches are available and for closer islands in which a higher immigration rate accounts for higher diversity.

Observations and experiments provide support for the island biogeography theory. In the late 1960s, Simberloff and Wilson conducted experiments in which all resident arthropods were killed on six small islands. Within a year, species diversity had reached equilibrium on each island, and the least diversity was found on the island most distant from the coast. Although the number of species on each island returned to the same level, the species composition was different. Chance events affected the species composition of these communities. Community structure and dynamics are much less predictable than ecologists once may have believed.

STRUCTURE YOUR KNOWLEDGE

1. Develop a concept map organizing your understanding of the important factors that structure a community.

2. Describe succession and the factors that contribute to this process.

TEST YOUR KNOWLEDGE

MULTIPLE CHOICE: *Choose the one best answer.*

1. Which of the following is not part of Gleason's concept of communities?
   a. Communities are chance collections of species that are in the same area because of similar environmental requirements.
   b. There should be no distinct boundaries between communities.
   c. The consistent composition of a community is based on interactions that cause it to function as an integrated unit.
   d. Species are distributed independently along environmental gradients.

2. Boundaries of communities
   a. occur along abrupt changes in abiotic factors.
   b. are located wherever community ecologists choose to make them.
   c. are not discrete or distinct.
   d. all of the above may be true.

3. The trophic structure of a community
   a. is characterized by the prevalent form of vegetation.
   b. is determined by competition.
   c. includes the feeding relationships within the community.
   d. is a measure of species diversity.

4. Co-evolution
   a. often is found in organisms competing for the same niche.
   b. describes the adaptations found in most community interactions.
   c. often can be seen in predator–prey and parasite–host interactions.
   d. all of the above are correct.

5. According to the principle of competitive exclusion,
   a. two species cannot co-exist in the same habitat.
   b. extinction or emigration are the only possible results of competitive interactions.
   c. two species cannot share the same realized niche in a habitat.
   d. intraspecific competition results in the success of the superior species.

6. Character displacement
   a. may allow competing species to co-exist.
   b. may change an organism's fundamental niche into its realized niche.
   c. allows a prey species to use confusing markings to escape predation.
   d. allows a predator to switch prey when one prey species dwindles.

7. The ability of some herbivores to eat plants that have toxic secondary compounds may be an example of
   a. co-evolution.
   b. Batesian mimicry.

    c. Mullerian mimicry.
    d. aposematic coloration.

8. Two equally good-tasting prey species may defend against predation by
    a. Mullerian mimicry.
    b. cryptic coloration.
    c. secondary compounds.
    d. aposematic coloration.

9. The role of predation within a community may be to
    a. maintain species diversity by decreasing competition between prey species.
    b. increase the relative abundance of the dominant species.
    c. stabilize host–parasite relationships.
    d. encourage character displacement and co-evolution.

10. Mutualistic relationships
    a. may have developed through co-evolution.
    b. may involve host–parasite relationships.
    c. are difficult to document.
    d. all of the above are correct.

11. The most important factor in determining community structure
    a. may change from one community to another.
    b. is predation.
    c. is competition.
    d. is history.

12. Primary succession
    a. involves the first colonists that arrive to an area.
    b. occurs when no previous organisms were present in an area.
    c. is a result of facilitation.
    d. occurs following a severe disruption of a community.

13. A climax community
    a. would always be the same for each type of habitat.
    b. may not develop in a periodically disturbed habitat.
    c. results from secondary succession.
    d. is the goal of succession.

14. Inhibition
    a. may prevent the achievement of a climax community.
    b. is evidence for the equilibrium theory of succession.

    c. is one of the factors that determines the most tolerant species in an area.
    d. eventually may be overcome by competitively successful species.

15. Which of the following is not part of the nonequilibrium model of succession?
    a. Chance events such as dispersal and disturbance play major roles in succession.
    b. Species diversity may be greatest when disturbances are intermediate in severity and frequency.
    c. When succession reaches a climax community, only extinctions make room for new colonists.
    d. Even when a community is mature, disturbances may prevent the achievement of a climax community.

16. The taxonomic similarity of species within biogeographic realms
    a. is due to the similarity in environmental characteristics.
    b. is mostly a reflection of past history of species' distributions.
    c. is greatest on larger islands.
    d. reflects the inability of species to live outside their ranges.

17. Transplant experiments
    a. are usually ecological disasters.
    b. give evidence for continental drift following the breakup of Pangaea.
    c. show whether a species can live outside its range.
    d. are used in the study of island biogeography.

18. An island that is small and far from the mainland
    a. would be expected to have a low species diversity.
    b. would be expected to be in an early successional stage.
    c. would be expected to have a small species diversity but a large abundance of organisms.
    d. would be expected to have a low rate of colonization and a high rate of extinction.

CHAPTER 49

ECOSYSTEMS

FRAMEWORK

This chapter describes the energy flow and chemical cycling that occur through the trophic structures found at the organizational level of ecosystems. Producers convert light energy into chemical energy, which is then passed, with a loss of energy at each level, from consumer to consumer, and to detritivores. Energy makes a one-way trip through ecosystems. Chemical elements, such as carbon, oxygen, nitrogen, and phosphorus, are cycled in the ecosystem from reservoirs in the atmosphere or soil, through producers, consumers, and detritivores, and back to the atmosphere or soil.

The burning of fossil fuels, deforestation, and the dumping of toxic chemicals into the environment and excess nutrients into lakes are human activities that have altered normal ecosystem dynamics and pose threats of future ecological disruption.

CHAPTER SUMMARY

An *ecosystem* is a community and its physical environment; it includes all the biotic and abiotic factors within an area. The two processes of energy flow and chemical cycling are associated with the ecosystem level of organization. Most ecosystems are powered by the input of sunlight energy, which is transformed to chemical energy by autotrophs, passed to a series of heterotrophs in the organic compounds of food, and continually dissipated in the form of heat. Energy flows through the system; it is not recycled. Chemical elements, such as carbon and nitrogen, are cycled between the abiotic and biotic portions of the ecosystem.

Trophic Levels and Food Webs

The *trophic structure* of an ecosystem determines the flow of energy and chemicals through different feeding (trophic) levels. The *producers* are autotrophs, which usually use light energy to photosynthesize sugars for use as fuel in respiration and as building materials for other organic compounds. The *primary consumers* are herbivores, which eat plants and algae. *Secondary consumers* are carnivores, which eat herbivores; *tertiary consumers* eat other carnivores. The *detritivores* consume organic wastes and dead organisms.

Plants are the main producers in terrestrial ecosystems; photosynthetic protoctists (algae) and cyanobacteria form the basis for most aquatic ecosystems; bacteria are the major producers in a few ecosystems. Multicellular algae and phytoplankton are producers within marine ecosystems. The chemosynthetic bacteria found around thermal vents deep in the seas depend on geothermal energy and transform chemical energy rather than solar energy.

Fungi and bacteria are the most important decomposers in most ecosystems. Earthworms and such scavengers as cockroaches and bald eagles are also detritivores.

A *food chain* shows the transfer of food between trophic levels. Most ecosystems have complex, branching food chains called *food webs*.

Energy Flow

Of the visible light that reaches leaves and algae, less than 1% is converted to chemical energy by photosynthesis. The world-wide photosynthetic production is 170 billion tons/yr of organic material.

Different ecosystems vary in productivity. The rate of

conversion of light to chemical energy in an ecosystem is called *primary productivity*. The *net primary productivity* (NPP) is the *gross primary productivity* (GPP) minus the amount used by plants in their own cellular respiration. For most plants, 50% to 90% of gross primary productivity remains as net primary productivity.

Primary productivity can be expressed as energy per unit area per unit time ($kcal/m^2/yr$), or as biomass ($g/m^2/yr$). *Biomass* is measured in terms of the dry weight of organic material. The *standing crop* is the total biomass of plants; the productivity is the rate at which new biomass is synthesized. Tropical rain forests are the most productive ecosystems; estuaries and coral reefs also have high rates of productivity.

Productivity in terrestrial ecosystems is influenced by precipitation, heat, light intensity, length of growing season, mineral content of the soil, and even CO_2 supply. Productivity in the seas is greatest in shallow waters and along coral reefs. The productivity of the open ocean is generally low because mineral nutrients are limited near the surface where light is available for photosynthesis. Phytoplankton communities are most productive where upwellings bring nitrogen, phosphorus, and other organic nutrients to the surface. In freshwater ecosystems, light intensity, temperature, and availability of minerals affect productivity.

Gross primary productivity can be measured in a pond by the comparison of oxygen concentrations in dark and transparent bottles in which, respectively, just respiration or both respiration and photosynthesis have been occurring.

Secondary productivity is the rate at which the heterotrophs of an ecosystem—herbivores, carnivores, and detritivores—produce new biomass from organic material. Each trophic level sees a decline in productivity, partly because of a loss of energy to heat as each consumer converts organic fuel into its own molecules or into energy in cellular respiration.

Herbivores cannot digest much of the cellulose that they eat, and two-thirds of the energy they do absorb is used for cellular respiration. In a grass field, insects may convert about 4% of the net primary productivity to secondary productivity. Carnivores can digest and absorb more of the organic compounds in their food, but they use over 90% for cellular respiration.

Approximately 10% of the chemical energy of one trophic level is transferred to the next trophic level, a loss of energy that can be represented as an *energy pyramid*. A *pyramid of biomass* illustrates the total dry weight of organisms at each trophic level, which also indicates the amount of chemical energy stored in organic compounds, and thus also steps steeply from the producers to the top trophic level. Some aquatic ecosystems may have inverted biomass pyramids, in which the zooplankton primary-consumer level may outlive and outweigh the heavily consumed phytoplankton primary-producer level. Although more biomass is stored in the zooplankton, the energy pyramid is normal in shape; more energy is converted to organic material by phytoplankton than reaches the zooplankton.

Because of the loss of energy at each trophic level, most food chains are limited to four or five links. Only about one-thousandth of the chemical energy produced in photosynthesis makes it to a tertiary consumer, such as a hawk feeding on sparrows. Eating grain-fed beef as a main source of calories is an inefficient means of obtaining the energy trapped by photosynthesis.

Chemical Cycling

Chemical elements are passed between the abiotic and biotic components of the ecosystems through *biogeochemical cycles*. Atoms of organisms are returned to the atmosphere or soil through respiration and the action of decomposers. Plants and other autotrophs use these inorganic nutrients to build new organic matter, which then is passed through the food chain.

The route of the chemical cycle depends on the element involved. Global cycles, which involve atmospheric reservoirs, occur for carbon, oxygen, and nitrogen. Less mobile elements, such as phosphorus, sulfur, potassium, calcium, and the trace elements, have a more localized cycle in which the soil is the main abiotic reservoir.

The actual movement of elements through biogeochemical cycles is quite complex, with influx and loss of nutrients from an ecosystem occurring in many ways. Ecologists have worked out general schemes in several ecosystems by adding radioactive tracers to chemical elements.

Carbon moves between the abiotic and biotic portions of ecosystems through the processes of photosynthesis and cellular respiration. CO_2 from the atmosphere or dissolved in aquatic environments is incorporated into organic matter, providing the carbon source for consumers. Respiration breaks down this organic matter and releases CO_2. The oxygen cycle is tied to the carbon cycle, because oxygen is both split and released from water and also incorporated from CO_2 into organic matter, during photosynthesis. Carbon generally cycles at a fast rate, but wood, coal, petroleum, and shells can store carbon for long periods.

Plants cannot assimilate atmospheric nitrogen. Only certain prokaryotes can fix or reduce N_2 into ammonia (NH_3), which can then be incorporated into amino acids and other nitrogenous organic compounds. Soil bacteria and symbiotic bacteria in root nodules fix nitrogen in terrestrial ecosystems; cyanobacteria do so in aquatic ecosystems. Industrial production of nitrogen fertilizers adds to the nitrogenous minerals in agricultural soils. Other bacteria, in a process called *nitrification*, convert ammonia to NO_2^- (nitrite) and NO_3^- (nitrate). Plants can absorb either ammonia or nitrate. Animals obtain their nitrogen in organic form from plants or other animals. Detritivores decompose nitrogenous compounds from organic wastes and dead organisms and return ammonia to the soil. Denitrifying bacteria convert ammonia to N_2, which returns to the atmosphere.

Phosphorus is an ingredient of ATP and nucleic acids, bones, and shells. Weathering of rock adds phosphorus to the soil, usually in the form of PO_4^-, which can be absorbed by plants. Organic phosphate is transferred from plants to consumers, and returned to the soil through the action of decomposers or by excretion by animals. Humus and soil particles usually bind phosphorus, keeping it available for recycling. The phosphorus that leaches into the water table eventually travels to the sea, where sediments become incorporated into rocks. These rocks may eventually return to terrestrial ecosystems as a result of geological processes. The sedimentary cycle works within geological time, whereas the local cycle between soil, plants, and consumers operates in ecological time.

Human Intrusions in Ecosystem Dynamics

The trophic structure, energy flow, and chemical cycling of most ecosystems have been influenced by human activities and technology. This widespread ecological impact is often harmful.

The Hubbard Brook Experimental Forest Study, coordinated by Bormann and Likens, has looked at nutrient cycling in a forest ecosystem for 25 years. The mineral budget for each of six valleys was determined by measuring the influx of key nutrients in rainfall and their outflux through the creek that drained each valley. About 60% of the precipitation exits through the stream; the rest is lost by transpiration and evaporation. Mineral influx and outflow were nearly balanced.

The effect of deforestation on nutrient cycling was measured for three years in a valley that was completely logged. Compared with a control valley, water runoff from the deforested valley increased 30% to 40%; net losses of minerals, such as Ca^{++}, K^+, and nitrate, were huge.

Lakes undergo a natural succession in which their primary productivity increases as mineral nutrients are added from runoff and recycled through the lake's food chain. Lakes eventually become *eutrophic*, or very productive and nutrient-rich. Sewage, factory wastes, and runoff of animal wastes or fertilizers from agricultural lands often accelerate this process. An increase in phosphorus can cause an explosive increase in producers, resulting in oxygen shortages, due to plant and algal respiration at night, and eventually from the metabolism of decomposers that work on the accumulating organic material. Reduced oxygen levels kill off many fishes and other lake organisms.

Toxic chemicals have been dumped into many ecosystems. Many of these are synthetic and nonbiodegradable; some may become more harmful as the result of reaction with abiotic and biotic environmental factors. One of the most serious environmental threats is *radioactive fallout*. Toxic substances may be retained within the tissues of organisms that absorb them. In a process known as *biological magnification*, the compounds become more and more concentrated in each succesive link in the food chain.

DDT is an example of biological magnification. Traces of DDT have been found in nearly every organism tested, and concentrations became high enough to interfere with successful reproduction in many birds.

The concentration of CO_2 in the atmosphere has been increasing since the Industrial Revolution as a result of the combustion of fossil fuels. The level has increased 7% in the past 30 years. Much of the CO_2 added to the atmosphere has been absorbed into the sea as bicarbonate.

The increase in CO_2 could affect species composition in communities as C_3 plants may be able to outcompete C_4 plants. An increase in temperature may also influence community structure. Through a phenomenon known as the *greenhouse effect*, CO_2 in the atmosphere absorbs infrared radiation that is escaping from Earth and causes an increase in temperature. The continued increase in atmospheric CO_2 concentration may have far-reaching consequences in weather changes, in the amount and distribution of precipitation, and in rising sea levels due to the melting of the ice caps.

The *Gaia hypothesis*, first proposed by Lovelock, maintains that the biosphere, as a sort of superorganism, has shaped the climate and atmosphere of Earth and is capable of regulating the environment. An example of such regulation is the increased uptake of CO_2 by plants, and the absorption into the oceans and incorporation into shell-building plankton, of approximately one-half of the CO_2 released from burning of fossil fuels and wood.

The concentration of O_2 has remained fairly constant for the past 200 million years, partly as a consequence of methane, produced by certain bacteria and termites, which reacts with O_2 to form CO_2 and H_2O.

Many scientists, including Lubchenco as quoted in the interview, question whether the biosphere operates as a superorganism with a self-regulated metabolism. The Gaia metaphor, however, does point out the interconnectedness of the trophic levels, energy cycling, and chemical cycles within and between ecosystems, and the inherent dangers of poisoning or disrupting these ecological balances.

STRUCTURE YOUR KNOWLEDGE

1. Two processes that emerge at the ecosystem level of organization are energy flow and chemical cycling. Develop a concept map that explains, compares, and contrasts these two processes.
2. Sketch out a generic food web, and then include examples of each link that could be found in an aquatic ecosystem.
3. Describe three human intrusions in ecosystem dynamics that have detrimental effects.

TEST YOUR KNOWLEDGE

MULTIPLE CHOICE: *Choose the one best answer.*

1. Which of the following organisms and trophic levels are mismatched?
   a. cyanobacteria — producer
   b. zooplankton — secondary consumer
   c. fungi — detritivore
   d. field mouse — primary consumer

2. Chemosynthetic bacteria found around deep sea vents are examples of
   a. producers.
   b. detritivores.
   c. chemical cycling.
   d. secondary productivity.

3. A food web
   a. shows the pyramid of biomass.
   b. must include tertiary consumers.
   c. may be inverted in some aquatic ecosystems.
   d. is a complex and branching food chain.

4. Primary productivity
   a. is measured by the standing crop.
   b. is greatest in marine ecosystems.
   c. is the rate of conversion of light to chemical energy in an ecosystem.
   d. is all of the above.

5. Net primary productivity
   a. is less than gross primary productivity.
   b. takes into account the energy used by plants in respiration.
   c. can be measured by the increase in biomass ($g/m^2/yr$).
   d. all of the above are correct.

6. The productivity in freshwater ecosystems is affected by
   a. temperature.
   b. light intensity.
   c. availability of nutrients.
   d. all of the above.

7. Secondary productivity
   a. is measured by the standing crop.
   b. is the rate of production of biomass in consumers.
   c. is greater than primary productivity.
   d. is only slightly less than primary productivity.

8. An energy pyramid shows that
   a. only one-half of the energy in one trophic level is passed on to the next level.
   b. most of the energy from one trophic level is incorporated into the biomass of the next level.
   c. the energy lost as heat or in cellular respiration is about 10% of the available energy of each trophic level.
   d. eating grain-fed beef is an inefficient means of obtaining the energy trapped by photosynthesis.

9. Biogeochemical cycles are global for
   a. elements that are found in the atmosphere.
   b. elements that are found mainly in the soil.
   c. carbon, nitrogen, and phosphorus.
   d. elements that have been marked with radioactive tracers.

10. The role of detritivores in the nitrogen cycle is to
    a. fix N_2 into ammonia.
    b. release ammonia to the soil.
    c. denitrify ammonia to return N_2 to the atmosphere.
    d. convert ammonia to nitrate, which can be absorbed by plants.

11. Carbon cycles relatively rapidly except when it is
    a. dissolved in aquatic ecosystems.
    b. released by respiration.
    c. converted into sugars.
    d. stored in coal, wood, or shells.

12. The geological time scale of phosphorus cycling involves
    a. its incorporation into shells.
    b. the weathering of rock to add PO_4^- to the soil.
    c. sedimentation to form rocks in the sea bed.
    d. the incorporation of phosphorus into fossils.

13. The Hubbard Brook Forest
    Study showed that
    a. most nutrients cycle out of watersheds.
    b. deforestation results in an increase in precipitation.
    c. mineral losses from a valley were great following deforestation.
    d. deforestation led to the drying up of the valley's stream.

14. Biological magnification shows that
    a. top predators may be most harmed by toxic environmental chemicals.
    b. DDT interferes with egg-shell production in birds.
    c. the greenhouse effect will be centered over the poles.
    d. the biosphere is able to expand beneficial processes to compensate for environmental disruptions.

15. The greenhouse effect
    a. could lead to the flooding of coastal areas.
    b. could result in more C_3 plants in plant communities.
    c. causes an increase in temperature when CO_2 absorbs more sunlight.
    d. can do all of the above.

16. The increase in CO_2 concentration in the atmosphere is a result of
    a. an increase in primary productivity.
    b. an increase in average temperature on Earth.
    c. the burning of fossil fuels and wood.
    d. an increase in methane production by termites.

17. Eutrophication
    a. results from the dumping of nonbiodegradable wastes into lakes.
    b. may result in fish kills due to a lack of oxygen caused by a reduction in photosynthesis.
    c. may involve a huge increase in producers due to an increase in phosphorus in a lake.
    d. involves a reduction in primary productivity.

18. The Gaia hypothesis
    a. shows that atmospheric O_2 concentration is regulated by methane.
    b. was developed by Bormann and Likens.
    c. has proved that the biosphere is a superorganism, able to regulate the atmospheric CO_2 concentration.
    d. may serve as a useful metaphor for the interactions between the biosphere and the physical parameters of Earth.

CHAPTER 50

BEHAVIOR

FRAMEWORK

This chapter introduces the complex and fascinating subject of animal behavior. The study of behavior involves the integration of the study of biochemistry, genetics, physiology, evolutionary theory, and ecology.

Behaviors can range from simple fixed action patterns in response to specific stimuli to insight learning in novel situations. Behaviors are the result of interactions between environmental stimuli, experience, and the genetic makeup of the individual. The parameters of behavior are controlled by genetics, and thus are acted on by natural selection. The ultimate cause of reproductive fitness can be used to interpret foraging behavior, agonistic interactions, mating patterns, and altruistic behavior. Sociobiology also extends evolutionary interpretations to human social interactions.

CHAPTER SUMMARY

Behavior can be defined as any observable movement or action of an animal in response to a stimulus. Animal behavior includes such activities as feeding, courtship, mating, and communication. *Ethology* is the study of how animals behave in their natural environments. The behavior of animals as they interact with their environments is related to all levels of ecological organization: individual, population, and community.

Studying Behavior

The mechanisms that produce behavior are complex and variable. *Anthropomorphism*—ascribing human emotions, thoughts, or reasoning to other animals—should not be a part of the study of behavior. We do not know whether other species have conscious feel-

ings or thoughts. The graylag goose example points out the need to look for the cues that are guiding animal behavior.

Proximate causes, occurring in ecological time, explain behavior in terms of present stimulations and responses. *Ultimate causes* of behavior are found on the scale of evolutionary time. Patterns of behavior are, to a greater or lesser degree, under genetic control. Behavior, thus, is inherited and subject to evolution by natural selection.

The age-old controversy of nature versus nurture extends to the determination of how much of an organism's behavior is innate or instinctive and how much is learned. Much of vertebrate behavior appears to result from the complex interaction between physiology and external cues. Although behavior often can be modified by experience, and thus can be phenotypically variable, it still has its basis in an organism's genes.

Innate Components of Behavior

The work of Lorenz and Tinbergen provided evidence for the innate components of behavior and established the foundation of ethology. The *fixed action pattern*, or *FAP*, was the fundamental concept to emerge from their work. An FAP is a highly stereotyped, innate behavior that, once begun, is carried through to completion. An FAP is triggered by a *sign stimulus*—some external sensory stimulus that, when perceived by the animal, "releases" the specific behavior.

The term *releaser* usually refers to sign signals that communicate between individuals of the same species. The releaser for attack behavior in stickleback fish is the red belly of the intruder, and may extend to any red colored object entering its territory. FAPs often are associated with interactions between parents and young, such as the begging behavior of young birds and the feeding behavior of their parents. A human infant's smile and grasp are FAPs, initiated respectively by the sight of a face and tactile stimuli.

The mechanistic aspect of FAPs is illustrated by the repeated stereotyped behavior seen in some animals, such as a digger wasp dragging a paralyzed cricket to a specific distance from its nest and then inspecting the nest. When the cricket is repeatedly moved a short distance away from the nest, the digger wasp performs the drag-and-inspect subroutine over and over.

The experiments of ethologists have shown that sign stimuli usually involve one or two simple characteristics of the object or organism that trigger the behavior. The pecking of a herring chick on its parent's beak releases feeding behavior by the adult. The pecking behavior of the chick can be released by a spot swung horizontally at the end of a long object, corresponding to the red spot on the parent's moving beak.

An animal's sensitivity to certain stimuli is closely correlated to the sign stimuli to which it responds. Frogs have retinal cells quite good at detecting movement, and the movement of objects releases the frog's feeding behavior.

Stronger responses often are elicited by exaggerated sign stimuli. The nestling with the widest-gaping beak will be most likely to be fed. The experimental use of *supernormal stimuli*, such as giant models of eggs, illustrates these behavioral tendencies.

The use of simple cues to release preprogrammed behavior ensures behavioral responses requiring the least amount of processing and integrating of input. Natural selection has resulted in innate programming of behavior that can be elicited by simple stimuli in the animal's environment.

Innate releasing mechanisms (IRM) were proposed by early ethologists as the link between stimuli and responses. Networks of neurons or cells were proposed as IRMs, but neurophysiologists have not been able to locate any such centers for specific behaviors. A giant neuron in the nudibranch *Tritonia* that triggers escape behavior when artificially stimulated has been identified. In vertebrates in particular, the concepts of releasers and IRMs are consistent with the observation of behaviors, but are not explanations of how these behaviors happen.

The idea of *drive* or motivation is used to explain the release of some behaviors. Feeding, reproduction, and other behaviors are not continuously performed. It may be said that animals have variable drives to feed or mate. Stronger tendencies toward a behavior may be correlated with a physiological change, such as seen by the increase in stereotyped pecking when chickens are hungry.

The concept of drive does not apply to all behaviors. Fleeing from a predator is tightly coupled with an external stimulus of the predator. Describing this behavior as a drive does not make sense. Again, the use of the concept of drives does not give any information about the mechanisms causing behaviors.

Learning and Behavior

Learning can result in the modification of innate components of behavior. *Habituation* refers to the loss of sensitivity to unimportant or repeated stimuli. Even the graylag goose has a limit to the number of eggs she will retrieve.

Practice may result in the more effective performance of innate behaviors. This improvement, however, may derive from the natural development and maturation of the animal, and thus may not be a function of the repetition of the task. Birds prevented from flying until older are able, when released, to fly without practice or learning.

The phenomenon of *imprinting* is a clear case of learning closely associated with innate behavior. Young ducks and geese instinctively follow the movement of an object, whether it be their mother, a ticking box, or Konrad Lorenz. Species recognition and reproductive behavior are tied to this early imprinting. The return of salmon to their home stream to spawn is an example of olfactory imprinting.

A *critical period* is a limited time during which some type of imprinting may occur. When Lorenz isolated geese totally during their first two days, they failed to imprint afterwards on anything. Adult birds appear to imprint on their young in the period following hatching. Thus, the young of cuckoos, and other bird parasites who lay their eggs in other species' nests, are accepted by their foster parents. Sexual imprinting may occur later in some species. Males of one species of finches, after initial rearing with their own species, were placed only with members of another closely related species during their critical period for sexual identity. Afterwards, they readily mated with females of the second species, but only reluctantly mated with their own species.

The close interaction between learning and innate behavior is illustrated by the learning of bird songs. Songbirds learn to sing normally only when they can hear the song of their species. If only the song of a related species is heard by a bird during its critical period, the bird does not learn that song. Genetics determine what song a bird can sing, but the bird can learn that song only by hearing it from other birds.

In *associative learning*, animals learn to associate one stimulus with another. *Classical conditioning*, described through the work with salivating dogs done by Pavlov in the early 1900s, illustrates the association of an irrelevant stimulus with a fixed behavioral and physiological response. The same type of conditioning may be associated with the broadening and sharpening of the perception of release stimuli in an animal's natural environment.

Operant conditioning refers to the trial-and-error learning through which an animal associates a behavior with a reward or punishment. Skinner's work with rats placed in "Skinner boxes" is the best-known laboratory study of operant conditioning. The association of good or bad tastes with food items is probably a common form of operant conditioning in animals in nature. Sometimes, animals are able to learn by watching the behavior of others. The spread of milk-bottle pecking through the English tit population was an example of this type of observational learning.

Insight learning, sometimes called reasoning, involves the ability to perform a behavior correctly in a novel situation. The stacking of boxes by a chimpanzee in order to reach a banana is commonly cited as an example of the insight learning that is often seen in primates and some other mammals. Most animals do not appear to have the ability to use insight.

The capacity to learn confers a survival and reproductive advantage and can be acted on by natural selection. The genetic base and internal mechanisms of learning are largely unknown. Some simple kinds of learning have been linked to biochemical or physiological changes. Most animal behaviors can be explained as relatively fixed patterns that may be modified by simple kinds of learning.

Behavioral Rhythms

Feeding, sleeping, reproducing, and migrating are all regularly repeated behaviors that show a temporal rhythm. Animals, as well as plants, show *circadian rhythms* of roughly 24-hour duration. To determine whether these rhythms are based on exogenous (external) cues or endogenous (internal) timers, researchers have placed animals in environments with no external cues and monitored the animals' rhythmic behaviors. A biological clock appears to be involved, although exogenous cues are necessary to keep the behavior timed with the real world. The exogenous cue, sometimes called a *zeitgeber*, is usually light. Humans living in free-running conditions with no external time cues have a biological clock with a period of about 25 hours.

In many species, *circannual* behaviors, such as reproducing and hibernation, are based on physiological and hormonal changes that are linked with changes in day length. Few studies of the endogenous control of these long-term behaviors have been done. Fat deposition in ground squirrels, associated with hibernation, has been shown to occur regularly, even in a constant environment.

The timing mechanism of the biological clock is unknown. Current hypotheses propose some sort of biochemical clock, perhaps based on regular molecular interactions.

Orientation and Navigation

Animals use a variety of cues to orient themselves in and navigate through their environments. A *taxis* is a more or less automatic movement directed toward or away from a stimulus. House-fly larvae are negatively phototactic after feeding; trout are positively rheotactic and automatically swim or orient into a current. A *kinesis* is a change in activity rate in response to a stimulus. Although kinetic movements are randomly directed, they still tend to maintain organisms in favorable environments. Often an animal's random movements will slow down in response to favorable stimuli.

Long-distance migration requires sophisticated mechanisms of orientation. Some species of birds and other animals navigate by the heavens, using the sun or stars as directional cues. Internal timing mechanisms are needed to compensate for the continuous daily movement of these reference bodies. The indigo bunting appears to navigate by using the unmoving North Star, but other organisms have the ability to adjust to the apparent movement of the sun and stars both throughout the day and during the movement of the animal over its migratory route.

Many bird species are able to continue migrating under clouds or through fog because of their ability to detect and orient to Earth's magnetic field. Experimentally manipulating the magnetic field changes the migratory orientation of some species. Magnetite, an iron-containing mineral, has been found in the heads of some birds, in the abdomens of bees, and in certain bacteria that orient to the magnetic field.

Foraging Behavior

Feeding often involves relatively simple behavior, including fixed action patterns released by the sight of prey. Some animals are generalists, feeding on a large variety of items, whereas others are specialists with quite restricted diets. Specialists usually have morphological and behavioral adaptations that are specific for their food item and make them extremely efficient at foraging. Generalists may not be as efficient, but they have more options should one food item become unavailable.

Generalists will often concentrate on a particular abundant food item, developing what we call a *search image* for that favored item. When that item becomes less abundant, the animal appears to change its search image and to concentrate on a new item. Search images may allow for short-term specialization while maintaining the advantages of being a generalist. The concept of search image describes what appears to happen but, once again, does little to explain the underlying mechanisms.

Behavioral ecologists study the foraging adaptations of animal, with the prediction of *optimal foraging*: that natural selection would favor animals that make efficient choices that maximize energy intake over expenditure. Tradeoffs in foraging involve choices such as taking a larger prey item, which may contain more food energy but require more energy to obtain if farther away, or to subdue and manipulate, or selecting closer or easier to subdue but smaller prey. Most studies have indicated that animals have surprising abilities to modify their foraging behavior in order to maximize overall energy intake. The ability appears to be innate, although experience and physical maturation are thought to increase foraging efficiency.

Competitive Social Interaction

Social behavior involves the interaction between two or more animals, usually of the same species. Mating behavior is mutually beneficial to the reproductive fitness of both individuals, but many types of interactions are competitive. Because members of the same species occupy the same niche, the potential for conflict over limited resources, especially in *K*-selected species, is quite high.

Agonistic behavior involves a contest that determines which competitor gains access to resources such as food or mates. The encounter may include a test of strength, or, more commonly, threat displays and *ritual* behavior, which serve to avoid actual physical conflicts. A submissive or appeasement display by one of the competitors inhibits further aggressive activity and indicates that the other animal has won the contest. Natural selection apparently favors the ending of a conflict without a violent combat that would reduce the reproductive fitness of the winner as well as the loser.

Dominance hierarchies, as illustrated by the "pecking order" in hens, establish which animals get first access to resources and prevent continual combats. In wolf packs, the top female controls the mating of the others. When food is abundant, all females are allowed to mate; when food is scarce, the top female restricts mating.

A *territory* is an area that is defended from other members of the same species. Territories typically are established for feeding, mating, rearing young, or a combination of these activities. Territory size varies with species, function, and abundance of resources. A home range is the area in which an animal may roam, but a territory is the area which the animal defends.

Agonistic behavior is used to establish and to defend territories. Vocal displays, scent marks, and patroling may be used to proclaim ownership continually. Usually only conspecifics are excluded from an animal's territory.

Both dominance systems and territories help to stabilize population density. They ensure that at least some individuals have a sufficient amount of a resource when the resource may be in limited supply. Animals low in the hierarchy or lacking territories usually are ready to step up should an established individual die.

Mating Behavior

Most animals are programmed to view conspecifics as threatening competitors to be driven off. This aversion must be overcome for mating to be accomplished. Complex courtship interactions, unique to each species, assure that individuals are not a threat and are of the proper species, sex, and physiological mating condition. These complex rituals often consist of a series of fixed action patterns, each released in sequence by the reciprocal behavior of the individuals involved. Appeasement gestures are often part of the mating ritual.

Ritualized acts probably evolved from actions that had a more direct meaning. The behaviors of various species in the predaceous fly family Empididae show a progression from a male carrying a dead insect for the female to eat while he mates with her, to carrying a dead insect inside a silk balloon (perhaps because the silk helped to subdue the insect or because it made the gift look larger), to carrying an empty silk balloon.

Mating relationships vary a great deal among species. Many species have *promiscuous* mating, with no strong pair bonds forming. Longer-lasting relationships may be *monogamous* or *polygamous*. Polygamous relationships are most often *polygynous* (one male and many females), although a few are *polyandrous* (one female and many males).

According to Darwinian considerations, an individual is valuable to a member of the opposite sex only as a vehicle to help his or her genes get into subsequent generations. Reproductive behavior can be analyzed on the basis of maximizing fitness, and may differ between the sexes based on their *sexual investment* in reproduction. Especially in vertebrates, females have a greater inherent investment in an offspring than do males, and it adds to the female's reproductive fitness to show discriminant selection of a mate. For males, sperm do not represent a large energy investment, and it may be reproductively advantageous to try to mate often and with a number of partners.

The needs of young offspring are reflected in the reproductive patterns of the parents. Newly hatched birds require more food than one parent may be able to supply. A male will ultimately leave more offspring, and thus increase his reproductive fitness, by helping to care for these young rather than by going off in search of more mates. With mammals, often only the female is needed as a source of food, and males are not monogamous.

Communication

Vision, hearing, and olfaction are common methods of communication between animals. *Pheromones* are chemical signals that often are involved in reproductive behavior, both to attract mates and to release specific courtship behaviors. The trailing behavior of ants involves scents released by scouts. Olfactory signals are also used to mark territories.

One of the most complex communication systems is that of honeybees, who use ritualized dances to communicate the location of food sources. The communication dances of bees were first described by von Frisch. Round dances are used when the food source is relatively close to the hive, and regurgitated nectar provides a scent to direct other bees to the food. The location of more distant food is communicated by waggle dances, in which the duration and vertical orientation of the dance on the comb respectively indicate the distance from the hive and the direction to the food relative to the horizontal angle to the sun.

Altruistic Behavior

Altruistic behavior is difficult to explain in Darwinian terms of improving the reproductive fitness of the individual, especially in examples such as warning calls by Belding's ground squirrels that may increase the caller's risk of being killed, or the attack behavior of honeybees that results in the death of the stinging bee. The cooperative raising of young grey-breasted jays is another example of altruistic behavior that does not seem to increase the reproductive fitness of the individual.

These examples have been explained by the fact that the cooperating individuals often are benefitting individuals that may be siblings or close relations, and thus are carrying copies of the altruistic individual's genes. *Kin selection*, first proposed as an explanation by Hamilton, suggests that altruistic behavior should be strongest between close relatives and less common as genetic relatedness decreases. Indeed, all worker bees are genetic sisters of the queen, and most flocks of jays represent an extended family unit of related individuals. Alarm calls usually are given by female ground squirrels. Males tend to disburse to other colonies after mating, whereas females are surrounded by offspring and other family relations. Thus, survival of female genes is enhanced by this altruistic behavior.

When altruistic behavior involves nonrelated animals, the explanation offered is *reciprocal altruism*; there is no immediate benefit for the altruistic individual, but some future benefit may occur when the helped animal may "return the favor." Reciprocal altruism often is used to explain altruism in humans.

Genetically maintained unselfish behavior, especially if it ended in harm to the altruistic animal, would be selected against. Altruistic behavior in animals must somehow enhance the survival of the individual's own genes. The extent to which human behavior can be interpreted in such terms is a controversial subject. Many researchers maintain that seeming examples of kin selection in other animals can be explained adequately as individual selection. Controversies over selfish genes, kin selection, and altruistic behavior continue.

Human Sociobiology

E. O. Wilson has elaborated the thesis that social behavior has an evolutionary basis, and that behavioral characteristics are expressions of genes that have been acted on by natural selection. Much of Wilson's work dealt with the application of evolutionary theory to social behavior of insects, but he also speculated on the evolutionary basis of certain social behaviors of humans.

The sociobiology debate continues the nature versus nurture controversy. In the example of human incest, the "nurture" side would argue that the aversion to incest must not be innate or else the taboos and laws against it would be unnecessary. The avoidance of incest must be a learned behavior, based on the experience that individuals who break the taboo are more likely to have defective children. The "nature" side would argue that the occurrence of such a behavior across many diverse cultures is evidence that it must have an innate component. Taboos and laws only reinforce a behavior that developed through natural selection as the result of decreased reproductive success of incestuous individuals.

Wilson presents the Israeli kibbutzim as evidence of the innate aversion to incest. Parents encourage their children to marry within the kibbutz, but marriages between these children, who essentially have been raised as siblings, is rare.

Some sociobiologists, including Wilson, explain that cultural and genetic components of social behavior are linked together in a cycle of reinforcement. The development of cultural regulations of innate behaviors serves as an additional environmental factor in the natural selection of that behavior. According to this view, human nature represents an integration of genes and culture.

Sociobiological considerations of human behavior do not mean we are totally controlled by our genotypes. The behavior of humans is probably more plastic than that of any other animal. The parameters of social behavior may be set by genetics, but the environment undoubtably shapes behavioral traits just as it influences the expression of physical traits. Genes involved with complex behavior probably have very broad "norms of reaction." An evolutionary interpretation of human social behavior must include environmental components.

STRUCTURE YOUR KNOWLEDGE

1. How does the nature versus nurture controversy apply to behavior?

2. Distinguish the following types of learning: imprinting, classical conditioning, operant conditioning, and insight learning.

3. Why are many interactions betweeen members of the same species agonistic? What phenomena reduce violent encounters?

4. What are the advantages and disadvantages of being a generalist or a specialist feeder?

5. How does the concept of Darwinian fitness apply to reproductive and altruistic behavior?

TEST YOUR KNOWLEDGE

MULTIPLE CHOICE: *Choose the one best answer.*

1. Ethology is the
   a. study of the behavior of animals in their natural environments.
   b. application of human emotions and thoughts to other animals.
   c. study of animal conditioning.
   d. study of the origin and ultimate causes of animal behavior.

2. Proximate causes
   a. explain the evolutionary significance of a behavior.
   b. are immediate causes of behavior, such as hunger or external cues.
   c. indicate that much of animal behavior is innate.
   d. all of the above are correct.

3. The controversy between nature and nurture
   a. tries to determine whether behavior has proximate or ultimate causes.
   b. relates to whether an animal's behavior is innate or learned.
   c. has shown that genetics is more important than environment in determining behavior.
   d. deals with whether animals have conscious feelings or thoughts.

4. Which of the following is not true of fixed action patterns?
   a. They are highly stereotyped, instinctive behaviors.
   b. They are triggered by sign stimuli in the environment.
   c. They include such things as bird songs and pecking and feeding behaviors.
   d. They are often released by one or two simple characteristics of an object or organism.

5. Supernormal stimuli
   a. are innate releasing mechanisms between individuals of the same species.
   b. may elicit stronger responses or FAPs.
   c. are illustrated by the removal of the cricket from near a digger wasp's nest.
   d. are illustrated by the bobbing of a stick with a red spot past herring chicks.

6. An animal's "drive" to perform a certain behavior
   a. may be related to its physiological condition.
   b. may depend on the size of the predator.
   c. will be greater with supernormal stimuli.
   d. is correlated with the sensitivity of its receptors to sign stimuli.

7. Habituation
   a. develops as a result of experience and practice.
   b. occurs during a critical period of development.
   c. is the loss of sensitivity to unimportant stimuli.
   d. is the development of a behavior habit or common type of response.

8. Learning
   a. cannot modify innate components of behavior.
   b. is a type of imprinting.
   c. involves the use of insight.
   d. is a change in behavior as a result of experience or practice.

9. A critical period
   a. is the time right after birth when sexual identity is developed.
   b. is a limited time during which imprinting or learning can occur.

c. is a case of learning closely associated with innate behavior.

d. is all of the above.

10. The return of salmon to their home streams to spawn is an example of
    a. habituation.
    b. associative learning.
    c. olfactory imprinting.
    d. operant conditioning.

11. In classical conditioning,
    a. an animal associates a behavior with a reward or punishment.
    b. an animal learns as a result of trial and error.
    c. an irrelevant stimulus can elicit a response because of its association with a normal stimulus.
    d. a bird can learn the song of a related species if it hears only that song.

12. Which of the following researchers is incorrectly paired with his work?
    a. Tinbergen—sign stimuli as releasers of FAPs
    b. Lorenz—imprinting with geese
    c. Pavlov—operant conditioning with a normal stimulus.
    d. Wilson—sociobiology

13. Insight learning
    a. only occurs in human beings.
    b. was studied by Skinner in his use of Skinner boxes.
    c. involves the ability to perform successfully in a novel situation.
    d. is associated with fixed action patterns.

14. Circadian rhythms
    a. may be timed to the real world by the use of a *zeitgeber*.
    b. have aproximately a 24-hour period.
    c. are based on a biological clock, which may rely on a biochemical timer.
    d. all of the above are correct.

15. Circannual behaviors
    a. rely solely on endogenous cues.
    b. are often linked to changes in day length.
    c. involve foraging, reproduction, and migration.
    d. do not occur in free-running conditions.

16. A kinesis
    a. is a randomly directed movement that is not caused by external stimuli.
    b. is a movement that is directed toward or away from a stimulus.

c. is a change in activity rate in response to a stimulus.

d. is illustrated by trout swimming upstream.

17. Which of the following is not true of long-distance migrations?
    a. Animals may use the stationary North Star as a point of reference.
    b. Navigation using celestial bodies requires an internal clock to compensate for the daily movement of these objects.
    c. Birds that are able to continue migrating during fog and overcast weather may be sensing Earth's magnetic field.
    d. They are circadian behaviors.

18. A change in search image
    a. would occur most often in specialists.
    b. may be associated with dwindling food supplies.
    c. replaces FAPs as foraging strategies of generalists.
    d. produces optimal foraging when some prey items are too large.

19. Agonistic behavior
    a. is most common between members of the same species.
    b. may include such things as threat displays and ritual behavior.
    c. may be used to establish dominance hierarchies and territories.
    d. all of the above are correct.

20. An animal's territory may
    a. be smaller than its home range.
    b. change in size with changing resource density.
    c. exclude only conspecifics of the same sex.
    d. all of the above are correct.

21. Pheromones are most likely to be involved in
    a. habituation.
    b. escape from predators.
    c. intraspecific communication.
    d. learning.

22. In a species in which females provide all the needed food for the young, but both young and females need protection,
    a. males are likely to be promiscuous.
    b. mating systems are likely to be polygynous.
    c. mating systems are likely to be polyandrous.
    d. males will have little sexual investment in reproduction.

23. The ability of honeybees to fly directly to a food source, after having to wait
several hours from the time of the
waggle dance, indicates
   a. that the waggle dance provided directions
   relative only to the position of the hive.
   b. that bees have an internal clock that
   compensates for the movement of the
   sun during the elapsed time.
   c. that these bees must have been
   to that food source before.
   d. that bees are directed more by
   olfactory cues than by directional cues.

24. Reciprocal altruism
   a. is never known to occur, because
   there is no immediate benefit
   to the altruistic organism.
   b. occurs only between
   closely related individuals.
   c. has been used to explain human altruism.
   d. is an excellent example of kin selection.

25. Sociobiology
   a. explains the evolutionary basis
   of behavioral characteristics
   within animal societies.
   b. applies evolutionary explanations
   to human social behaviors.
   c. studies the roles of culture and
   genetics in social behavior.
   d. does all of the above.

ANSWER SECTION

CHAPTER 46: THE PHYSICAL ENVIRONMENT

Suggested Answers to Structure Your Knowledge

1. Ecology is the study of the relationships of organisms to their abiotic and biotic environments, focusing on the abundance and distribution of species. Questions cover a range that includes the consideration of individuals and their physiological adaptations, the growth and regulation of populations, the interactions (such as predator–prey and competition) within communities, and the flow of energy and chemicals through ecosystems. Experimental manipulation of variables is done both in the laboratory and in the field, along with observational measurements of various physical factors and kinds and numbers of organisms. Evolution, with the principles of natural selection and adaptation, is the key organizing theory.

2. Biomes include the eight major world communities, which are determined on the basis of abiotic and biotic similarities. Temperature and precipitation, which are related to latitude, altitude, and location (as in coast or interior of continent), are the factors that most influence the plant and animal life forms found in each community. Convergent evolution, as organisms adapt to similar environments, may account for the similarities in life forms in each biome.

Answers to Test Your Knowledge

Multiple Choice:

1. b		6. c	
2. d		7. a	
3. d		8. c	
4. c		9. c	
5. d		10. d	

Matching:

1. C	3. D	5. A	7. B
2. H	4. E	6. G	8. F

CHAPTER 47: POPULATION ECOLOGY

Suggested Answers to Structure Your Knowledge

1.

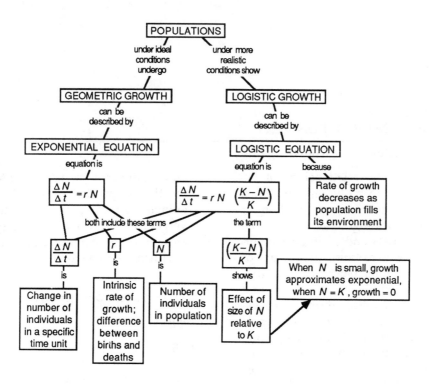

2. The age structure of a population affects population growth as a result of the differential reproductive capabilities of various age groups. The larger the proportion of reproductively active individuals, the higher the growth rate. Fertility tables report birth rates for various age groups of females. Combining fertility tables with life tables (which report age-specific mortality) yields a net reproductive rate of the number of female offspring each female of the population can be expected to produce. This rate, along with the generation time (age at which females begin reproduction), determines population growth rates.

Answers to Test Your Knowledge

Matching:

1. J	6. I
2. G	7. F
3. H	8. L
4. C	9. B
5. A	10. D

Multiple Choice:

1. b	4. a	7. c	10. d
2. c	5. a	8. b	11. a
3. d	6. b	9. d	12. b

CHAPTER 48: COMMUNITIES

Suggested Answers to Structure Your Knowledge

1.

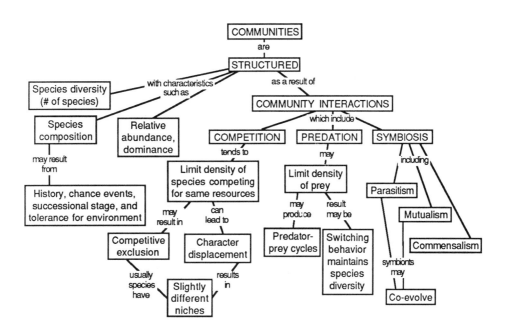

2. Succession is the gradual transition in species composition in a community, following a disturbance or creation of a new habitat. Three factors may be involved in this transition. Each stage of vegetation alters the environment somewhat and may *facilitate* the transition to the next stage. Early pioneers in the habitat may also *inhibit* the establishment of other species, until more competitive species are able to become established. The change in species composition also may be due to the gradual success of the species that are most *tolerant* of the environment found in that habitat.

Answers to Test Your Knowledge

Multiple Choice:

1. c	7. a	13. b
2. d	8. b	14. d
3. c	9. a	15. c
4. c	10. a	16. b
5. c	11. a	17. c
6. a	12. b	18. a

CHAPTER 49: ECOSYSTEMS

Suggested Answers to Structure Your Knowledge

1.

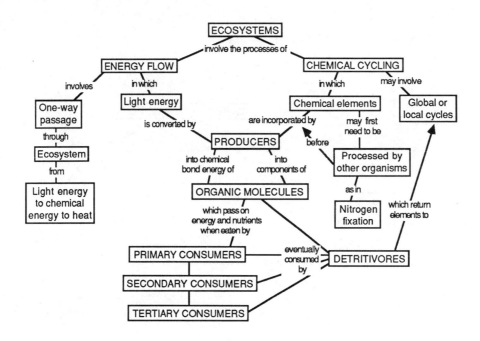

2.

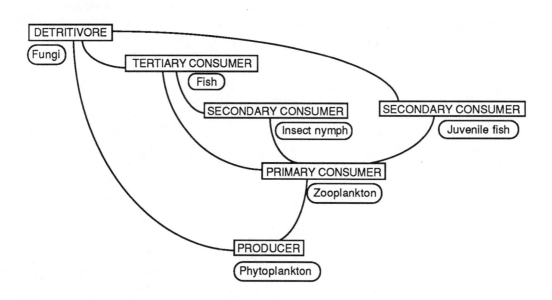

3. a. Deforestation causes an increase in water runoff with an accompanying loss of soil and minerals. The cutting of tropical forests results in a loss of species diversity, a reduction in productivity of the area, and weather changes.

b. Dumping of wastes and runoff from agricultural lands has produced eutrophication in many lakes, killing off many fishes and other organisms.

c. The introduction of toxic chemicals into the environment has resulted in their being incorporated into the food chain. As a result of biological magnification, these substances pose added threats to top-level consumers.

d. An increase in CO_2 levels in the atmosphere as a result of the combustion of fossil fuels and wood adds to the greenhouse effect. The resulting increase in temperature may have far-reaching effects on climate and sea level.

Answers to Test Your Knowledge

Multiple Choice:

1. b	7. b	13. c
2. a	8. d	14. a
3. d	9. a	15. a
4. c	10. b	16. c
5. d	11. d	17. c
6. d	12. c	18. d

CHAPTER 50: BEHAVIOR

Suggested Answers to Structure Your Knowledge

1. There is controversy over how much of animal behavior is innate and genetically programmed and how much is a product of experience and learning. Fixed action patterns clearly are preprogrammed behavior, and many seemingly complex animal behaviors can be isolated into a series of FAPs. In other cases, genetics may set the parameters for an organism's behavior, but experience can modify behavior and learning is clearly evident.

2. Imprinting is a type of learning that occurs during a critical time in an organism's development. The animal appears to be genetically programmed to identify with whatever species it is placed with at that time, to remember a smell as a guide to return "home" to spawn, or to follow whatever moves as "mother." Innate behavior and learning are tightly coupled in imprinting.

   The concept of classical conditioning was developed by Pavlov. As the concept is used in a laboratory study, an animal learns to associate an unrelated stimulus with a stimulus that elicits a behavioral or physiological response. Later, the unrelated stimulus can cause the behavior. In nature, classical conditioning is the type of learning in which an organism adds to its collection of stimuli that may be associated with particular cues or releasers.

   Operant conditioning, as studied by Skinner, is learning in which an animal comes to associate a positive or negative reward with a particular behavior. In a laboratory or circus, it can be used to teach animals all sorts of "tricks." In nature, it is a way in which an organism can learn the good or bad tastes of food or similar lessons.

   Insight learning involves the ability of an animal to reason and to figure out an appropriate behavior in a novel situation.

3. Members of the same species occupy the same niche and are thus competitors for resources such as food, territory, and mates. In agonistic encounters, one animal may establish its right to these resources. Threat displays, appeasement gestures, dominance hierarchies, and territories all serve to reduce violent encounters between conspecifics.

4. A generalist is not as efficient a forager but is able to use more food items. The development of search images may improve the efficiency of generalists. A specialist often has anatomical and behavioral adaptations to feeding efficiently on one type of food item. Should this item become scarce, however, the specialist is at a disadvantage.

5. According to the concept of Darwinian fitness, an animal's behavior should help to increase its reproductive success. The best strategy for reproductive behavior will be determined on the basis of the probability of producing the most offspring to carry one's genes into the next generation. Females that have a high sexual investment in each offspring will maximize their fitness by a discriminate choice of a mate. A male's reproductive fitness may be maximized by frequent mating, or by helping to rear young if two parents are needed for offspring to survive. Altruistic behavior has been explained on the basis of kin selection; animals will show altruistic behavior if their efforts will benefit related animals who may be carrying duplicates of their genes.

Answers to Test Your Knowledge

Multiple Choice:

1. a	6. a	11. c	16. c	21. c
2. b	7. c	12. c	17. d	22. b
3. b	8. d	13. c	18. b	23. b
4. c	9. b	14. d	19. d	24. c
5. b	10. c	15. b	20. d	25. d